刘亦师

清华大学

近代校园规划与建筑

编著

清华大学出版社
北京

图书在版编目（CIP）数据

清华大学近代校园规划与建筑 / 刘亦师编著. — 北京：清华大学出版社，2021.4
ISBN 978-7-302-57948-9

Ⅰ.①清… Ⅱ.①刘… Ⅲ.①清华大学－校园规划②清华大学－教育建筑 Ⅳ.①TU244.3

中国版本图书馆CIP数据核字(2021)第063098号

责任编辑：刘一琳
装帧设计：陈国熙
责任校对：刘玉霞
责任印制：杨 艳

出版发行：清华大学出版社
　　　　网　　　址：http://www.tup.com.cn，http://www.wqbook.com
　　　　地　　　址：北京清华大学学研大厦 A 座　　　　邮　　编：100084
　　　　社 总 机：010-62770175　　　　　　　　　　　邮　　购：010-62786544
　　　　投稿与读者服务：010-62776969，c-service@tup.tsinghua.edu.cn
　　　　质量反馈：010-62772015，zhiliang@tup.tsinghua.edu.cn
印 装 者：北京博海升彩色印刷有限公司
经　　销：全国新华书店
开　　本：210mm×285mm　　　　印　张：38.25　　　插　页：1　　字　数：1018 千字
版　　次：2021 年 6 月第 1 版　　　　　　　　　　　印　次：2021 年 6 月第 1 次印刷
定　　价：298.00 元（全两册）

产品编号：091075-01

清华大学

近代校园观划与建筑

庚子大寒 刘亢晓题

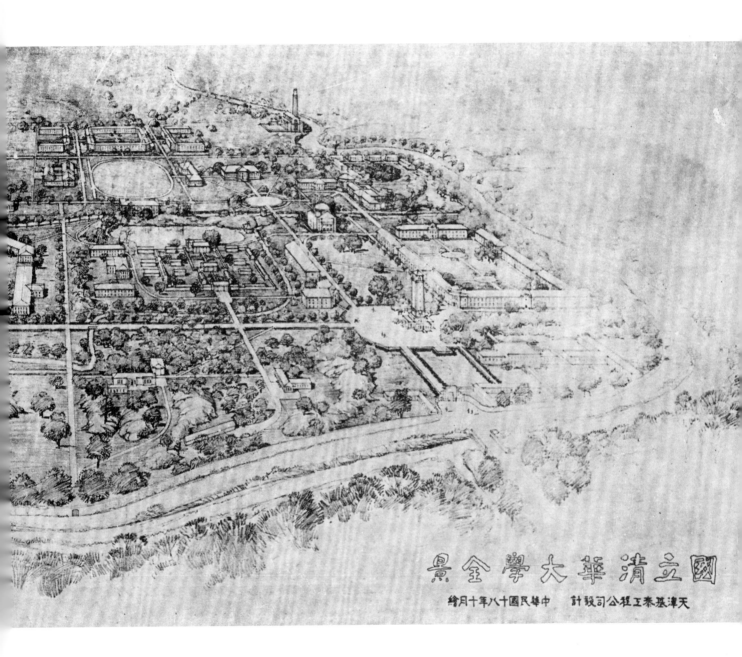

國立清華大學全景

天津基泰工程司公設計　中華民國十八年十月繪

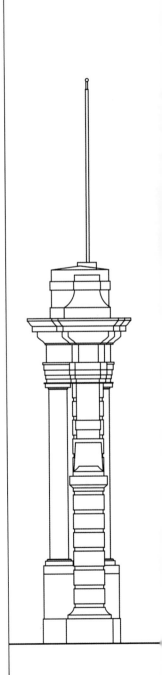

序（一）PREFACE

　　大学是功能独特的文化机构，肩负着文化传承创新和文化育人的使命。大学校园不仅是大学生学习、工作、生活的物质环境，也是体现大学理念、蕴育大学精神的精神家园，其本质是文化空间。所谓文化可以理解为人为了生存和发展，使外在世界按照人追求的真善美的境界"人化"；而当人构建起人化了的世界时，它又通过"人文化成"而造化人自身，即"化人"。文化是人化和化人的过程及其产物。大学校园是历代大学生依据大学理念开物成境的成果，是提高师生文化素质、提升学校文化品位的重要环境，由此，做好大学校园的规划与建设意义重大。

　　本书作者刘亦师作为清华大学建筑学院建筑历史与理论研究所的青年教师，自2001年邂逅清华老校区后，便为其厚重的建筑文化内涵所吸引。2003年，刘亦师到清华建筑学院攻读中国近代建筑史，2014年清华大学建筑学院博士后出站留校，在撰写《中国近代建筑史概论》的同时，考虑到自1909年开始的清华校园规划和建设是中国近代建筑史的重要研究对象，着手编著图文并茂、内涵丰富的《清华大学近代校园规划建设历史研究》。学校的校园规划和建设史，是校史的重要组成部分，能够起到存史、资政、育人的作用。

　　刘亦师在清华大学近代校园规划建设研究中，继承了清华务实会通的学术风格，着力"务博综、尚实证"，注重中西融会，古今贯通，理实结合，在综合中提升认识境界。他从清华学校曾是一所留美预备学校，规划与建筑风格深受欧美影响的史实出发，首先实地考察了弗吉尼亚大学的大草坪区与清华礼堂前草坪区的类似与差别，并对与清华大礼堂多具相似性的弗吉尼亚大学圆厅图书馆穹顶结构进行了深入比较研究；继而又专程赴耶鲁大学，对该校校友、为清华学校校园规划和建筑设计做出重大贡献的建筑师墨菲（Henry Killam Murphy，1877—1954，一译茂飞）进行了档案发掘，幸运地发现了墨菲关于清华学校的设计资料，包括大礼堂的平面、立面和剖面设计图。回到清华大学后，他在学校房管处支持下对大礼堂进行了测绘，通过比较，对其与弗吉尼亚大学圆厅图书馆的异同有了具体而深入的认识。这一经历使他体会到，爬梳原始档案、考察历史文献，并积极组织测绘取得第一手资料，有着广阔的前景，并以此为基本思路，拓展了对清华校园规划与建设的系统研究。

清华是享誉瀛寰的建筑规划和设计师的摇篮，梁思成在这里开拓了中国建筑史学术研究的先河。针对中国建筑史罕人问津的境况，他不禁发问："我们中华民族有着数千年的灿烂文化，为何独独没有自己的建筑史？"他发誓："中国的建筑史将一定由中国人自己来撰写！"为此他放弃了国外优越的生活和工作条件，回到了当时民不聊生的中国，自1931年开始他从实物调查入手，对北京地区的古建筑进行测绘、分析和鉴定。1932年春，他走出京城，"以测量绘图摄影各法将各种典型建筑实物做有系统秩序的记录"，与妻子林徽因及同仁通过整整5年艰苦卓绝的实地考察与测绘，用现代科学的方法开辟了中国建筑史的研究道路。1944年，《中国建筑史》编写完成，后梁思成主持的"中国古代建筑理论及文物建筑保护"研究被授予国家自然科学奖一等奖，梁思成也被英国学者李约瑟称为研究"中国建筑历史的宗师"。同时，我们也看到，对于中国近代建筑史的研究和教学活动曾经有过一个时期的停顿，直至1985年在清华大学汪坦教授和张复合教授的推动下才重新启动。青年副教授刘亦师于2019年9月出版的《中国近代建筑史概论》和即将出版的《清华大学近代校园规划与建筑》，可以看成清华大学建筑学院群体继承梁思成先生开拓的中国建筑史研究事业的一个最新环节。

我们注意到，刘亦师在上述两本著作所涉及的时域局限在"近代"，即从史学上定义的"从1840年的鸦片战争到1949年中华人民共和国成立的110年历史"，对清华校园规划与建设研究则局限在1909年留美学务处成立到1966年不到一个甲子的时间。实际上，清华大学校园建设大发展是在改革开放之后的40年里，其间对清华校园建设乃至现当代中国建筑事业产生重大影响的大师级人物，如梁思成先生的弟子吴良镛、关肇邺、李道增等的业绩和学术轨迹，都需要全面梳理总结，希望清华大学建筑学院建筑历史与理论研究所的新秀们，包括刘亦师，能够对现当代的清华大学校园规划与建设事业做出持续的研究！

2020年7月于清华园

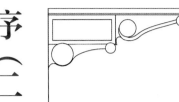

序（二）PREFACE

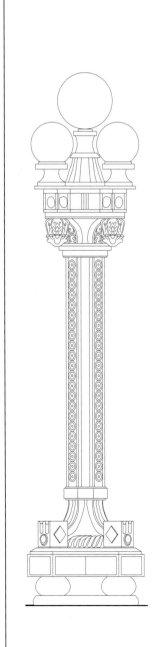

20世纪80年代以来，关于清华大学的研究广泛开展，在基于校史、校志编研以及从中国近代教育史角度进行探讨的诸多著述中，都或多或少涉及清华大学的近代校园规划和建设。但是，最早把清华大学近代校园规划和建设作为专门课题进行研究的，则首推清华大学建筑学院教授罗森先生；他发表在《新建筑》1984年第4期的《清华大学校园建筑规划沿革（1911—1981）》一文，为关于清华大学近代校园规划和建设研究的开山之作。

1985年8月，由汪坦先生发起和组织的"中国近代建筑史研究座谈会"召开之后，中国近代建筑史研究在全国范围内普遍进行，有力地推动了中国近代建筑史研究学科的形成和发展，使近代学校校园规划和建设研究成为广受关注的一个重要课题，清华大学的近代校园规划和建设研究亦随之得以进一步深入。

2003—2006年，亦师在清华大学攻读中国近代建筑史硕士学位，三年期间，他对清华大学的校园建筑有着切身的感受。2012年，他从美国加利福尼亚大学伯克利分校取得博士学位回到清华大学，进博士后工作站，随之留校任教，承担中国近代建筑史研究工作并讲授"中国近代建筑史"课程，推出专著《中国近代建筑史概论》（2019年9月，商务印书馆出版）。

正如他所说，"我在清华大学从事近代建筑史的研究和教学，自感把清华近代校园和建筑的研究做好、做深责无旁贷"。正是出于这种责任心和使命感，亦师"务博综、尚实证"，从2013年开始，奔走于国内海外，爬梳原始档案和未经广泛使用的历史文献，访谈健在的亲历者，实地考察相关校园规划和建设，并组织清华大学校园重要近代建筑测绘，取得第一手资料。历经八年，大体上掌握了从清华大学建校到20世纪60年代校园规划和建设的发展线索及基本史实，得以在清华大学即将迎来建校110年庆典之际，结集其阶段性研究成果及其所依据的海外史料与清华大学校园重要近代建筑测绘成果成书。

值得注意的是，清华大学的创建源自美国退还超索的部分庚子赔款，在建校后的很长一段时间里与西方各国外交部门（尤其美国）、洛克菲勒基金会等官方与民间机构均有所牵涉，一些外国建筑师亦曾参与清华大学近代校园规划和建设且发挥重要作用。这决定了对清华大学近代校园规划和建设的研究，应该投置在全球时空的框架之中。针对以往研究中的不足，亦师"拓展研究视野"、抱定"搜罗资料的决心"，想方设法，做出了锲而不舍的努力，搜集到"与清华建设密切相关，但此前学者罕少使用的海外文献"，为他的研究提供了坚实有力的支撑点，得以回转时空、"置身其中"，取得具有创新性的成果。

此外，档案史料中所见的相关设计图纸与最终落成的建筑实体之间有所差别，甚至相去甚远。查找、分析其间的异同，抽丝剥茧，觅踪寻源，才能建立起物质空间演变与社会学以及思想史、教育史方面的联系，看出校园规划和建设折射出的治校风格、教育理念以及校务管理等诸多背景。正是在清华大学校园重要近代建筑测绘中所付出的辛劳，积累了"大量技术图纸和分析图"，使得亦师对清华大学校园规划和建设"有了很多新的认识"，才能使档案史料在亦师手里物尽所用，为他的研究搭建起跨越学科、具有拓展性的平台，使研究变得"富有张力"。

基于以上两点，对于清华大学近代校园规划和建设的研究来说，此书是具有创新性的力作，会对中国近代建筑史研究学科的发展产生影响，推进近代学校校园规划和建设研究的深入；对于清华大学的校史、校志编研以及从中国近代教育史角度对清华大学进行探讨来说，此书亦是具有重要价值的文献，会为清华大学的相关研究提供学术支撑。

同时，此书对清华大学校园重要近代建筑的保护，对清华大学现代校园的规划和建设，也必将发挥积极的作用！

2020年7月9日于学清苑

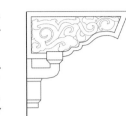

序（三）PREFACE

　　对一所大学的校史研究中，关于校园规划和校园建筑发展历史的梳理与研究，应当是其重要的组成部分。校园的建筑风貌，在一定程度上体现了学校的文化传统与办学特色。研究校园规划与建筑的历史，对于全面了解学校的发展变迁和传统特色，有着十分重要的意义。

　　清华园，被誉为全球最美的大学校园之一。长期以来，清华大学高度重视校园规划和建设，在校史研究中也一直有所体现。从1991年起，校史研究室陆续编纂了"清华大学史料选编"丛书，目前已出版6卷10余册，各卷都有专门章节收录不同时期建筑、校产等方面的重要史料。校史研究室黄延复老师曾对清华大学建筑风物进行过专门研究，编著了《清华园风物志》一书，连出三版，并曾在校报《新清华》上连载。当时我正担任《新清华》的主编，连载的文章引起校内外读者广泛关注。校史研究室特约研究员苗日新老师也曾出版《熙春园·清华园考》一书，考证了熙春园、清华园的历史沿革。由清华大学校史馆编辑、百年校庆后出版的《清华大学图史》中，也刊发了不同时期的校园地图和建筑图片等。此外，建筑学院的一些专家学者和学生曾以清华校园建设为对象，进行过专题研究，发表了相关的论文。在清华大学校史展览中，校园建设一直是各个历史时期不可缺少的展示内容。

　　为迎接清华大学建校110周年，学校决定对校史馆进行全面提升改造。讨论改造方案时，大家一致提出要在新的校史展览中，以现代化、多媒体的新方式，对清华校园变迁进行专门的展示，反映清华园百余年的发展变化。在我担任校史研究室主任和档案馆馆长工作后，校史研究室副主任金富军老师向我介绍，建筑学院刘亦师副教授对清华校园规划的历史进行了较为系统的研究，并邀请他来校史馆作过专题讲座。因此，校史研究室与刘亦师副教授商定，请他以委托课题的形式，在过去已有研究的基础上，承担校史展览中"校园变迁"专题展示的专业指导工作。

如今，焕然一新的"清华大学110年校史展览"在校庆110周年之际如期开幕，其中"校园变迁"电子沙盘成为吸引师生校友和各界观众的一大亮点。与此同时，刘亦师老师多年耕耘的成果《清华大学近代校园规划与建筑》也将出版发行。这都是非常值得庆贺的事情。

翻阅完数百页的书稿，我深感这是一本在众多前辈学人研究的基础上，对清华近代校园规划与建筑的历史进行系统梳理和深入研究的著作。我从未学过规划与建筑专业，更没有对这方面进行过深入研究，因此没有资格对刘亦师老师的著作进行专业的评论。但从一个校史工作者和档案工作者的视角来看，刘亦师老师在研究过程中，发掘和使用了大量的历史档案文献，比如在美国耶鲁大学和洛克菲勒基金会等查阅和复制的史料和图纸，以及当时在海内外发行的各种期刊、报纸上的资料等。这些材料提供了认识了解不同历史时期校园规划和建筑的原始素材，也为读者还原出一幅幅当年的生动画面。这就使得本书在具备较强专业性的同时，又具有了很强的史料性和可读性。此外，刘亦师老师还将他的研究团队8年来在清华校内实测或根据实测复原的十余幢建筑的几百张图纸集中起来，真实地反映了清华校园近代建筑的建设情况，这也为其他研究者进行相关研究提供了很大的便利。

我相信，《清华大学近代校园规划与建筑》既是一部建筑史领域研究的新作品，也一定是大学校史研究中独具特色的新成果。它将有益于清华大学校史研究的不断深化，也将对近代大学校园研究的广泛开展起到引领作用。希望刘亦师老师进一步推进这一研究课题，比如将研究范围扩及现代部分，使我们对清华校园建设历史的认识更加完整和深入。衷心期待和祝愿刘亦师老师取得更新的研究成果。

范宝龙

2021年4月于清华荷清苑

下编 | 校园近代重要建筑测绘图集 / 375

中编 有关近代清华校园规划、建设之海外资料汇纂

清華学堂
80.5'/.

高崇生贈之
2020年4.25. 时年83岁.

第①章　清华创建及其早期建设的相关档案资料

1.1 Arthur H. Smith. China and America To-day: A Study of Conditions and Relations. New York: F.H. Revell Company, 1907. 明恩溥著《今日之中国和美国》封面与目录页影印，及有关退还超索庚子赔款用于在中国兴办教育论述之节录（pp.210-224）

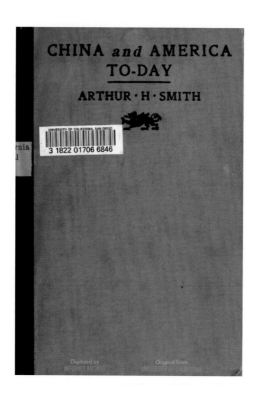

编者注：

　　1906年3月6日，明恩溥到白宫拜谒西奥多·罗斯福总统，并建议将中国清政府的庚子赔款超索部分退还，专门开办和补贴在中国的学校。1907年明恩溥出版《今日之中国和美国》一书，分析了当时美国工业产品在中国遭到抵制的原因（"排华法案"）与迫切改变美国形象的需要，论述应该多让一些中国知识分子来美国留学，"以免日本和欧洲国家占领先机"，敦促美国政府采取措施，通过吸引中国留学生来打造一批从知识和精神上为美国支配中国的新的领袖。此议深得罗斯福总统的赞同。后通过驻美大使梁诚的积极活动，最终促成退还部分庚子赔款用于兴建清华学堂。下文为该书第八章《美国在中国的机遇及责任》相关部分之节录。

Washington was strongest at just those points where the Oriental is weakest, and the Oriental recognises that fact at a glance. Views like this of our greatest men have been fermenting in the minds of Japanese, Chinese, and Koreans for a long time. A recent paper mentions that in a popular vote of the scholars in a Japanese school as to their favourite hero, Washington received a few more than sixty votes, and Lincoln almost as many, while the great Japanese war-Admiral Togo did not rise to forty ballots! The new China is to be officered and piloted by new men. All the impulses which have brought about the renaissance of Japan, and those which are yet to do the same for China, are impulses from without, and not from within. China is now turning to other nations for guidance and for help in educating her young men. It is but a few years since she sent her first students to Japan; but during the past two years the hegira of Chinese youth to the Island Empire is probably without historic parallel. Japan no doubt expects to pay back her age-long debt to China by exerting there a dominating influence as a step toward her anticipated hegemony of Asia. Even in the stress and strain of the Russian war she set apart numbers of her army and naval officers, as well as civilians, for the instruction of the Chinese students. Friction arose between these scholars and the Chinese Minister to Japan, who was a Manchu, and the rising spirit of Chinese patriotism renders the whole Manchu race especially obnoxious to young China.

Freed from the wonted restrictions of home and of the Confucian training in which they were born, the Chinese students resented Japanese control, and several thousands of them returned to China, sometimes abusing the opportunity afforded them by their travel to write and to speak in a way to excite anti-dynastic feeling, already far too strong for safety. At the present time it is estimated that there are about 15,000 Chinese in Japan, nearly all in Tokyo, representing almost every Province of the Empire.

The public vice which is so conspicuous a feature of the capital of Japan has never been known in China. It has demoralised very many of the Chinese students. Some of them have even thrown off the trammels of Confucianism, and are openly adopting an attitude of contempt for the ancient Sages. One such remarked to a foreigner: "It was old K'ung [Confucius] who ruined China! " The only creed: (aside from Christianity) available to replace the teaching of China's hereditary masters, is Epicureanism, which has hitherto never been in China a recognised cult. For China itself such a state of things is an alarming symptom and a menace to her relations not with Japan only, but with other nations as well. The Court in Peking has with excellent reason long looked with disfavour upon this unbalanced influence of Japan, fearing especially its anti-Manchu tendencies, but the Government is apparently quite helpless to stem the swelling tide.

Under circumstances such as these, is it not the part of wisdom for us to put forth our best exertions to deflect this stream of students to our own shores, not for the good of China alone, but also for the welfare of America and of the world? Our former ill-treatment of those who in the past have desired to come is the greater reason for the adoption of this policy upon a large scale. A Chinese gentleman once said to the writer that he would much have preferred to have his son study in the United States, but having vainly spent six months of time and much money in the effort to get him into the country, he had sent him to more hospitable England. The unmitigated folly of our course of action is now becoming manifest even to ourselves. It only requires an educated public opinion not merely to remove restrictions, but to extend a welcome to Chinese students to our educational institutions all over the land.

It is interesting to know that the international committee of the Y. M. C. A., with characteristic foresight and energy, has undertaken a work of broad range among these students, from which large results are sure to flow.

As an excellent specimen of various papers which have been indited upon this subject of national and international importance, the reader may be glad to have the opportunity of perusing one written early in 1906 by a distinguished American educator, submitted to the President of the United States, and privately circulated.

Memorandum concerning the sending of an Educational Commission to China, by Edmund J. James, President of the University of Illinois.

The recent developments in the Orient have made it apparent that China and the United States are destined to come into ever more intimate relations, social, intellectual, and commercial. The Chinese will come to this country for the purpose of studying our institutions and our industry. A striking evidence of this fact is afforded by the work of the Chinese Commission now or lately in the United States. Our own people will go to China for the purpose of studying Chinese institutions and industry. Anything which will stimulate this mutual intercourse and increase mutual knowledge must redound to the benefit of both nations.

A great service would be done to both countries if the Government of the United States would at the present juncture send an educational commission to China, whose chief function should be to visit the Imperial Government, and with its consent each of the provincial governments of the Empire, for the purpose of extending through the authority of these Provinces to the young Chinese who may go abroad to study, a formal invitation on the part of our American institutions of learning to avail themselves of the facilities of such institutions. The appointment of such a commission would draw still closer the bonds which unite these two great nations in sympathy and friendship.

China is upon the verge of a revolution. It will not, of course, be as rapid as was the revolution in Japan, if for no other reason, because of the vast numbers of the nation and the enormous extent of its territory. But it is not believed that this revolution which has already begun can ever again suffer more than a temporary backset and reaction.

Every great nation in the world will inevitably be drawn into more or less intimate relations with this gigantic development. It is for them to determine, each for itself, what these relations shall be,—whether those of amity and friendship and kindness, or those of brute force and 'the mailed fist.' The United States ought not to hesitate as to its choice in this matter. The nation which succeeds in educating the young Chinese of the present generation will be the nation which for a given expenditure of effort will reap the largest possible returns in moral, intellectual, and commercial influence. If the United States had succeeded thirty- five years ago, as it looked at one time as if it might, in turning the current of Chinese students to this country, and had succeeded in keeping that current large, we should to-day be controlling the development of China in that most satisfactory and subtle of all ways,—through the intellectual and spiritual domination of its leaders.

China has already sent hundreds, indeed thousands, of its young men into foreign countries to study. It is said that there are more than five thousand Chinese studying in Japan, while there are many hundreds in Europe—three hundred in the little state of Belgium alone. This means that when these Chinese return from Europe they will advise China to imitate Europe rather than America,—England, France, and Germany, instead of the United States. It means that they will recommend English and French and German teachers and engineers for employment in China in positions of trust and responsibility rather than American. It means that English, French, and German goods will be bought instead of American, and that industrial concessions of all kinds will be made to Europe instcad of to the United States. Now it is natural, of course, that the vast majority of Chinese youth should go to Japan to study rather than to European countries or the United States, owing to its proximity, to racial affinity, and to the smaller cost of travel and living. On the other hand, the Chinese are in many points jealous of the Japanese, and, other things being equal, would often prefer to send their young people to other countries. Among all these countries the United States would be the most natural one to choose, if it had not been for our anti Chinese legislation, and still more for the unfriendly spirit in which we have administered this legislation, for the Chinese Government at any rate never really objected to our legislation directed toward preventing the immigration of Chinese labourers, but only to the manner in which we passed such laws and the way in which we administered them.

We are the natural friends of the Chinese. We have been their real political friends. We have stood between the Chinese Empire and dismemberment; we have come more nearly giving them the square deal in all our relations in the East than any other nation. They are consequently less suspicious of us, as far as our politics are concerned, than of any other people. Their justly sore feeling over our treatment of Chinese gentlemen in our custom-houses will yield quickly to fair and decent conduct on our part. It is believed that by a very small effort the good-will of the Chinese may now be won over in a large and satisfactory way. We may not admit the Chinese labourer, but we can treat the Chinese student decently, and extend to him the facilities of our institutions of learning. Our colleges and universities are to-day far better adapted for giving the average Chinese student what he desires in the way of European civilisation, than the schools and colleges of any European country. We need but to bring these facts to their attention in order to secure their attendance here, with all the beneficial results which would flow from such an opportunity to influence the entire current of their thought and feelings.

If a commissioner with one or two assistants were sent to China representing the American Government in a formal way in the field of education, and should extend to the Chinese people, through the Government at Peking and through the provincial governments, a cordial invitation from the United States, and from the institutions of higher learning in the United States to avail themselves of these advantages exactly as they would if they were their own institutions, it is apparent that a great impression might be produced upon the Chinese people. The Chinese appreciate, as well as we, the compliment implied in sending a formal commission of this sort to another country. It is a recognition such as any country might be proud of, and the Chinese are a singularly proud and sensitive people in everything that concerns their own dignity.

Such a commission going to each of the Provinces would have an opportunity to give the Chinese Government much information about the United States and its educational institutions; and as the inquiries of such governments would not be limited, of course, to education and educational institutions, so the information spread abroad throughout China would not relate simply to educational matters, but to industrial and commercial as well. It would be possible, through this method of coming in contact with influential Chinese, to recommend directly to them in response to their requests, American teachers, engineers, and other people whose services they might like to obtain. I mention this point especially because I know that the leading Chinese statesmen are anxious to get just the right kind of men from America and Europe as assistants in all sorts of business and governmental enterprises, having had myself, during the last year, four inquiries from different Chinese governments for young men who would be willing to spend five or six years in the Chinese public service in responsible and influential positions.

In a word, the visit of such a commission would exert a manifold and far-reaching influence, exceeding greatly in value any possible cost of the enterprise. It would have results in many unexpected directions outrunning all our present anticipations, and showing new and surprising possibilities of usefulness in the fields of education, business, and statesmanship. The extension of such moral influence as this would, even in a purely material sense, mean a larger return for a given outlay than could be obtained in any other manner. Trade follows moral and spiritual domination far more inevitably than it follows the flag.

If this wise and statesmanlike proposal of President James has not thus far resulted in action, it must be due to inertia on the American side of the Pacific, and not to the probability of opposition in China. The matter should by no means be suffered to rest until something is accomplished. As soon as the importance of welcoming Chinese students to America under existing conditions was brought to the attention of the Overseers of Harvard University they at once voted to extend through the Chinese Imperial Commissioners then in the country to the Chinese Government an invitation for ten Chinese students to attend that institution. The same step was soon after taken by Yale University; and on behalf of Chinese women, to whom three scholarships were offered, by the trustees of Wellesley College, an institution which the Imperial Commissioners visited at the special command of the Empress Dowager, who had become greatly interested in what she had heard of American education for women. When the immense influence which has been exerted in Japan by the comparatively small number of her daughters who have been educated in America is remembered, the importance of this small beginning for her sister empire may be faintly forecast. But all these movements, and many others like them, are utterly inadequate to cope with the present opportunity and emergency. It is well known that after all public and private claims arising from the Boxer disturbances of 1900 have been satisfied, there will eventually remain in the hands of the American Government a sum of perhaps $20,000,000 (gold), a part of the indemnity of 450,000,000 taels of silver arranged by all the Powers in the peace protocol of 1901.

Upon two previous occasions, once with China and once with Japan, the American Government has established a precedent (so far as appears unique among nations) of returning unexpended balances of indemnities.

The suggestion is often made that this money should be treated in the same way as its predecessors. Many Americans, however, intimately acquainted with China's condition, are profoundly convinced that if such a sum were handed back to China without conditions, it would at once be applied to purposes which would distinctly endanger the peace of the world, and make more difficult and insoluble a problem already taxing the ingenuity of the Occident to deal with. It is of course easy to say that if this money is ours we should keep it; if it belongs to China, to China it should go. But is it not perfectly reasonable to claim, as many do claim, that this sum represents not merely replacing value of fixed capital destroyed, but that it should be considered as a *punitive* indemnity for a great criminal act of Chinese officials, and in reality of the Chinese Government, against the American Government in the person of its Legation? *We are under as much obligation to see that this money is so used as to make similar outbreaks in future more difficult as we are to return it at all.* Ought we not, acting upon the wise suggestion of President James, upon two previous occasions, once with China and once with Japan, the American Government has established a precedent (so far as appears unique among nations) of returning unexpended balances of indemnities.

The suggestion is often made that this money should be treated in the same way as its predecessors. Many Americans, however, intimately acquainted with China's condition, are profoundly convinced that if such a sum were handed back to China without conditions, it would at once be applied to purposes which would distinctly endanger the peace of the world, and make more difficult and insoluble a problem already taxing the ingenuity of the Occident to deal with. It is of course easy to say that if this money is ours we should keep it; if it belongs to China, it should go. But is it not perfectly reasonable to claim, as many do claim, that this sum represents not merely replacing value of fixed capital destroyed, but that it should be considered as a punitive indemnity for a great criminal act of Chinese officials, and in reality of the Chinese Government, against the American Government in the person of its Legation? We are under as much obligation to see that this money is so used as to make similar outbreaks in future more difficult as we are to return it at all. Ought we not, acting upon the wise suggestion of President James, to propose to the Chinese Government to use this sum (which will fall due annually for a generation to come), or at least a part of it, in educating Chinese students in the United States?

During the preceding hundred years there has been a mighty collision between the civilisation of the West and the civilisation of the East. We have had commerce, followed by war, and war succeeded by diplomacy. The Western nations have established Legations at Peking, and consulates at the ports, while the Chinese have been persuaded to establish Legations in Western lands and consulates in foreign ports to look after the interests of

Chinese subjects. Thus times have vastly changed since 1858, when one of the Chinese plenipotentiaries, in response to a suggestion that his Government should appoint consuls abroad to look after the interests of the Emperor's subjects settled in foreign lands, said: "When the Emperor rules over so many millions, what does he care for a few waifs that have drifted away to a foreign land?" It was stated that some of those in the United States were growing rich from the gold mines and that they might be worth looking after on that account. "The Emperor's wealth," he replied, "is beyond computation; why should he care for those of his subjects who have left their home, or for the sands they have scooped together?" It is not so long ago that diplomacy was counted upon to settle all the issues between the East and the West as soon as China should have been beguiled into the "sisterhood of nations"; but the ultimate outcome of this process, deftly mingled with perpetual Western aggression and outrage, was the Boxer movement, and the siege of the Legations in Peking. The climax of this diplomacy, was exhibited in 1901, when the Powers found it difficult to agree upon anything; and when at last they did agree, the net result of their elaborate specifications (except only the indemnity) was, after a few years had elapsed, as nearly as possible nothing at all. The world is slowly and with difficulty becoming disabused of its obsession that commerce is in itself an elevating agency. On the contrary, when unregulated by conscience, it furnishes fire-water and fire-arms to savages, engages in the slave trade and the coolie traffic, and in the "red rubber" atrocities on the Congo, at which the civilised world is aghast, "Commerce, like the rainbow, bends toward the pot of gold." Neither is moral renovation to be expected from such industrial revolution as is taking place in Japan, and will within a few decades wholly transform China. Listen to "The Bitter Cry of the Children," and see how even in our own Christian land we are barely able (if indeed we are as yet able) to check the downward tendencies of unregulated industrialism which wrecks the lives of women and destroys more children and youth than an army of Minotaurs. A critic of our civilisation, writing under the guise of a Chinese, bitterly complains of the persistent attempts of the Occident to substitute for the old Chinese "moral order" Western "economic chaos." There is much to be said for this contention, for it is as good as certain that when China shall have been quite drawn into the modern commercial and industrial maelstrom, while she will be financially richer, she will be morally poorer.

Much light has come to China from many sources, unwilling as she has been to receive it. The foreign-controlled Imperial Maritime Customs has been a standing object lesson in Occidental methods of honestly administering great public trusts, but the Chinese would be glad to be rid of the foreign element, when, without higher motives than rule at present, "Chaos and Old Night" would soon set in again. An able and intelligent foreign press, the large body of foreign residents in Chinese ports, and especially the Chinese students who have been educated abroad, have all had an important though widely different part in the gradual leavening of a small portion of China. Yet these have only touched the fringes of the Empire, or the banks of its chief river. But there has been in China another force incomparably more influential than all of these combined. It is the originally small, but always steadily growing body of Protestant missionaries, beginning a century ago with a single Englishman, and now numbering more than 3,800, from six different countries of Europe, and from all quarters of the British Empire, the United States being (at the end of 1906) represented by 1,562 persons. These men and women instead of living beside the Chinese, as do residents of the ports, live among them in cities, towns, and hamlets in every Province of the Empire, speaking every dialect, going everywhere, inquiring into everything, constantly meeting and mingling with all classes of Chinese, from officials in their yamens to coolies and beggars on the street. Much knowledge of China has, indeed, come to the outside world from other than missionary sources; but for many decades nearly all trustworthy information of outside countries which filtered into the minds of the bulk of the Chinese people came through missionary channels.

Upon the spiritual aspect of their work (the most important because fundamental) it is aside from our purpose to dwell further than to remark that universal experience has shown that the introduction of Christianity into any land is the most powerful moral force in human history. (pp.210-224)

（邓可、王睿智转录）

1.2　Richard Arthur Bolt. "The Tsing Hua College, Peking." The Far-Eastern Review, 1914, 10（09）：363–369. 全文影印

February, 1914　　THE FAR EASTERN REVIEW　　363

THE TSING HUA COLLEGE, PEKING

With special reference to the Bureau of Educational Mission to the United States of America.

[BY RICHARD ARTHUR BOLT, A.B., M.D., PHYSICIAN OF TSING HUA COLLEGE.]

In the summer of 1881 the first Chinese Educational Mission to the United States of America, under Yung Wing, came to its untimely end. The fate of the Mission was sealed in 1876, when Yung Wing accepted the post of Associate Minister to the U.S.A., leaving his pet scheme open to the machinations of the Commissioner of the Educational Mission, Wu Tzu-ih. Mr. Tong Kaison, one of the original Yung Wing students, has aptly summarized the different accusations which were brought against the students as follows: "The boys had, by imbibing the spirit of American institutions, lost all sense of duty and obligation towards their Sovereign and their country; they had adopted American customs, sentiments, habits of life and thought; they entertained, by reason of their republican education and training, the most radical and subversive opinions and the most partisan feelings; consequently, to patronize such men was to sow the seed of future discord and rebellion; they had all rejected the doctrines of Confucius, and some of them had embraced the tenets of an alien faith; they were impatient of authority and cherished contemptuous opinions of their government; and, lastly many of them had shown a desire to become naturalized and to marry and settle down as citizens of the United States."

Whether true or not, it is easy to understand how Chinese official circles of thirty years ago would accept such accusations and make capital of them. The bringing of the boys back to China at this juncture, and the breakup of the Mission, produced the very results which the Commissioner and his colleagues had hoped to avoid. It is interesting, therefore, to study the causes which have led to the formation of another Educational Mission and the safeguards placed about it in order to escape the pitfalls of the former one.

The relations between the United States and China have, on the whole, been most friendly, despite the fact that a considerable labor element on the Pacific slope forced the Anti-alien Laws, and have succeeded in keeping them in force ever since.

As the people of the United States have come into more intimate contact with the better merchant class and students of China a high regard for their integrity and ability has developed. The United States Government itself has shown its friendly feelings by numerous exchanges of diplomatic courtesies, by its decided stand for the "Open Door" in the Far East, by its insistence upon the "Integrity of the Chinese Nation" and by its early recognition of the Chinese Republic. The most notable act of friendship of recent years was the remission of a portion of the Boxer Indemnity Fund. According to the Protocol of September 9, 1901, the United States was to receive as indemnity from China the sum of $24,400,000 gold. After making arrangements for the payment of all just claims, and adjusting some differences which were not forseen at the time the original indemnity was decided, Congress of the United States passed a bill authorizing the President to arrange to remit to China the remainder of the indemnity. The original indemnity bond was modified from $24,400,000 gold to 13,655,492.29 gold with interest

at 4% per annum. Of this latter amount $2,000,000 gold were held pending hearings on private claims within one year. In round numbers, then, this gave the Chinese Government $11,000,000 gold, to be remitted from the total indemnity—collected quarterly—during thirty years.

An envoy was dispatched from China to the United States to thank the President for this "act of friendship." In a note from Prince Ch'ing to Minister Rockhill, dated July 14, 1908, the purpose for which the money was to be expended is stated as follows:

"The Imperial Governement wishing to give expression to the high value it places on the friendship of the United States finds in its present action a favorable opportunity for doing so. Mindful of the desire recently expressed by the President of the United States to promote the coming of Chinese students to the United States to take courses in the schools and higher educational institutions of the country, and convinced by the happy results of past experience of the great value to China of education in American schools, the Imperial Government has the honor to state that it is its intention to send henceforth yearly to the United States a considerable number of students there to receive their education. The Board of Foreign Affairs will confer with the American Minister at Peking concerning the elaboration of plans for the carrying out of the intention of the Imperial Government."

The plans finally evolved resulted in the formation of the "Bureau of Educational Mission to the United States of America." This Bureau was under the joint control of the old Wai Wu Pu (Board of Foreign Affairs) and the Hsüe Pu (Board of Education.) All affairs pertaining to the selection and sending of Chinese Government students to the United States, and their entire support, was to be under the jurisdiction of this Bureau. The Wai Wu Pu thereupon appointed H. E. Chow Tze-chi as Director, and H. E. Tong Kaison Co-Director, to confer with Mr. Fan Yuen-lien as Co-Director from the Hsüe Pu. Dr. C. D. Tenney, then Chinese Secretary of the American Legation at Peking, was delegated by the American Minister to represent him in an advisory capacity to the Bureau. Comprehensive "Rules for the Selection and Sending of Students to America" were drawn up by the Bureau and sanctioned by an Imperial Decree of July 10, 1909.

It was originally deemed advisable to select students on competitive examination from any school in the different Provinces. In this case the Bureau simply acted as an examining body with considerable discretionary powers. According to this plan the first competitive examination was held in Peking at the Yamen of the Board of Education in September, 1909. Over 600 students came up to the Capital from practically every Province to try for the coveted honor of being chosen to go to the United States. After a preliminary test in Chinese and English Literature and Composition, the number of students considered eligible to proceed with the final examination was reduced to 80. These were further examined in Algebra, Plain and Solid

Main Entrance to Tsing Hua College.　Photo by R. A. Bolt.

清华大学近代校园规划与建筑

Geometry, Trigonometry, German and French, Latin, Physics, American and English History, History of Greece and Rome, and Chemistry. A final selection of 47 students was made. It was discovered from this examination that the greatest difficulty was shown in the proper use of the English language. After a medical examination by the Physician to the American Legation Guard, and suitably equipping themselves at Shanghai, they sailed for America on October 12, 1909, under the care of H. E. Tong Kaison. The students were distributed among a number of the leading Academies and Colleges of the United States.

A second examination was held in the Law College, Peking, from July 21st to July 29th 1910. This time 400 students from all over the Empire participated, the Provinces of Kiangsi and Kwangtung being most largely represented. The list of examination questions was practically the same as at the first examination. After the preliminary test 172 boys were allowed to proceed with the finals. Of this number 70 were chosen to go to America. It was stated that a larger number of these students were able to enter American colleges without additional preparatory work, but it was forcibly brought home to the examiners that more thorough preparatory work would have to be done in China before the students could do full justice to their advanced studies in the United States. The original plan contemplated that 100 students would be sent each year for the first four years and 50 per year for a period of 29 years. It was found impossible to select so large a number at first well enough prepared in English to enter the higher educational institutions. Then came the Revolution which crippled the finances so much that the original plan had to be held in abeyance.

In order that the students should be better equipped to enter American universities without further loss of time in preparatory schools in the United States it was finally decided to establish under the Bureau a school in which all future Government students, to be sent to the States on the Indemnity Fund, should be trained. Strong arguments were advanced for the establishment of such a school in the vicinity of Peking. It was urged that while the munificent sum of $150,000 gold was available annually it would be a good investment to build a model institution capable of developing into the best in China. For such an institution the Chinese Government, at the expiration of the "indemnity fund," would certainly find other means of support. In the outline for this new school it was stated that "American teachers, both for the advanced and the elementary classes shall be engaged, and the methods shall be those of American schools, so that the students by their familiarity with American methods may suffer no inconvenience in entering American schools. This school shall be solely for the purpose of giving temporary instruction to the students chosen from the different provinces in order to test their character and ability."

The manner of selecting the American teachers for this school is worthy of note, as showing the confidence the Directors

placed in a man representing avowed Christian ideals and principles. After most careful consideration the Directors, acting upon the suggestion of Mr. Tong Kaison, decided to delegate to Mr. John R. Mott the securing of suitable teachers for the new "Indemnity School." Knowing that Mr. Mott was in close touch with organisations in America which could indicate well qualified teachers, and having the necessary office machinery at his disposal, proved the farsightedness of the Directors in placing in his hands the selection of the teachers. After considerable deliberation an American faculty, consisting of nine ladies and eight men, was selected. This number included a Physician and a Physical Director for the College. These teachers, with the exception of the Physical Director, sailed from San Francisco on the s.s. "Tenyo Maru," January 18, 1911. They were officially received and royally entertained at Honolulu, throughout Japan and in Shanghai, arriving at Peking February 21st. Here they were detained some six weeks awaiting the completion of the new buildings at Tsing Hua Yuen. Since then the Physical Director has arrived; five men and one lady from the U. S. A. have been added to the teaching staff; while three ladies and two men have left the school.

The preparatory school, now known as Tsing Hua College, is situated about four miles outside the northwest city gate (Hsichihmen) of Peking on the line of the Peking-Kalgan Railroad. It is beautifully environed amidst rural surroundings in sight of the new Summer Palace, and almost adjacent to the magnificent ruins of the Old Summer Palace (Yuen Ming Yuen). It has been stated that this site was originally occupied as the country residence of the notorious Prince Tuan of "Boxer" fame, and that after his downfall the old Empress-Dowager confiscated his property and turned it over to the Wai Wu Pu, for the use to which it is now being put. Others question this and affirm that the Park was originally the property of Prince Tun, the fifth son of Tao Kuang. In fact the local name for the place has long been Wu Ye Yuan, which means the "Park of the Fifth Prince."

Hospital and Dispensary Building, Tsing Hua College

When the Park was handed over to the Bureau of Educational Mission to the United States by Imperial Edict issued at the instance of the Prince Regent in 1911 the Tsing Hua College was at once laid under the direction of the then Vice-Minister of Foreign Affairs, Mr. Chow Tze-chi, who was once Charge d'Affaires in Washington. The late Mr. Tong Kaison, who was eventually President of the College, and Mr. Y. L. Fang, afterwards Minister of Education, co-operated with Mr. Chow Tze-chi in this work, and as a result of their efforts a number of buildings were erected in 1910 and 1911 sufficient to accommodate 600 students, the contract being carried out by Mr. Emil S. Fischer, formerly of New York and now head of Fischer and Co. of Tientsin and Peking. The cost of the work was over half a million Taels, or about $350,000 gold dollars.

Photo by C. B. Malone.

Regular Daily Exercise at Tsing Hua College. Main High School Building in Background, Assembly Hall to right in picture.

A well built stone wall completely surrounds the 60 odd acres which make up the College Compound, which will be improved as soon as necessary funds are available. The College has recently acquired 60 more acres to the west. The Compound is attractively landscaped with artificial tree-topped hills, lotus ponds, and a small stream which runs directly through it. A considerable number of trees have been planted, a greensward laid out, and many flowers placed on the hillsides. The buildings of foreign, semi-foreign and Chinese architecture are conveniently grouped in t h e Campus. T h e offices of the Bureau of Educational Mission to U. S. A. were originally housed in the reconstructed Yamen, strictly Chinese in every detail, picturesquely surrounded by low hills and opening at the rear on to a marble porch which overlooks a lotus pond. There are substantial residences for the President of the College, the Dean, and a small Compound for the Chinese teachers, now numbering sixteen. The American teachers have had nine double bungalows allotted for their use, one of which has been comfortably fitted up as a Club House.

The High School department occupies a large two-wing double story grey brick building. This contains the educational office, offices for the President and Dean, an attractive Faculty room, the Chemical and Physical Laboratories, and various class rooms. Behind this building is an Assembly Hall which will accommodate about 400. Adjoining this at the rear are the High School study-halls, dormitories and dining room—all one story brick buildings connected by covered corridors. At the west end of the dining room a workshop for Manual Training has been fitted up, the boys of the third and fourth year Middle School and first year High being required to take this subject.

The Middle School is housed in a group of buildings separated from the High School by a small stream bridged in three places. It comprises classrooms, Proctors' and Secretaries' Offices, the beginning of a Museum collection, study-halls, dormitories, dining-room, kitchen, and out-buildings. These are all connected by covered passage-ways. Suitable bath and toilet arrangements have also been provided. Just west of the Middle School is a well laid out athletic field, tennis courts, basket ball grounds, football and baseball field, running track and space for field events, and an archery area. Athletics have now become compulsory for every student. He must take regular calesthenic exercises each morning as well as spend one hour daily in some out-of-door sport. Careful account is kept

Entrance to "Yamen" of Bureau of Educational Mission to U. S. A. and in background the Compound for the Chinese teachers, Tsing Hua College

of attendance a n d credit allowed for the athletic work. Beyond the athletic field is a Hospital and Dispensary building.

Every student is given a complete medical inspection and physical examination upon entrance to the College, and further examinations are required throughout the year. Special attention is given to Preventive Medicine and Hygiene. Plans have been considered for a Gymnasium, a Library and a Science Building, and these will probably be erected in due time. The entire plant is well lighted with electricity, and telephone connections have been established with Peking.

After considerable delay and uncertainty Tsing Hua College was formally opened with Chinese ceremonies on April 1st, 1911. The High School Department was begun with 128 advanced students who had failed in their previous examinations to be selected to go to the U. S. A. 307 younger students were selected by the Provincial examiners, ostensibly upon competitive examination to enter the Middle School. Practically every Province in China, besides Manchuria and Mongolia, was represented in the School. After a complete physical examination the boys were allowed to take up residence. The teaching of the younger students was at first almost completely in the hands of Chinese teachers who then numbered about twenty. The American teachers were assigned, rather arbitrarily, to teach the older boys, from among whom it was understood that fifty would be selected that year to proceed to the United States.

The first Dean was a young man recently graduated from an American College of high standing. His record as a student had been exceptionally good, especially in mathematics, but he lacked executive ability and had no real experience in school organization. It was evident from the first that this inexperienced dean was not going to coöperate with the American Faculty, nor plan the College work in such a manner as to inspire their confidence. In the outline for the first curriculum mathematics overbalanced everything else; in fact it crowded the study of English almost to second place. A foolish attempt was made to introduce a course in Esperanto; in fact the Dean at first succeeded in dubbing the new institution, "La Kolegio de Juna Hinujo." After trying in all tactful and courteous ways to obtain from the Dean some idea of what he was attempting to accomplish the American teachers felt it their duty to offer the following suggestions:—

"1st. That the curriculum be adapted to meet the requirements for

Photo by C. B. Malone
Group of High School Students at Tsing Hua College.

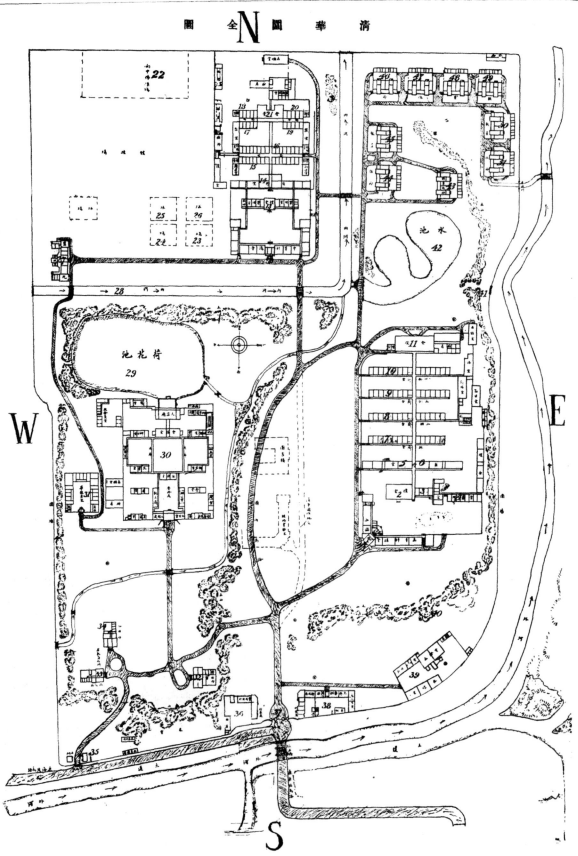

Plan of Tsing Hua College (See Key on page 369.)

entrance to American Colleges." (Up to this time there had been but little indication that this was being done).

"2nd. That the students be informed as soon as possible in regard to the above requirements." (Heretofore there had been great doubt and uncertainty).

"3rd. That immediate steps be taken to thoroughly prepare as many as possible of the advanced students for entrance to American Colleges next fall." (The policy of the Dean had seemed to thwart this very purpose).

"4th. That the teachers be kept in close touch with the educational plans and workings of the School." (Up to that time no faculty meetings had been held. The Regulations of the school were made out in Chinese largely by the Dean himself, and it was only after considerable pressure that translations were given to the American teachers. The Dean undertook to organize all the teaching work without consultation with the teachers, and assigned courses which were very illy balanced.)

Following closely upon these recommendations the Dean "resigned of his own accord." The Directors then realized that the welfare of the new school depended upon the selection of a capable successor who could command the respect of the American faculty. Great tact and wisdom was shown in their next move. Mr. Chang Pe-ling of Tientsin, a man of rare personal traits, an avowed Christian gentleman with marked ability for educational work, was persuaded to give a portion of his time to help reorganize the educational scheme. He was formally appointed Dean by the Directors, and immediately began by making a thorough survey of the whole College. He met all the teachers personally, and secured from them suggestions as to necessary steps to be taken to place the work upon a sound educational basis. In many ways pains were taken to inspire confidence that the new regimen would conserve the purpose for which the school was established.

The American teachers were requested to present recommendations as to the best method of selecting students to go to the United States. They replied, "We have come to the con-

The Central Court of "Yamen" of Bureau of Educational Mission to U. S. A., Tsing Hua College.

clusion that the most important thing now before us is the formulation of proper and just plans for determining upon the students to be sent to the United States this summer. We would suggest that in considering the qualifications of students for entrance into American colleges the following features are of the greatest importance, and will result in the best selection, namely, 1. Past records of scholarship. 2. Present class standing, and 3. A fair general examination. After due consideration the Directors announced that the advanced students would be chosen on the basis of 50% being allowed for their class standings and 50% on a final examination in college entrance requirements. The examinations took place at Tsing Hua College from June 23rd to 29th inclusive. The examination questions were set by the Faculty, and the examination conducted by Bureau officials in a very thorough manner. The questions, on the whole, were about the same as those set for entrance into the best American colleges. The papers were graded by Chinese scholars, most of whom were returned students from America who held important positions of trust in Peking.

From the 134 advanced students taking this examination, 63 were chosen to go to the Unites States. With few exceptions these were the best prepared students in the College. The American teachers were satisfied that every effort was made to pick students best qualified to go, although it was felt considerable improvement could have been made in the preparation of some of them had they remained longer in the school. Three of these boys were detained on account of their physical condition; each student being required to pass a rigid physical examination before being allowed to proceed to the States. The successful ones sailed from Shanghai on the s.s. "Persia," August 5th, 1911. They were placed in the leading colleges of the Middle and Eastern States.

At the opening of the College Mr. Tong Kaison was away on a Mission to America and Europe and was compelled to be away during the uncertain months of preliminary organization. Mr. Chow Tze-chi was commissioned to go to the Coronation in England at a most critical time in the initial period of organization.

Photo by R. A. Bolt.

Middle School Group of Buildings, Tsing Hua College.

Under the able leadership of Dr. W. W. Yen, then Acting-Director, and Mr. Chang Po-ling as Dean the policy of the College began to assume definite shape and gave promise of conserving the ideals for which it had been founded. Shortly after Mr. Tong Kaison and Mr. Chow Tze-chi returned a reorganization of the Bureau of Educational Mission to the

U.S.A. took place. This was further effected after the Revolution. Mr. Chow Tze-chi was appointed to the Governorship of Shantung; Mr. Fan Yuen-lien became Minister of Education; Mr. Chang Po-ling withdrew to concentrate upon his work as principal of Nan Kai Middle School; Mr. Tong Kaison was elevated to Director and President of Tsing Hua College, and in April, 1912, Mr. Y. T. Tsur received the appointment of Vice-Director and Dean. Mr. Tsur came to his new work with high ideals and a determination to place the educational scheme of the College on a firm basis. Mr. Tong Kaison worked heroically for the school during the trying period of the Revolution and afterwards in getting sufficient funds to keep it going. Even when confined to his bed with a fatal heart disease his interest in the College never fagged, and until the last he was planning for its welfare. After his death in August, 1913, Mr. Y. T. Tsur, then in the United States with the 20 students he had taken to place in the colleges, was appointed President of the College, and recently Mr. G. T. Chao, a graduate of Wisconsin University, has been appointed Dean. Everything now points the way for a bright future for Tsing Hua College. The new officials give promise of conserving what was best in the fundamental policy of the institution and of adding necessary new blood to meet new conditions.

Tsing Hua College as now reorganized consists of two departments—a High School and a Middle School. The Middle School receives boys from 11 to 13 years who have been selected by the Provincial Educational authorities. In this school they are early led into studies taught in the English language, and by the time their four years are finished they are able to take practically all work in English in the High School. The High School gives a College preparatory course leading either into scientific or general literary work. It aims to so thoroughly prepare the students that they may enter the American colleges and universities without conditions, or without any further preparatory studies. Some of the recent graduates have been able to enter the Sophomore year in several of the American colleges. Practically the entire last batch of 20 students were able to enter without condition the best institutions of the Middle and Eastern States. In the United States they are allowed to pick the college suited to their purpose and are given from four to six years to complete their course. All the expenses while in the Tsing Hua College, and, in the United States, are met from the returned "Indemnity Fund." In the United States a Director and a Secretary for keeping in intimate touch with all the Chinese students are maintained at Washington. At present Mr. T. T. Wong is Director and Mr. Chung Wen Ough Secretary for the Chinese students in the United States.

This educational scheme, and similar ones, force us to the question, "What will be the outcome of modern scientific education in China, and especially the influence of American education upon the new generation of students?" Confused answers are already being given to this question. A vigorous writer in the *Edinburgh Review*, just before the Revolution, surmised that, "America was desirous, for motives that can be guessed, to attract Chinese students, and the Chinese Government had a particular reason for sending them to a country that was going so fast ahead in the van of progress. The result, however, has been to stimulate the same kind of unrest that is troubling India............. Thus the constitution of the Celestial Empire, the oldest in the world, is disordered by a few doses of the new learning." After a few more well-turned

sentences this same writer bubbles over with, "We cannot wonder that the venerable despotism of China should have been painfully discomposed by the rapid absorption of American stimulants." This critic certainly had some ground for his fears, but in all fairness he should have differentiated those students returning to China from American colleges and those educated elsewhere. The returned students from the Yung Wing Educational Mission brought back to China new and vigorous life and introduced here many of the modern reforms. We need not plead for such men as Y. C. Tong, Jeme Tien-yu, Tong Shao-yi, Tong Kaison, Tsai Ting-kan, Liang Tun-yen and Liang Ju-hao.

Many of the younger generation of returned students have also acquitted themselves creditably. A Chinese gentleman who really knows the facts states the case thus: "In various ways our American educated men and women are leaving their impression on our national life. They are giving a good account of themselves, they are reflecting credit on the country from which they drank inspiration, and they are setting us a worthy example of serious and efficient service." In another place Y. T. Tsur has compiled a list of above 100 returned students from the United States together with their colleges in the States and occupations in China. This shows them to be largely influential on Government Boards, as educators, editors, physicians, engineers, etc., etc. Whatever may be the ultimate outcome of this Western education, it is at present true that the educated Chinese youth are earnestly seeking opportunities to go abroad and find out for themselves. We cannot at present stem the tide. Our real responsibility is to give them the best we have—the most fundamental moral, mental, and physial education we have, and not a superficial gloss of Western materialism. While we cannot pass any final judgment upon this recent educational movement, it is certain that if the new learning is introduced into China with moral earnestness it will be the greatest factor in the uplift of this virile people.

Photo by C. B. Malone.

Front View of Main High School Building, Tsing Hua College.

OUTLINE OF COURSES OF STUDY AT TSING HUA COLLEGE
FOR 1913-14.

Middle School.		High School.	
1st Year.	Hours per week.	1st Year.	Hours per week.
Chinese	10	Chinese	3
Reading	5	Reading	4
Grammar & Sentence	3	Rhetoric and Composition	3
Dictation	2	General History	3
Penmanship	1	Elementary Chemistry	3
Nature Study	3	Physiology	3
Arithmetic	3	Plane Geometry	3
Drawing	2	Ethics (Lectures)	1
Music	2	Music	1
Health Drill	1	Health Drill	1
		Manual Training	2
		Elementary Physics	2
	32 hours.		29 hours.

One hour daily required of all students on Athletic Field.

2nd year.		General Course.	Scientific.
Chinese	10	Chinese	3	3
Nature Study	3	English Lit.	6	6
Reading	5	Modern Language	5	5
Grammar & Sentences	5	Ancient History	4	4
Dictation	2	Ethics (Lectures)	1	1

February, 1914   **THE FAR EASTERN REVIEW**   369

Conversation I	Health Drill .. I I	
Arithmetic 3	Music .. .. .. I Elective	
Drawing 2	subject .. I	
Music 2	Solid Geometry and	
Health Drill I	Trigonometry .. 4 Required .. 4	
	Elective.	
	Latin (Elective) .. 5. Mechani-cal Drawing 2	
	Second Modern	
	Language .. .. 5. Manual	
	Training .. 2	

       32 hours.

3rd Year.

	General Course.	Scientific
Chinese 6	Chinese 3 3	
Reading 5	English Literature 3 6	
Grammar & Sentence 4	Modern Language 5 5	
Dictation 2	Mediæval & Modern	
Conversation .. I	History .. 4 4	
Geography 3	Higher Arithmetic I I	
Algebra 3	Ethics (Lectures) .. I I	
Drawing 2	Health Drill .. .. I I	
Manual Training .. 2	Elective Subjects. Advanced	
Music 2	Algebra .. 4	
Health Drill I	Latin (2nd Year). 5. Chemistry 5	
	2nd Modern Elective subject.	

       31 hours.   Language .. .. 5 Music .. I

Advanced Algebra. 4.

Advanced Physiology 3.

Middle School	High School	
4th Year.	General Course.	Scientific
Chinese 6	Chinese 3 3	
Reading 4	English Literature 5 5	
Grammar & Rhetoric 3	American History .. 4 4	
English Composition. 2	Civil Government .. 2 2	
Conversation I	Political Economy 2 2	
General Geography 3	Ethics (Lectures) .. I I	
Elementary Chemis-try 2	Health Drill .. .. I I	
Hygiene I	Theme Writing .. 2 Physics .. 5	
Algebra 3	Public Speaking .. I	
Manual Training .. 2	Electives .. .. .. 3 Electives	
Drawing I	Psychology. Logic .. 2 Chemistry 5	
Music I	Second Modern	
Health Drill I	Language. 3rd year 5 5	
	Analytical Geometry 3 3	
	Mathematical Survey 2 2	
	Music .. I	

       30 hours.

Statistics of the Bureau of Educational Mission to U.S.A. for 1913.

Tsing Hua College

Number of Chinese Teachers 16	
Number of American Teachers 18	

Number of Students in Tsing Hua College November 20, 1913.

High School	Middle School
Senior Class 36	4th Year 56
Junior ,, ... 45	3rd ,, ... 50
Sophomore Class .. 37	2nd ,, ... 44
Freshman ,, .. 56	1st ,, ... 72
————	————
174	222

Total number of students .. .. 396.

Number of Students in the U.S.A. on the "Indemnity Fund"

1st Batch 47	
2nd ,, ... 70	
3rd ,, ... 60	
4th ,, ... 19	
————	
196	

Students taken over on the "Indemnity Fund" as Specials.

Chinese Legation Students 3	
Sent by Board of War 2	
Sent by the Customs College 2	
Sent by the Peiyang College 14	

Special Scholarship Students 9	
————	
30	
$500 gold—Scholarship Students 32	
Girl students taken on to the Fund 4	
————	
36	

Total Number of Indemnity Students .. .. .. 272.

Total Number of Chinese Students now in the U.S.A. is about 1,000.

Key to Plan of the Tsing Hua College Buildings

No. 1:—The Main College Building with upper story, of imposing architectural construction, with class rooms, lecture rooms, chemical studies, presidential and secretarial reception and office rooms, all laid out for over 300 students.

No. 2:—The Li Pai Tang, or festival hall, as well as music study for the students, seating capacity about 400, good acoustic, and airy hall.

No. 3:—Adjoining the Festival Hall the quarters for the Business Manager of the College and Proctor.

No. 4:—The temporary Library.

No. 5:—The Students' rooms in which the students do their preparatory work.

No. 6:—Ditto (both 5 and 6 are for the High School only).

No. 7, 8, 9 and 10:—Dormitories for 300 High School students.

No. 11:—The Large Dining Hall for the east section of buildings with its kitchen, and pantry; part of this dining room is now used as the study room for manual labour.

No. 12:—The Grammar School class rooms.

No. 13:—A small hall with adjoining offices.

No. 14:—The Studies.

No. 15 to 20:—The dormitories for 300.

No. 21:—The Dining Hall of the Grammar School students and adjoining kitchen and other buildings.

No. 22 to 26:—The Sport and Athletic field with tennis courts, basket ball grounds, etc.

No. 27:—The Hospital building in charge of Dr. Bolt.

No. 28:—The "Inner River."

No. 29 to 42:—Lotus Ponds.

No. 30:—The old Princely Residence, now generally called "the Yamen" all reconstructed and with fine Great Halls and living quarters of the officials of the College; here is also the central telephone station which connects the three million square feet of extended property of the Tsing Hua College and its various buildings, as well as offices.

No. 31:—The Chinese Teachers' residential quarters.

No. 32, 33, 34:—The Residential Houses for the President and Vice-Presidents of the College.

No. 35:—The South West Entrance (closed).

No. 36:—The College Grounds Police Head Quarters.

No. 37:—The Main Entrance Gate (Archway) with road to the Peking-Kalgan Railway Station, which is ten minutes off.

No. 38:—The Post-office and Porter Quarters, as well as stables.

No. 39:—The Electric Light station giving current, for 2,000 electric incandescent lamps.

No. 40 and 41:—The wooded hills in the old Princely Park.

No. 42:—Lotus Pond.

No. 43:—The Club of the American Professors and Teachers.

No. 44 to 51:—The Residential Buildings and section of the homes of the American teachers engaged by the Tsing Hua authorities directly from the States. Each building has a double residential section with sunny rooms and verandahs, Dining and Sleeping Rooms and Parlor. Most of these buildings face the pretty panorama of the Western Hills, with the Great Summer Palace in front, the Precious Stone Pagoda, and so on.

PHILIPPINE LUMBER CONCESSION

The Governor-General of the Philippines has granted to the Kolambugan Lumber Company a concession to cut two billion feet of merchantable lumber in the Moro province, the concession to run for twenty years. The area comprises 100 square miles and the capital is furnished by the Millars Karri Lumber Company, the largest firm of lumber dealers in the United Kingdom, which has for many years had huge lumber concessions and holdings in the south-western part of Australia.

1.3 Richard Arthur Bolt "The Tsing Hua College, Peking" 一文中有关清华学校之校园景观及建筑部分节译

The preparatory school, now known as Tsing Hua College, is situated about four miles outside the northwest city gate (Hsichihmen) of Peking on the line of the Peking-Kalgan Railroad. It is beautifully environed amidst rural surroundings in sight of the new Summer Palace, and almost adjacent to the magnificient ruins of the Old Summer Palace (Yuen Ming Yuen). It has been stated that this site was originally occupied as the country residence of the notorious Prince Tuan of "Boxer" fame, and that after his downfall the old Empress-Dowoger confiscated his property and turned it over to the Wai Wu Pu, for the use to which it is now being put. Others question this and affirm that the Park was originally the property of Prince Tun, the fifth son of Tao Kuang. In fact the local name for the place has long been Wu Ye Yuan, which means the "Park of the Fifth Prince."

When the Park was handed over to the Bureau of Educational Mission to the United States by Imperial Edict issued at the instance of the Prince Regent in 1911 the Tsing Hua College was at once laid under the direction of the then Vice-Minister of Foreign Affairs, Mr. Chow Tze-chi, who was once Charge d'Affaires in Washington, The late Mr. Tong Kaison, who was eventually President of the College, and Mr. Y. L, Fang, afterwards Minister of Education, co-operated with Mr. Chow Tze-chi in this work, and as a result of their efforts a number of buildings were erected in 1910 and 1911 sufficient to accommodate 600 students, the contract being carried out by Mr. Emil S. Fischer, formerly of New York and now head of Fischer and Co. of Tientsin and Peking. The cost of the work was over half a million Taels, or about $350,000 gold dollars.

A well built stone wall completely surrounds the 60 odd acres which make up the College Compound, which will be improved as soon as necessary funds are available. The College has recently acquired 60 more acres to the west. The Compound is attractively landscaped with artificial tree-topped hills, lotus ponds, and a small stream which runs directly through it. A considerable number of trees have been planted, a greensward laid out, and many flowers placed on the hillsides. The buildings of foreign, semi-foreign and Chinese architecture are conveniently grouped in the Campus. The offices of the Bureau of Educational Mission to U.S.A. were originally housed in the reconstructed Yamen, strictly Chinese in every detail, picturesquely surrounded by low hills and opening at the rear on to a marble porch which overlooks a lotus pond. There are substantial residences for the President of the College, the Dean, and a small Compound for the Chinese teachers, now numbering sixteen. The American teachers have had nine double bungalows allotted for their use, one of which has been comfortably fitted up as a Club House.

The High School department occupies a large two-wing double story grey brick building. This contains the educational office, offices for the President and Dean, an attractive Faculty room, the Chemical and Physical Laboratories, and various class rooms, Behind this building is an Assembly Hall which will accommodate about 400. Adjoining this at the rear are the High School study-halls, dormitories and dining room—all one story brick buildings connected by covered corridors. At the west end of the dining room a workshop for Manual T raining has been fitted up, the boys of the third and fourth year Middle School and first year High being required to take this subject.

The Middle School is housed in a group of buildings separated from the High School by a small stream bridged in three places. It comprises classrooms, Proctors' and Secretaries' Offices, the beginning of a Museum collection, study-halls, dormitories, dining-room, kitchen, and out-buildings. These are all connected by covered passage-ways. Suitable bath and toilet arrangements have also been provided. Just west of the Middle School is a well laid out athletic field, tennis courts, basket ball grounds, football and baseball field, running track and space for field events, and an archery area. Athletics have now become compulsory for every student. He must take regular callisthenic exercises each morning as well as spend one hour daily in some out-of-door sport. Careful account is kept of attendance and credit allowed for the athletic work. Beyond the athletic field is a Hospital and Dispensary building.

Every student is given a complete medical inspection and physical examination upon entrance to the College, and further examinations are required throughout the year. Special attention is given to Preventive Medicine and Hygiene. Plans have been considered for a Gymnasium, a Library and a Science Building, and these will probably be erected in due time. The entire plant is well lighted with electricity, and telephone connections have been established with Peking.

After considerable delay and uncertainty Tsing Hua College was formally opened with Chinese ceremonies on April 1st, 1911. The High School Department was begun with 128 advanced students who had failed in their previous examinations to be selected to go to the U.S.A. 307 younger students were selected by the Provincial examiners, ostensibly upon competitive examination to enter the Middle School. Practically every Province in China, besides Manchuria and Mongolia, was represented in the School. After a complete physical examination the boys were allowed to take up residence. The teaching of the younger students was at first almost completely in the hands of Chinese teachers who then

① 编者注：即端郡王载漪。

② 编者注：即奕誴。

③ 编者注：清华园之渊源及俗称详见本书上编第2章。

④ 译者注：载沣。

⑤ 编者注：手工劳动室或工艺馆为美国大学的基本建设内容之一，为学生实习之所。清华工艺馆建成于20世纪20年代初，此前曾利用同方部等建筑作此用途。

numbered about twenty. The American teachers were assigned, rather arbitrarily, to teach the older boys, from among whom it was understood that fifty would be selected that year to proceed to the United States.

* * *

现名为"清华学校"的预科学校，坐落于京张铁路沿线，距北京西北的城门——西直门约6.5千米。学校处于可以看到颐和园的优美田园环境中，且与圆明园遗址相邻。据说，这里曾是"义和团事变"祸首之一端郡王①的城外居所。在他垮台之后，慈禧太后将其财产交给外务部，如今作学校之用。也有人质疑称其主人是惇郡王②。当地人长期以来一直称这里为"五爷园"，即"五皇子的园子"③。

1911年摄政王④下令将园子移交美国教育代表团时，清华学校第一时间交由彼时的外务部副部长，曾任华盛顿临时代办的周自齐领导。已故的清华学校校长唐国安，与后来的教育部部长范源濂同周自齐合作，在他们的努力下，于1910年至1911年建造了一批足以容纳600名学生的建筑物，建设合同系与曾工作于纽约，现斐士公司（天津、北京）（Fischer and Co. of Tientsin and Peking）负责人埃米尔·S.斐士（Emil S. Fischer）签署。工程费总计超过50万两白银，约合35万美金。

精致的石墙环绕着60余英亩的校园，一旦有充足的资金，这里还将被继续修缮。学校近期又收购了西侧60余英亩的土地。校园中风景宜人，有种满树木的人工山丘和一片荷塘，一条小溪流淌其中。这里植满绿树，铺有草坪，山丘种满花朵。西式、折衷式、中式建筑以便捷联系的方式组织在校园中。美国教育代表团的办公室最初被安排在重修过的衙门内，这片建筑完全为中式风格，被小山环绕，风景优美，后方还有可以俯瞰荷塘的大理石门廊。学校里有提供给校长与教务长的宽敞居所，还有一个小院给现有的16名中国教师居住。美国教师被安排在九栋联排单层别墅中，其中一栋被改造成了舒适的俱乐部。

高等科位于一座较大的两翼双层灰砖建筑中。这里包括教育办公室、校长与主任办公室、一间舒适的教员室以及化学、物理实验室和各类教室。在这座建筑后方是可容纳400人的礼堂。其后是高等科的自修室、宿舍和食堂，这些都是单层砖砌建筑，之间有敞廊连接。食堂西侧已建成手工劳动室⑤，初等科三、四年级和高等科一年级的学生需要修习这项内容。

初等科的建筑群与高等科相隔一条小溪，两岸有三座桥相连。这些建筑中包括教室、监督员和秘书办公室、一座简易的陈列馆、自修室、宿舍、食堂、厨房及室外建筑物，由敞廊相连。这里提供配套的洗浴和厕所设施。初等科西侧是布局合理的运动场、网球场、篮球场、足球场、棒球场、跑道以及田径项目场所和射击场。体育现在是所有学生的必修项目。学生必须按时参加每天的早操，同时保证每天一小时的室外活动时长。体育活动会详细记录出勤并仔细计算学分。运动场另一端是医院和药房。

每位学生入学时，都会接受一次完整的体检和体质测试，在之后的一年中也会持续进行测试。预防医学和卫生工作得到了额外的重视。学校有计划在合适的时机加建体育馆、图书馆和科学楼。整个校园有良好的照明系统，电话已接入北京市线网。

在经历长期的延误与不确定事件后，清华学校于1911年4月1日举行中式庆典，正式开幕。高等科最初迎来128位学生，由此前在赴美选拔考试中落选的人组成。初等科的307位学生表面上由选拔考试选出，事实上是由各省考官挑选而来。除了东三省和内蒙古，几乎所有省都有代表学生入学。这些学生在经过全面体质测试后获准入住校园。低年级课程最初几乎全部由20名左右的中国教师负责。美国教师在自愿基础上被指派教授高年级学生，据了解，这些学生中有50人在当年被派往美国。

（王睿智转录，梁曼辰译，杜林东校）

清华古月堂
87.5.

第 2 章　墨菲档案之清华建设相关文件及相关研究

2.1 Memorandum Report of Interview of June 13, 14, 15 at Tsing Hua, Peking, China, between President TSUR & H. K. Murphy (June 26, 1914). 墨菲与周诒春第一次见面《会谈录》首页影印与全文转录及翻译

June, 26 1914.

TSING HUA COLLEGE.

Memorandum Report of Interviews of June
13, 14, 15, 1914, at Tsing Hua,   Peking
China, between President TSUR & H.K.Murphy.

At the invitation of Mr.Y.T.Tsur, Yale, President
of Tsing Hua College, H.K.M. made a trip out from Peking
on the afternoon of June 13, to discuss with him and with
Dean CHAO, the future architectural developments of the College; and after a short interview, President Tsur engaged
H.K.M's firm, Messrs. Murphy & Dana, as architects for the
new buildings to be built immediately; and to lay out, as
soon as possible, the scheme of ultimate development for
the future University. For their services as Architects,
Murphy & Dana are to render bills, at such intervals/ as may
be convenient to them, for double the total costs incurred
by them, on the work up to the date of each bill, less the
amount of previous payments on account.- On Aug.I,1914,
Murphy & Dana are to draw on Mr.Jenkins, Sec'y of the International Y.M.C.A., New York City, for a preliminary retaining
fee of Five Hundred ($500.00.) Dollars gold; and Pres. Tsur
will arrange that subsequent payments to the Architects

At the invitation of Mr. Y.T.Tsur, Yale, President of Tsing Hua College, H.K.M. made a trip out from Peking on the afternoon of June 13, to discuss with him and with Dean CHAO, the future architectural developments of the College; and after a short interview, President Tsur engaged H.K.M's firm, Messrs. Murphy & Dana, as architects for the new buildings to be built immediately; and to lay out, as soon as possible, the scheme of ultimate development for the future University. For their services as Architects, Murphy & Dana are to render bills, at such intervals, as may be convenient to them, for double the total costs incurred by them, on the work up to the date of each bill, less the amount of previous payments on account. On Aug.1, 1914, Murphy & Dana are to draw on Mr. Jenkins, Sec'y of the International Y.M.C.A., New York City, for a preliminary retaining fee of Five Hundred ($500.00) Dollars gold; and Pres. Tsur will arrange that subsequent payments to the Architects will be made either through Mr. Jenkins or through Pres. Tsur's representative at Washington, D.C.

Immediately after H.K.M's arrival in New York (probably July 17) preliminary drawings are to be started as follows:

1）Plan of entire present group of buildings, showing proposed location of the following new buildings:

    a）Auditorium ⎫
               ⎬ (possibly combined in one building)
    b）Library ⎭

    c）Science Building

    d）Gymnasium

2）Plans and Perspective of Auditorium.

3）Plans and Perspective of Library.

4）Plans and Perspective of Science Building.

5）Plans and Perspective of Gymnasium.

6）Group-plan of future University buildings to be built on properly recently acquired, adjoining present group on West.

To assist in this work, H. K. M has obtained from Pres. Tsur the following data:

1）Book of "Views of Tsing Hua College, Peking".

2）Large mounted photographs of the Principal Buildings.

3）Tsing Hua College Bulletin of Information No.1.

4）Reprint from "Far Eastern Review" of Dr. Bolt's article on Tsing Hua.

5）April 15, 1914 issue of the "Useful Knowledge" published at College.

6）Memorandum by Pres. Tsur of costs of Present Buildings.

7）Blue Print of Survey at 1″ =100′0″ of Original Property.

8）Blue Print of Survey at 1″ =100′0″ of New Property.

9）Report of Committee on Location of Buildings (3/28/14) with plans.

10）Minority Report of Committee on Location of Buildings sq. ft. (Heinz) (3/12/14) with plans.

11）Two tracings of Original Property.

12）One tracings of New Property.

13）Eight Blue-prints of small-scale drawings of Present Buildings.

14）Four card-memorandum Programs for New York (by Pres. Tsur).

15）Memorandum Program for Science Building by Prof. Pierlé (chemistry).

16）Memorandum Program for Science Building by Prof. Wold (physics).

17）Memorandum Program for Science Building by Gymnasium Mr. Shoemaker (Physical Director).

About Aug.1, Pres. Tsur will telegraph Murphy & Dana whether they are to engage a man to go out from New York, in the Fall of 1914, to supervise the letting of the contracts and the actual construction. (This H.K.M. strongly advised, in view of the danger of great financial losses to the College, unless this precaution is taken, through "squeeze" and through ignorance of modern scientific American building methods; and Pres. Tsur expects to be able to cable affirmatively).

On Aug.1, Murphy & Dana are to send to C.A. Pierlé, Pomona, Missouri, sketches of Chemistry Section of Science Building; Mr. Pierlé will then send back to N.Y. his criticisms and suggestions, before leaving his home for China again.

On Aug.18, Murphy & Dana are to send to R.D. Whitmore, Sunderland, Mass, sketches of Science Building, showing provision (in Basement?) for Manual Training Dept.; these will be at once returned to N.Y. with criticisms, as above.

On Aug.29, Murphy & Dana are to mail to Y.T. TSUR, Esq., c/o Chinese Consul, San Francisco, Cal., all sketches ready at that time; so that Pres. Tsur may study them on his way East across the United States from San Francisco, where he expects to arrive on Sept.5. (He will leave Peking on July 24, to remain at Shanghai until Aug.15; Shanghai address c/o Y.M.C.A. 3 Quinsan Road).

On Sept. 31 Pres. Tsur expects to reach N.Y. and will go at once to Murphy & Dana's office for a full conference with Mr. Murphy and Mr. Dana on the preliminary drawings.

About Oct.10 Pres. Tsur will return to N.Y. for a final conference with Mr. Murphy and Mr. Dana, on revisions made meanwhile in their drawings as a result of the Sept. 21-25 conferences. Pres. Tsur will then leave for China and will take with him the man selected by Murphy & Dana to supervise the work. (See above)

About Nov. 10, Murphy & Dana's Superintendent will reach China and will devote the next four months to a careful study, on the ground, of all the conditions affecting the letting of contracts, methods of construction, materials etc.; and will interview as many as possible of the best Engineers and Architects in Peking and other cities of China, in order that the Tsing Hua Work may be done in the best way at the lowest total cost consistent with good work. He will also collect during this period data to enable him later to amplify the outline specifications to be prepared in New York.

About Feb.15, Murphy & Dana are to send to Peking the finished working drawings and outline specifications for all of the buildings to be built immediately, so that the Superintendent in Peking may send the plans and full specifications out by March 15, for competitive bids by several of the best Contractors in and about Peking.

As soon as satisfactory estimates are received, construction work is to start on all the immediate new work, probably in April 1, 1915; and it is hoped that the new buildings will be ready for occupancy by the opening of College in Sept.1916.

The matter of the expansion of the present "College" (consisting of a Middle School and a High School) into a University, is one on which no definite assurances can be given at this time. However, there is at present no great National University in China; and Pres. Tsur feels that Tsing Hua, with its exceptional natural advantages in location (seven miles out of Peking;) in beauty of grounds (formerly a park;) in the fact that it is a government institution; in its intimate connection with America (its foreign faculty are all Americans, and its students are all trying for the honor of being sent to America, on the Boxer Indemnity Fund;) and in the good architectural start it will have when the new buildings about to be built are completed; will be more than likely to receive from the China Government the financial support necessary to carry out Pres. Tsur's plans for expansion. Certainly the need for such an institution is great; and in Pres. Tsur, Tsing Hua College has a man combining the enthusiasm, energy and high ideals necessary for carrying out such a project.

The University, for which Murphy & Dana are to prepare, immediately, a tentative block plan, will follow in general the plan of the American University, rather than the English plan of separate small Colleges, and will comprise:

1）College of Arts and Letters.

　　Literature, History, Economics, Education, Languages etc.

2）Administrative Building.

3）School of Engineering.

Civil Eng., Mechanical Eng., Mining Eng., Electrical Eng., Sanitary Eng.

4）School of Applied Sciences.

5）School of Political and Social Sciences.

6）School of Medicine.

7）School of Agriculture.

Forestry, Experiment Stations.

8）School of Architecture.

9）School of Chinese Classics.

10）School of Astronomy.

Observatory.

11）Dormitories.

Faculty Residences.

12）University Library.

13）Amphitheatre.

14）Infirmary.

15）Athletic Field.

Gymnasium.

16）Central Plant for supplying Steam Heat and Power (to be used as Work Shop by Engineering Students).

It is not possible at this time to say how many students will have to be provided for in each department of the University; in general, however, the group-plan should be made to conform to the average requirements of an American University with the College of Arts and Letters (leading to the degree of B.A.) as the principal feature. On account of this uncertainty in the matter of future developments, the buildings for each Department should be laid out with a special view to individual expansion, separately from the general expansion of the whole University.

It will be necessary to keep a large Middle School (perhaps 1000 students ultimately) properly to feed the University itself; in order to have the boys well-prepared, it is necessary to start with them while quite young.

In addition to the buildings to be erected immediately, the Superintendent sent to China in the Fall of 1914 will have to take up for immediate consideration, Sewage Disposal and Water Supply. For the latter, Pres. Tsur will have artesian (or other) wells driven at once, near the present Power House; and the Architect's Superintendent will have to arrange for pumping apparatus, tanks, piping, etc. Electric light current is already available; but an extension of the telephone system will have to be arranged to include more of the buildings than are at present equipped.

Another matter for consideration is the possibility of securing for use in the new buildings, some of the stone from the ruins of the old Summer Palace, about a mile from the College Grounds. Much of this is beautifully carved—moulded plinth-stones, columns, brackets, balustrades etc. which could be used in the approaches to the buildings; the majority of the available material is in the form of large blocks of stone, from $1' \times 3' \times 1'$ up to $2' \times 3' \times 8'$ in size, especially adapted for foundations and lintel-stones. There is also a great deal of small material, which would break up well for concrete. The Old Summer Palace ruins are the property of the recently deposed Manchu royal family; and owing to their present somewhat impoverished state, Pres. Tsur think it may be possible to get permission to cart away whatever could be used for a comparatively small payment. H.K.M. advised Pres. Tsur to get possession of the material, if possible to do so economically.

The matter of the architectural style to be adopted for the new buildings was discussed at considerable length. At first H.K.M. assumed that it was to be some form of Chinese; but on close study of the existing buildings, he found that none of them (with the exception of the "YAMEN" built by the Prince who formerly occupied the property) was really Chinese at all. The Chinese effect of the group comes almost solely from the gray brick of the walls, the gray Chinese tie of the roofs and the fact that all but one of the buildings are of only one story. The blunted gables of the roofs in the High School Dormitories and the Chinese Faculty Houses, are English; the windows are casements of the French type; the Middle School gables have a strongly Spanish character, while the large High-School Building is of a non-descript style of architecture, with a red French tile roof, treated at the corners with mansards. The Entrance to the latter building (which at present dominates the group) is effective and dignifies, though the detail and proportions are not really good. As Pres. Tsur has no preference for the Chinese style, the only strong reasons for its adoption, so far as the individual characteristics of Tsing Hua is concerned, are the presence, as a permanent feature, of the really beautiful "Yamen", and the curious small artificial hills which wind about through the grounds, like miniature ranges of mountains. Although Pres. Tsur recognizes the educational value, to the Chinese, of a group of modern buildings in the Chinese style, (as in the case of the new Yale Mission buildings at Changsha;) he feels that the style, if at all well carried out imposes many restrictions and limitations, from the utilitarian point of view, on the design of buildings intended for class-room and dormitory purposes. H.K.M. agreed with him in this, and advised that no attempt be made, in the new buildings, to carry out Chinese forms, except that the buildings should be of the same gray brick as the present buildings, and should be kept as low as would be consistent with economy of construction. Pres. Tsur and H.K.M. both felt that certain features, about the grounds, might be made purely Chinese; and that a fine Chinese Entrance Gate might be made a feature of the New University Campus. The decision not to use the Chinese style for the new buildings was also partly due to considerations of cost construction.

The detailed programs for the immediate new buildings, so far as at present determined, are as follows:

(A) AUDITORIUM (possibly combined with Library building) :

· to seat one thousand persons, all with good view of stage. Impressive Entrance, and easy exit facilities. Large Stage (for graduation exercises, concerts etc.) with convenient separate entrance form outside.

(B) LIBRARY:

· Stock Room.

· General Reading Room.

· Reference Library for various Departments.

· Physics, Chemistry, History, Economics etc.

· Museum and Picture Gallery.

· Administrative Offices.

(C) GYMNASIUN: (consult J.H. McCurdy, Springfield, Mass. Y.M.C.A.)

· Abundant light and air for all rooms.

· Two separate floor spaces, each about $60' \times 80''$; separated by moveable partition rolling up to give one floor $80' \times 120'$.

· Indoor Running Track around whole building, with corners devoted to gallery seats for spectators. At least $12'-0''$ clear height under running track.

· Swimming Pool $30' \times 75'$, tiled throughout, with Gallery for spectators of Water, polo games, etc. At least $15'-0''$ clear space at one end of pool, and $10'-0''$ at one side; $5'-0''$ to $8'-0''$ on other side and end.

· Locker Rooms: 1000 lockers, each about $12'' \times 16'' \times 5'0''$ high (single tier), separate locker rooms for Faculty & visiting Athletic Teams.

· Drying Room ($10' \times 12'$): for drying clothing by steam heat.

· Shower Room: 25 showers, hot and cold water, arranged around walls of two rooms; showers to come out from walls, with no partition between; gutters around walls, and dry floor in centre.

• Steam Room (9′ × 12′?): for Turkish bath.

• Toilet accommodations (proportionate to Locker Room): basins and W.C's, Towel Closet.

• Servants' Rooms: 3 rooms (to sleep two servants in each.).

• Fencing Room.

• Boxing Room.

• Handball Room.

• Trophy Room (20′ × 40′?).

• Physical Directors' Office:

Three adjoining rooms; one at least having outside window(20′×15′. 12′×15′. 12′×15′).

Large double-deck Porch facing athletic field, to serve as grand-stand; second story sloped.

(D) SCIENCE BUILDING:

• For requirements of Chemistry Dept., see sketch plan and notes by Mr. Pierlé;

• For requirements of Physics Dept., see sketch plan and notes by Mr. Wold.

• Allow for future expansion of each department on upper floor.

• Provide accommodations in Basement of Manual Training Dept. and Mechanical Drawing Dept. (Mr. Whitmore).

• Probably raise level of 1st floor of Science Building 1′–0″ to 1′–6″ above level of 1st floor of High School, to allow better light in Basement for Mr. Whitmore's work.

COST OF IMMEDIATE WORK:

The Chinese Government have appropriated One Hundred and Fifty Thousand ($150,000) Dollars gold to cover the cost of the immediate work, including payments to Architects, etc. H.K.M. estimated that Murphy & Dana's bill for work to be done by them up to the time of Pres. Tsur's departure for China in Oct. 1914, and including tentative layout for future University, as well as preliminary drawings for Immediate Work as outlined above, would be about Two Thousand ($2,000) Dollars gold; that the next Stage of the work, including the preparation of actual working drawings and outline specifications for the immediate work, would be about Five Thousand ($5,000) Dollars gold; and that the salary of Murphy & Dana's Superintendent for two years (Sept.1914 to Sept. 1916) would be about Eight Thousand ($8,000) Dollars gold. In other words, the total payments to the Architects up to the time of opening the new buildings included in the immediate work, would probably be about Fifteen Thousand ($15,000) Dollars gold. H.K.M. pointed out to Pres. Tsur that, while the cost would be kept down as low as possible, it would be safe to allow one or two thousand dollars gold, more than this, to cover possible complications arising in the preparation of the drawings.

Note:

It is expected that the College will be able to provide, each year, living guarantee for the (?) Superintendent in one of the unoccupied houses built for foreign (married) Professors.

--------------------- * * * ---------------------

6月13日，墨菲受其耶鲁大学校友、清华学校校长周诒春邀请，从北京出发前往美国，与周诒春及其副手赵国材讨论清华校园规划和未来建设一事。简短的会谈之后，周校长聘请墨菲的公司——茂旦洋行（Murphy & Dana），希望他们能尽快制定未来大学建设的总体计划。方便起见，茂旦洋行将会分期开出业务账单，其总额将为其实际开销的两倍。这些费用需在每笔账目到期前付清。1914年8月1日，茂旦洋行将通过纽约市国际基督教青年会干事詹金斯先生收取500美元（黄金）作为预付费用。周校长将通过他在华盛顿的代表人或詹金斯先生向建筑商支付后续费用。

在墨菲抵达纽约（大约在7月17日）不久后，如下内容的初步方案便应马上开始绘制，包括：

1）现有建筑的总平面图以及下列建筑的拟定位置

a）大礼堂

b）图书馆（a、b或可合并于同一座建筑）

c）科学楼

d）体育馆

2）大礼堂平面图与透视图

3）图书馆平面图与透视图

4）科学楼平面图与透视图

5）体育馆平面图与透视图

6）满足近期需求的未来大学部建筑组团平面图，与现有建筑组团的西侧相接

为帮助工作进行，墨菲请周校长提供了以下资料：

1）《北京清华学校风貌》一书

2） 现有主要建筑的相关照片

3）《清华学校公报》第一期

4）《远东评论》中博尔特博士关于清华文章的复印件[①]

5）1914年4月15日清华学校出版的《有用的知识》

6）周校长关于现有建筑建造花销的记录

7）原始场地1英寸：100英尺的测量图

8）新场地1英寸：100英尺的测量图

9）委员会对建筑选址的报告（1914年3月28日）以及平面图

10）少数派（亨氏）对建筑选址的报告（1914年3月12日）以及平面图

11）两份清华园的界址图

12）一份近春园的界址图

13）八份现有建筑的小比例尺图纸

14）四份周校长的方案备忘录，寄送纽约

15）化学系教授皮尔勒的科学馆功能布局备忘录

16）物理系教授沃尔德的科学馆功能布局备忘录

17）体育系主任休梅克的体育馆功能布局备忘录

周校长将在8月1日左右电报联络茂旦洋行，询问他们是否能从纽约派一人于1914年秋季前往北京，监督合同的订立和施工。（考虑到学校有遭受重大经济损失的可能，墨菲强烈建议这么做。只有这项预防措施得以实施，才能避免经济损失，也能保证美国现代科学建筑手段的实施。周校长希望通过电报确认此事。）

8月1日，茂旦洋行将科学楼化学部分的草图寄给密苏里州波莫纳学院的C.A.皮尔勒（C.A. Pierlé）。皮尔勒会在他再次前往中国之前将对于图纸的批评和建议寄回纽约。

8月18日，茂旦洋行将科学馆的草图寄给桑德兰的R.D.惠特莫尔，草图应包括工艺馆（可能位于地下室）。相关建议也将立刻寄回纽约，同上述。

8月29日，茂旦洋行将致信身在旧金山中国领事馆的周诒春，并附所有的草图。周校长可以在从旧金山前往美国东部的路上研究这些草图。他计划于9月5日到达纽约。（周校长将在7月24日离开北京，在上海停留至8月15日。他在上海的地址是昆山路3号基督教青年会。）

9月31日，周校长将抵达纽约，并将立即前往茂旦洋行，与二位就初期图纸进行深入的会议交流。

10月10日左右，周校长将回到纽约，与墨菲和丹纳进行最终会议，讨论9月21—25日会议中图纸问题的修改结果。之后周校长将与茂旦洋行挑选的驻场建筑师一同回中国。

11月10日左右，代表茂旦洋行的监理将抵达中国，并将在接下来的4个月中对场地情况、合同设立条件、建设手段、建筑材料等方面进行深入研究。同时，为了

①
编者注：全文及节录分别详见本书中编1.3节、1.4节.

用最低的开销将清华的工程做到最好，他将会面试大量来自北京和中国其他城市的优秀工程师和建筑师。于此期间，他还会收集更多的资料，以便在回到纽约之后充实简要标准规范。

2月15日左右，茂旦洋行会将完成的工作图纸与所有将开工建筑的简要标准规范寄往北京，以保证在北京的驻场建筑师能在3月15日发布全部图纸和全面标准规范，向北京及周边最好的建筑承包商发出投标邀请。

一旦取得满意的评估结果，建设项目将在1915年4月1日左右立刻开始实施。新建筑预期将在1916年9月开学时投入使用。

关于将现在的"学校"（包括中等科和高等科）扩建为大学一事，目前无法保证其可以成功。然而，中国眼下尚无一流国立大学；并且周校长认为，清华坐拥优越的地理位置（离北京7英里），优美的场地环境（曾经是一座花园）；同时还是一处政府机构；与美国关系紧密（由于庚子赔款，这里的外国教员都来自美国，这里的学生都以争得前往美国留学的机会为荣）；并且一旦新校舍落成，还将拥有良好的建筑基础，因此很可能得到中国政府的经济支持，以实现周校长的扩建计划。确实需要扩建这样一处学府。同时，周校长对于此扩建计划投入的热情、精力及崇高的理想，皆为实现扩建计划所必需。

茂旦洋行需要立即制定的大学部方案，布局应遵循美国大学的总体设计规划，而不是英国大学那种分立的小型学院布局。这个方案中将包括：

1）文学与艺术学院

　　文学、历史学、经济学、教育学、语言学等

2）行政楼

3）工程学院

　　土木工程学、机械工程学、矿业工程学、电子工程学、卫生工程学

4）应用科学学院

5）政治与社会科学学院

6）医学院

7）农业学院

　　林业学、农业实验站

8）建筑学院

9）经学院

10）天文学院

　　天文台

11）宿舍

　　教职工住宅

12）大学图书馆

13）阶梯教室

14）医务所

15）田径场

  体育馆

16）供热、供电的中央工厂（也将作为工程学院学生的工作站）

现在还无法确定将来各院系招收学生的数量。总体而言，建筑组团的设计应当参考一个中等规模、包括文理学院在内的一所美国大学来制定。考虑到将来未知的发展规划，每个院系的建筑应当与整个大学的总体扩建分开，各自考虑设计。

为了保证大学生源，同时经营一所较大的中学（大致1000名学生）很有必要，在男生们年纪较小时即招收到校学习，使他们在进入大学前便已完全适应这里的校园生活。

除了负责建筑建设，在1914年秋天被派至中国的驻场建筑师还要尽早考虑污水处理及供水问题。对于供水问题，周校长将在问题解决之前使用现发电厂旁的自流井或其他井的水。监督人需要设计水泵、水箱、管道等设施。电灯系统已可以使用。但电话系统需要重新设计以覆盖更大区域，联通更多建筑。

另外一件需要考虑的事是，使用距校园1英里的圆明园遗址中的石材用于新建筑的建设，是否存在可能性。这些精雕细琢的基石、柱子、支撑和栏杆，很多都可以再次利用，其中大部分是尺寸在1英尺×3英尺×1英尺到2英尺×3英尺×8英尺的大石块，很适合用作地基和楣石。废墟中也有很多小块的材料，可以用来制作混凝土。圆明园是近年被废黜的满清政府的皇家财产。由于清朝遗族当下似乎处境困窘，周校长认为可以用相对较少的报酬征得他们的许可，从圆明园里运来可以使用的各种材料。墨菲从经济性的角度建议周校长尽量多从这里获取所需的材料。

关于新建筑应当采取何种建筑风格的讨论持续了很久。起初墨菲认为应当使用一些中国形式，但在仔细研究现有建筑之后，他发现除了这片地曾经的皇子主人在这里建的"衙门"（即工字厅）外，其他建筑中并没有真正的中式建筑。这些建筑的中国元素主要体现在灰色的砖墙和灰色的屋顶瓦上，还有除了一栋建筑外，其他建筑都只有一层。清华学堂和中国教职工住宅①的山墙是英式的，窗框是法式的；中等科②的山墙有很强的西班牙风格；而高等科教学楼③则是一种难以定义的建筑风格——它的屋顶覆以法式红瓦，转角处施用孟莎屋顶。该座建筑目前统率着现有的建筑群落，其主入口虽然细节和比例处理得不是很好，但却非常具有威严和统率力。由于周校长并非特别偏好中国传统样式，目前看来唯一影响清华采取中式风格的是古色古香的"衙门"，以及遍布校园各处的假山。尽管周校长认为，像长沙的雅礼学校建筑群那样在现代建筑中融入中国元素，对于学生们是有教育意义的。但他同时认为，从实用的角度考虑，如果严格按照中国样式设计，对于教室、宿舍等将造成很多限制。墨菲表示赞同这一观点，并建议除了使用灰砖与现有建筑保持一

①
编者注：似应指刚竣工的二院一带的宿舍，非指古月堂。

②
编者注：指三院。

③
编者注：指清华学堂（一院）。

致性，以及尽可能地保持低矮，以符合建设的经济性之外，不应当花费精力去体现中式风格。周校长和墨菲一致认为，在场地规划等一些方面应当做到纯粹的中式；同时，新的大学校园将会有一个特征鲜明的中式大门。以上关于不在建筑中使用中式元素的考虑，一定程度上也考虑到了建设开销问题。

根据目前的意见，新建筑的详细方案如下：

（A）大礼堂（可能与图书馆结合）

· 可以容纳1000名观众，并保证所有位置都有很好的舞台视野。拥有一个特征鲜明的入口；可以快速疏散的设施；设有独立室外出口的大型舞台（为毕业典礼，音乐会等准备）。

（B）图书馆

· 书库
· 主阅览室
· 不同院系（物理、化学、历史学、经济学等）的参考书阅览室
· 博物馆和展廊
· 行政办公室

（C）体育馆（由来自马萨诸塞州斯普林菲尔德市的基督教青年会成员J.H.麦卡迪作为顾问）

· 每间屋子都有充足的光照和新鲜的空气
· 两个分别约为60英尺×80英尺独立空间，由活动隔板分隔，当隔板升起时，总面积可达到80英尺×120英尺
· 围绕着整个建筑的室内跑道，在角落处有为旁观者准备的展廊座椅。跑道下净高至少12英尺
· 30英尺×75英尺的游泳池，瓷砖铺砌，配有为旁观者准备的展廊，可举行水球比赛等。泳池一端至少有15英尺净宽，对侧为10英尺；另一组对边分别宽5英尺、8英尺
· 更衣室：1000个单层更衣柜，每个大约12英寸×16英寸×5英尺。教职工和来访的运动团队有独立的更衣室
· 烘干室（10英尺×12英尺）：用于蒸汽烘干衣物
· 淋浴室：25个冷热水淋浴头，排布在两间房间的墙壁上。淋浴头架在墙上，之间没有隔板。下水沟沿墙排布，房间中心是干燥区
· 蒸汽室（9英尺×12英尺）：蒸汽浴
· 卫生间（数量与更衣室的设置成比例配置）：洗手池、厕位和毛巾柜

- 服务人员用房：3间房（每间住两名服务人员）
- 击剑室
- 拳击室
- 手球室
- 奖杯陈列室（20英尺×40英尺）
- 体育教师办公室：

三间连续的房屋，至少一间有外窗（20英尺×15英尺，12英尺×15英尺，12英尺×15英尺）。

大型双层门厅面向田径场，作为正面大看台；二层倾斜。

（D）科学馆

- 关于化学系的需求，请看皮尔勒先生的平面图和备忘录
- 关于物理系的需求，请看伍德先生的平面图和备忘录
- 允许各院系在未来向上加建
- 在地下室为手工实践和机械绘图留出教室（按惠特莫尔先生建议）
- 将科学楼的一层抬高至比高等科一层高1英尺到1英尺6英寸的高度，以便给位于地下室的惠特莫尔先生办公室提供更好的光照

近期工作开销：

中国政府已拨款15万美元（黄金）来支付近期的工作开销，其中包括建筑师的佣金等。墨菲估计，茂旦洋行从周校长自1914年10月前往中国起至目前的所有工作，总费用大约为2千美元（黄金）。这期间的工作包括未来大学的试验性设计图纸，还有上述近期工程的初期图纸。下一阶段的工作报酬预计约为5000美元（黄金）。这部分工作将包括实际施工图纸和近期建设的设计说明。茂旦洋行派出的驻场建筑师在此后两年（1914年9月至1916年9月）的报酬应为8000美元（黄金）。综上所述，在新建筑投入使用前，支付建筑商的总费用大约为1.5万美元（黄金）。墨菲向周校长提出，尽管开销应尽可能压缩，但最好能多支付一两千美元（黄金）或更多，用以解决初期图纸阶段可能出现的问题。

注：

学校最好能为驻场建筑师提供一套为外籍已婚教授准备的住房作为其居所。

（梁曼辰转录、翻译）

2.2 Murphy & Dana Architects. Analysis of Buildings Shown on General Plan Dated October 30, 1914, of Future University for Tsing Hua, Peking, China. 墨菲方案《规划说明书》首页影印与全文转录及翻译

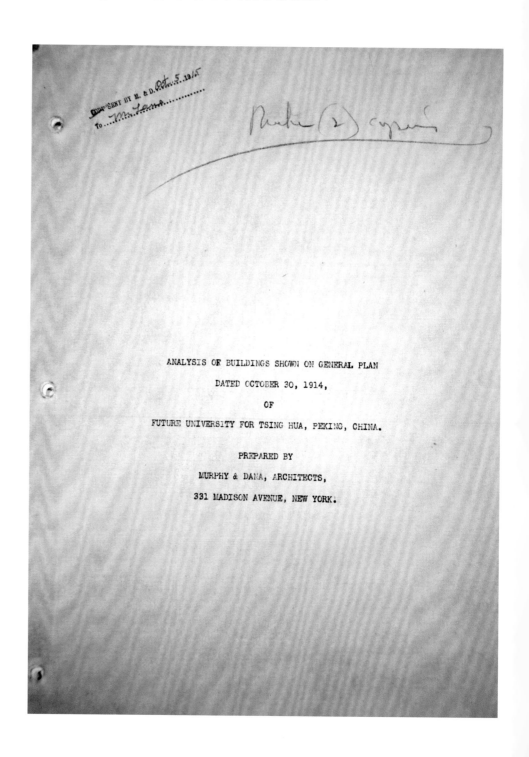

Tsing Hua University

ACADEMIC GROUP

ARTS & LETTERS

500 Students (41 square feet per student) ------------------------------------20500 sq. ft.

- Literature
- History
- Economics
- Languages
- Political and Social Sciences

Classroom accommodation for 500 Student – 30 to 40 in class (15 square feet per student)

--- 7500 sq. ft.

1 Lecture Room seating 250 / 10 square feet per student -------------------- 2500 sq. ft.

Reference Room 10000 Volumes -- 1500 sq. ft.

8 Professors (8 Rooms / 175 square feet each) -------------------------------- 1400 sq. ft.

20 Instructors (10 Rooms / 160 square feet each) --------------------------- 1600 sq. ft.

Janitors (4 Rooms / 150 square feet each) --------------------------------- 600 sq. ft.

Miscellaneous Small Rooms --- 1000 sq. ft.

Toilet, Lavatory & Coat Room accommodation ----------------------------- 1000 sq. ft.

17100 sq. ft.

Stairs, Corridors, etc. -- 3400 sq. ft.

Total for this Department --- 20500 sq. ft.

Note:

The Classroom unit corresponds closely to the unit used at the college of the City of New York.

CHINESE CLASSICS

200 Students (30 square feet per student) -------------------------------- 6000 sq. ft.

Classroom accommodation for 200 Students – 30 to 40 in class (15 square feet per student)

--- 3000 sq. ft.

4 Professors (4 rooms / 170 square feet each) --------------------------------- 680 sq. ft.

4 Lecturers (2 Rooms / 170 square feet each) ---------------------------------- 340 sq. ft.

Reference & Study Rooms, 5000 Volumes ------------------------------------- 500 sq. ft.

Toilet, Lavatory and Coat Room accommodation ---------------------------- 500 sq. ft.

5020 sq. ft.

Stairs, Corridors, etc. -- 980 sq. ft.

Total for this Department -- 6000 sq. ft.

Actual Floor Area for Two Arts Buildings ---------------------------------- 43000 sq. ft.

Floor Area required for Arts & Letters Departments ------------------------ 20500 sq. ft.

Floor Area required for Chinese Classics ----------------------------------- 6000 sq. ft.

26500 sq. ft.

Surplus -- 16500 sq. ft.

Floor Area required for 300 Students from Other Departments who may be

taking Classes in These Buildings (300 Students / 41 sq. ft. per student) ------- 12300 sq. ft.

Excess available for book storage and other purposes ----------------------- 4200 sq. ft.

EDUCATION

400 Students (45 square feet per student) ----------------------------------- 18000 sq. ft.

Classroom accommodation for 300 students – 24 to 40 in a class

(15 square feet per student) -- 4500 sq. ft.

1 Large Lecture Theatre, seating 200 (10 square feet per student) --------- 2000 sq. ft.

Kindergarten --- 900 sq. ft.

Model School Room -- 500 sq. ft.

Reference and Study Rooms --- 3000 sq. ft.

Other Small Rooms --- 800 sq. ft.

6 Professors (6 Rooms / 170 square feet each) ------------------------------ 1020 sq. ft.

20 Instructors (10 Rooms / 170 square feet each) --------------------------- 1700 sq. ft.

Janitors (2 Rooms / 150 square feet) -- 300 sq. ft.

Toilet, Lavatory & Coat Room accommodation ------------------------------ 900 sq. ft.

15620 sq. ft.

Stairs, Corridors, etc. --- 2380 sq. ft.

Total for this Department -- 18000 sq. ft.

Note:

On account of local conditions it is impossible to make a close comparison with buildings of a similar character in Western Universities. The accommodation provided above, however, corresponds generally with the Teachers' College at Columbia University, New York.

It is assumed that 100 or more students of this Department will be taking classes in the Arts & Letters and other Departments during every period, leaving only 300 students for whom provision must-be made in this building.

Total Floor area of this Building --- 19500 sq. ft.

Floor Area required for School of Education -------------------------------- 18000 sq. ft.

Excess available for book storage and other purposed ----------------------- 1500 sq. ft.

LAW & JOURNALISM

These two Departments will be housed in one building at preset; and in the event of expansion either department could be given one of the "Future" buildings indicated on plan.

LAW – 300 Students (41 square feet per student) ---------------------------- 12300 sq. ft.

Classroom accommodation for 225 Students – 24 to 40 in a class
(15 square feet per student) --- 3475 sq. ft.

Lecture Room seating 200 (10 square feet per student) ---------------------- 2000 sq. ft.

Debating Room -- 900 sq. ft.

Reference and Study Rooms, 9000 Volumes ---------------------------------- 1000 sq. ft.

Miscellaneous Small Rooms -- 750 sq. ft.

3 Professors (3 Rooms / 170 square feet each) --------------------------------- 510 sq. ft.

8 Instructors (4 Rooms / 170 square feet each) -------------------------------- 680 sq. ft.

2 Janitors (2 Rooms / 150 square feet each) ------------------------------------ 300 sq. ft.

Toilet, Lavatory and Coat Room accommodation ---------------------------- 600 sq. ft.

10215 sq. ft.

Stairs, Corridors, etc. --- 2085 sq. ft.

12300 sq. ft.

Note:

The total unit corresponds closely to the unit used in the Law School of the University of Pennsylvania.

It is assumed that 75 or more of the students of this Department will be taking classes in the Arts and Letters and other Departments during every period, leaving only 225 students for whom provision must be made in this building.

JOURNALISM

200 Students (30 square feet per student) ------------------------------------- 6100 sq. ft.

Classroom accommodation for 150 students at an average of 15 square feet per student

-- 2250 sq. ft.

Library and Study Rooms, 14000 volumes ------------------------------------ 2500 sq. ft.

2 Professors (2 Rooms / 170 sq. ft. each) ----------------------------------- 340 sq. ft.

4 Instructors (2 rooms / 170 sq. ft. each) ----------------------------------- 340 sq. ft.

Toilet, Lavatory & Coat Room accommodation ------------------------------ 500 sq. ft.

5930 sq. ft.

Stairs, Corridors, Etc. -- 1170 sq. ft.

Total for this Department --- 7100 sq. ft.

It is assumed that 50 or more of the students of this Department will be taking classes in the Arts and Letters and other Departments during every period, leaving only 150 students for whom provision must be made in this building.

Total Floor area of this building --- 19500 sq. ft.

Floor Area required for Law Dept. -- 12300 sq. ft.

Floor Area required for Journalism Dept. ----------------------------------- 7100 sq. ft.

19400 sq. ft.

Excess available for book storage --- 100 sq. ft.

MUSIC

50 Students (160 square feet per student) ------------------------------------- 8200 sq. ft.

Recital Hall, (seating 225, 8 square feet per person), with organ; and stage 40'–0" wide by 30'–0" deep -- 3000 sq. ft.

15 Rooms for individual practice / 100 square feet each -------------------- 1500 sq. ft.

3 Classrooms for 25 at an average of 15 square feet per student ----------- 1125 sq. ft.

Glee Club Room -- 500 sq. ft.

Directors' Room -- 200 sq. ft.

Assistant Directors' Room -- 170 sq. ft.

Janitors Room -- 150 sq. ft.

Toilet, Lavatory and Coat Room accommodation ----------------------------- 200 sq. ft.

6845 sq. ft.

Stairs, Corridors, etc. --- 1355 sq. ft.

8200 sq. ft.

Total Floor Area of this building -- 9000 sq. ft.

Floor Area required for this Department -------------------------------------- 8200 sq. ft.

Excess available for storage and other purposes ---------------------------- 800 sq. ft.

ART & ARCHITECTURE

50 Students (172 square feet per student) -------------------------------------- 8600 sq. ft.

Drafting Room -- 2400 sq. ft.

Studio for Cast Work --- 1200 sq. ft.

Lecture Room seating 50 (10 square feet per student) ------------------------ 500 sq. ft.

2 Small Rooms for free hand work, (to be used by students from other Departments)

800 sq. ft. in each --- 1600 sq. ft.

Reference and Study Room, 6000 volumes ------------------------------------ 500 sq. ft.

Directors' Room -- 170 sq. ft.

2 Instructors (1 room) --- 170 sq. ft.

Janitors (2 rooms / 150 sq. ft. each) -- 300 sq. ft.

Toilet and Cloak Room accommodation --------------------------------------- 300 sq. ft.

7140 sq. ft.

Stairs, Corridors, Etc. --- 1460 sq. ft.

8600 sq. ft.

The conditions governing a department of this character are subject to wide variation. Mr. Dana's experience as head of the Architectural Department at Yale University has been made the basis of our calculations as given above.

Total Floor Area of this building -- 9000 sq. ft.

Floor area required for this Department -------------------------------------- 8600 sq. ft.

Excess available for storage --- 400 sq. ft.

ENGINEERING

600 Students (63 square feet per student) ----------------------------------- 38000 sq. ft.

Civil Engineering -- 200 students

Sanitary Engineering --- 50 students

Mechanical Engineering --- 150 students

Electrical Engineering -- 100 students

Total -- 600 students

Drafting Room accommodation for 100 students (50 square sq. ft. per student)

--5000 sq. ft.

Lecture Room seating 200 / 10 sq. ft. per student -------------------------- 2000 sq. ft.

Lecture Room seating 100 / 10 sq. ft. per student -------------------------- 1000 sq. ft.

Classroom accommodation for 150 students – 25 to 40 in a class (15 square feet per student)

--- 2250 sq. ft.

General Engineering Laboratories, accommodation for 175 students, 20 to 25 in a Laboratory (40 sq. ft. per student) --- 3000 sq. ft.

Apparatus and other rooms in connection ---------------------------------- 1500 sq. ft.

Two small Laboratories for individual work ------------------------------- 600 sq. ft.

Sanitary Engineering Laboratories, accommodation for 25 students (40 sq. ft. per student)

--- 1000 sq. ft.

Apparatus and other rooms in connection --------------------------------- 500 sq. ft.

16850 sq. ft.

ENGINEERING, Continued

Forwarded --- 16850 sq. ft.

Small Laboratory for individual work ------------------------------------- 300 sq. ft.

Electrical Laboratories, accommodating 25 students (40 sq. ft. per student)

--- 1000 sq. ft.

Apparatus and other rooms in connection --------------------------------- 500 sq. ft.

Small Laboratory for individual work ------------------------------------- 300 sq. ft.

Mining Laboratories, accommodation for 40 students (75 sq. ft. per student)

--- 2800 sq. ft.

Museum and Reference Room --- 2500 sq. ft.

Miscellaneous Small Rooms -- 1500 sq. ft.

6 Professors (6 rooms / 150 sq. ft. each) -------------------------------- 900 sq. ft.

6 Professors' Research Laboratories, 170 sq. ft. each ---------------------- 1020 sq. ft.

20 Instructors (6 rooms / 150 sq. ft. each) ------------------------------------- 1700 sq. ft.

6 Janitors (6 rooms / 150 sq. ft. each) --------------------------------------- 900 sq. ft.

Toilet, Lavatory and Coat Room accommodation --------------------------- 1000 sq. ft.

31270 sq. ft.

Stairs, Corridors, etc. -- 6730 sq. ft.

38000 sq. ft.

Note:

The total unit compares closely with the unit used in the Engineering Building, Columbia University.

It is assumed that 100 or more students of this Department will be taking classed in the Physics and Chemistry Buildings, and in the workshops, situated in the Power House, during every period; leaving only 500 students for whom provision must be made in this building.

Total Floor Area of two Engineering Buildings ------------------------------ 44500 sq. ft.

Floor Area required for Department of Engineering ------------------------ 38000 sq. ft.

Excess available for storage of Apparatus, etc. -------------------------------- 6500 sq. ft.

AGRICULTURE & FORESTRY

300 Students (41 sq. ft. per student) ------------------------------------- 12300 sq. ft.

Agriculture --- 200 students

Forestry --- 100 students

Total --- 300 students

Agricultural & Forestry experimental laboratories, accommodations for 75 students (40 sq. ft. per student) --- 3000 sq. ft.

Apparatus and other rooms in connection -------------------------------------- 1000 sq. ft.

Lecture Room for 100 students (10 sq. ft. per student) ---------------------- 1000 sq. ft.

Classroom accommodation for 80 students (15 sq. ft. per student) --------- 1200 sq. ft.

Museum and Exhibit Room --- 2000 sq. ft.

3 Professors (2 rooms / 170 sq. ft. each) --------------------------------------- 340 sq. ft.

2 Professors' Research Laboratories / 150 sq. ft. each ------------------------- 300 sq. ft.

5 Instructors (3 rooms / 170 sq. ft. each) --------------------------------------- 510 sq. ft.

Janitor (2 rooms / 150 sq. ft. each) --- 300 sq. ft.

Toilet, Lavatory and Coat Room accommodation ------------------------------ 600 sq. ft.

10250 sq. ft.

Stairs, Corridors, etc. --- 2050 sq. ft.

12300 sq. ft.

It is assumed that 100 or more of the students of this Department will be working in the Physics, Chemistry and other buildings during every period; leaving only 200 students for whom provision must be made in this building.

Total Floor Area of this building -- 14000 sq. ft.

Floor Area required for Agricultural Dept. ----------------------------------- 12300 sq. ft.

Excess available for storage, etc. --- 1700 sq. ft.

An experimental station consisting of

· Model Dairy

· Stable

· Poultry Farm

· Piggery

· Kennels

· Conservatory

and other miscellaneous buildings pertaining to farming; to occupy a large treat of land (about 10 areas at least) indicated on the General Plan to be purchased.

PHYSICS & BOTANICAL LABORATORIES

--- 18150 sq. ft.

Physics Laboratories accommodation for 75 students (40 sq. ft. per student)

--- 3000 sq. ft.

Apparatus Rooms, Stock Rooms and other Rooms in connection ----------- 1500 sq. ft.

Botanical Laboratories, accommodation for 50 students (50 sq. ft. per student)

--- 2500 sq. ft.

Apparatus Rooms, Seale and Stock Rooms, and other rooms in connection --- 1000 sq. ft.

Lecture Room for 200 students (10 sq. ft. per student) ---------------------- 2000 sq. ft.

2 Classrooms for 50 (15 sq. ft.) --- 1500 sq. ft.

4 Professor (4 Rooms / 170 sq. ft.) --- 680 sq. ft.

2 Professors' Research Laboratories / 160 sq. ft. ------------------------------- 320 sq. ft.

```
10 Instructors (6 rooms / 170 sq. ft.) ----------------------------------------- 1700 sq. ft.

Janitors (3 Rooms / 150 sq. ft. each) ----------------------------------------- 450 sq. ft.

Toilet, Lavatory & Cloak Room accommodation ----------------------------- 500 sq. ft.
                                                                          15150 sq. ft.


Stairs, Corridors, etc. -------------------------------------------------------- 3000 sq. ft.
                                                                          18150 sq. ft.
```

Note:

The Laboratory unit used compares closely with the unit at Fayerweather Hall, Columbia University.

As this building will be used entirely by students of other departments, such as Engineering, Agriculture, etc., it is not practical to calculate a unit basis; the building as shown on the General Plan would hold 400 or more, if all rooms were occupied at one time.

```
Total floor area of this building ----------------------------------------- 19500 sq. ft.

Floor Area required for this Department ------------------------------- 18100 sq. ft.

Excess available for storage and other purposes ----------------------- 1400 sq. ft.
```

CHEMISTRY

```
-------------------------------------------------------------------------------- 30300 sq. ft.
```

To be used by Engineering, Medical, Agricultural and Architectural Students.

General Chemistry Laboratories, including Quantitative, Qualitative, Metallurgical, Analytical and other Laboratories, accommodating 200 students at an average of 50 sq. ft.

```
per student -------------------------------------------------------------------- 10000 sq. ft.

Apparatus and other Rooms in connection with Laboratories --------------- 4500 sq. ft.

Small Laboratories for individual work at 200 sq. ft. each ------------------- 1600 sq. ft.

Large Lecture Rooms for 250 (10 sq. ft.) ------------------------------------- 2500 sq. ft.

Small Lecture Room for 100 (10 sq. ft.) -------------------------------------- 1000 sq. ft.

Classroom accommodation for 150 students (15 sq. ft. per student) -------- 2250 sq. ft.

6 Professors' Rooms (170 sq. ft.) ------------------------------------------- 1020 sq. ft.

2 Professors' Research (160 sq. ft.) -----------------------------------------320 sq. ft.

12 Instructors' Rooms (6 Rooms / 170 sq. ft.) ------------------------------ 1020 sq. ft.

Janitors (3 rooms / 150 sq. ft.) --------------------------------------------- 450 sq. ft.

Lavatory and toilet accommodation ----------------------------------------- 600 sq. ft.
                                                                          25260 sq. ft.
```

Stairs, Corridors Etc. --- 5040 sq. ft.

30300 sq. ft.

Note:

The unit for Laboratories used compares closely with the unit in the Chemical Laboratories at Columbia University, New York.

Total Floor space in this building about -------------------------------------- 32000 sq. ft.

Amount required as per above schedule ------------------------------------- 30300 sq. ft.

Excess available for storage and other purposes ----------------------------- 1700 sq. ft.

MEDICAL GROUP

SCHOOL OF MEDICINE
SCHOOL OF DENTISTRY

200 students. --- 33000 sq. ft.

This Department is divided into two sections; one Building being assigned entirely to the classrooms and Lecture Rooms; and the other to the Medical and Biological Laboratories.

As it is impossible properly to analyze the requirements for a Department calling for such special study as a Medical School, without actually working them out in plan, we have followed the figures used in the new Medical School of the Western University of Pennsylvania. This provides for 200 students, with a total floor area of 32,000 sq. ft.; or 160 sq. ft. per student.

DORMITORY GROUP

DORMITORIES

120 students in a Dormitory (160 sq. ft.) -------------------------------------- 19200 sq. ft.

Each Dormitory is a full three-story building; the First and Second Floors being composed of suites of two Bedrooms and a Study (each Bedroom accommodating two students). Total 120 students. Allowance is also made for one bachelor instructor on each floor; and each floor would contain one General Lavatory.

The 12 Dormitories shown in color in the General Plan would accommodate 1440 students; and the S.W. Dormitory in which the Co-Operative Store occupies the First Floor would accommodate 90 students; making a total of 1530 students without building any of the buildings shown on the General Plan in the central portion of the Dormitory space, or in the tract indicated "to be purchased". In order to bring the total accommodations up to the maximum of 2800 students, it would be necessary to build only the Dormitories shown on the treat "to be purchased"; and it may, therefore, be possible always to preserve the fine architectural effect of the large open campus shown in the "Dormitory Group".

DORMITORIES, Continued

Each Dormitory contains

30 suites (2 students in a suite) 320 sq. ft. each ------------------------------- 9600 sq. ft.

30 Rooms (2 students in a room) 150 sq. ft. each ---------------------------- 4500 sq. ft.

3 suites for Bachelor Instructor (one on each floor / 300 sq. ft. each) -------- 900 sq. ft.

Toilet and Lavatory accommodation -- 1000 sq. ft.

16000 sq. ft.

Stairs, Corridors, etc. --- 3300 sq. ft.

19300 sq. ft.

Total floor area of each Dormitory -- 21000 sq. ft.

Floor area required as per above schedule ------------------------------------ 19300 sq. ft.

Excess available for storage, etc. -- 1700 sq. ft.

DINING HALLS

1200 students in each building (19 sq. ft. per student) ----------------------- 23500 sq. ft.

In order to avoid excessively large single Halls, and in order to make it possible to add moderate-sized units, as required for expansion, Halls seating 600 students each have been adopted. Each completed Dining Hall building contains two of these units of 600 each, with a central Kitchen and Pantry, capable of serving 1200 students. The large Halls are a single high story; with servants' quarters over the Kitchen.

Seating accommodation for 1200 students (14 sq. ft. per student) --------- 16800 sq. ft.

Kitchen and Pantry accommodation for 1200 (2 sq. ft. per student) -------- 2400 sq. ft.

19200 sq. ft.

Halls, Corridors, etc. --- 4300 sq. ft.

Total -- 23500 sq. ft.

Note:

The unit used for seating accommodation compares closely to the unit used at the College of New Rochelle, New Rochelle, New York; also in the Restaurant of the Vanderbilt Hotel, New York.

INFIRMARY

66 Beds – 275 sq. ft. per patient	18300 sq. ft.
60 Beds, 150 sq. ft. per bed. (Each patient having a single room)	9000 sq. ft.
Emergency Ward (6 beds, 80 sq. ft. per bed)	480 sq. ft.
Diet Kitchen	500 sq. ft.
Pharmacy	300 sq. ft.
2 Doctors' Suites / 300 sq. ft. each	600 sq. ft.
6 Nurses' Rooms (two rooms on each floor / 150 sq. ft. each)	900 sq. ft.
Convalescents' Lounging Room on First Floor	3500 sq. ft.
	15280 sq. ft.
Stairs, Corridors, etc.	3020 sq. ft.
Total	18300 sq. ft.

Note:

The unit used for bed accommodation compares closely with the unit used at the French Hospital, New York City.

AUDITORIUM

	seating 4000.
Main Floor	seating 3000.
Balcony	seating 1000.
Total	seating 4000.
Seating 3000 on Main Floor, including aisles (8 sq. ft.)	24000 sq. ft.
Stairs, Corridors	6000 sq. ft.
Stage, 70′–0″ wide × 40′–0″ deep	2800 sq. ft.
Dressing Rooms	800 sq. ft.
Toilet accommodation	1600 sq. ft.
Total	35200 sq. ft.

Note:

The unit used for seating accommodation, compares closely with the unit used in the Hill Memorial Auditorium, University of Michigan.

LIBRARY

225000 Volumes --- 10800 sq. ft.

Ground Floor contains Administrative Offices as follows:

- President's Office
- Dean's Office
- Registrar's Office
- Secretary's Office
- Business Manager's Office
- Miscellaneous Offices

MAIN FLOOR:

Reading Room accommodation for 250 students (22 sq. ft. per student), and providing shelving for 25000 volumes --------------------------------------- 5500 sq. ft.

Stack accommodation for 200000 volumes (in 4 tiers) / 55 sq. ft. per 1000 volumes

--- 2750 sq. ft.

Librarian's Office -- 250 sq. ft.

Assistant Librarian's Office -- 150 sq. ft.

8650 sq. ft.

Halls, Stairs, & Corridors -- 2150 sq. ft.

Total --- 10800 sq. ft.

Note:

The units used in figuring this building correspond with similar units in the College Library (which has been worked out in detail on the 1/16″ scale drawings) and with the figures given by "The Library Bureau" of New York.

ALUMNI HALL

--- 11525 sq. ft.

①
编者注: 原文如此, 疑应为fiat (指令)。

It is intended that the funds for this building should be contributed by the Alumni of Tsing Hua, as a figt[①] to their College.

- Auditorium on main floor, seating 1000.
- Club & Faculty Rooms on Second Floor.
- Toilet accommodations, storage, etc. in Basement.

Seating accommodation for 1000 (8 sq. ft. per person) ---------------------- 8000 sq. ft.

Stairs, Corridors, etc. --- 2000 sq. ft.

Stage 50′−0″ wide × 25′−0″ deep --- 1025 sq. ft.

Dressing Rooms, etc. -- 500 sq. ft.

Total -- 11525 sq. ft.

Note:

The unit used for seating compares closely with the unit used at the Hill Memorial Auditorium, University of Michigan.

OBSERVATORY

-- 1400 sq. ft.

Observatory (with dome) --- 600 sq. ft.

Apparatus Room -- 300 sq. ft.

Record Room and Library --- 200 sq. ft.

Director's Room -- 170 sq. ft.

1270 sq. ft.

Halls, Corridors, etc. --- 130 sq. ft.

Total -- 1400 sq. ft.

Note:

As this Building will be used by the students of other Departments, no definite number of students has been assumed; the Building being about the same dimensions as the Observatory at Columbia University, New York.

GYMNASIUM

1500 Students --- 20850 sq. ft.

MAIN FLOOR:

Gymnasium 90′-0″ × 125′-0″ providing a running track, 20 laps to the mile

---11250 sq. ft.

Locker Room on Second Floor, on one wing, accommodation for 1500 ------ 4800 sq. ft.

First Floor below this Locker Room will be occupied by showers, Faculty Locker Rooms, Hand Ball Court, etc.

Swimming Pool and accessories, will occupy the other wing ---------------- 4800 sq. ft.

20850 sq. ft.

Note:

The unit used in figuring this building, correspond with similar unit used in the College Gymnasium (which has been worked out in detail on the 1/16″ scale drawings and with the figures given by Mr. A.G. Spalding, New York City).

Tsing Hua High School

CLASSROOM GROUP

CLASSROOMS BUILDINGS
- N.W. Classroom Building: 400
- S.W. Classroom Building: 400
- Total: 800

N.W. Classroom Building – 400students (34 sq. ft. per student) ------------13600 sq. ft.

16 Classrooms (24 students in each / 23 sq. ft. per student) ------------------ 8812 sq. ft.

16 Teachers (5 Rooms, 3 in a room / 300 sq. ft. each) ------------------------ 1500 sq. ft.

Toilet and Lavatory accommodation --- 1000 sq. ft.

11312 sq. ft.

Stairs, Corridors, etc. --- 2288 sq. ft.

Total -- 13600 sq. ft.

Total floor space in this building --15215 sq. ft.

Floor space required as per above schedule --------------------------------13600 sq. ft.

Excess available for storage, and other purposes ---------------------------- 1615 sq. ft.

SOUTH-WEST CLASSROOM BUILDING

400 students (49 sq. ft. per student) -- 19000 sq. ft.

16 Classrooms (24 students in each / 28 sq. ft. per student) ------------------ 9832 sq. ft.

Assembly Hall seating 300 (10 sq. ft. per student) --------------------------- 3000 sq. ft.

16 Teachers (5 rooms 3 in a room / 300 sq. ft. per student) ------------------ 1500 sq. ft.

Toilet, Lavatory and Coat Room accommodation --------------------------- 1000 sq. ft.

15332 sq. ft.

Stairs, Corridors, etc. -- 3668 sq. ft.

19000 sq. ft.

Total Floor space in this building -- 22000 sq. ft.

Floor space required as per above schedule -------------------------------- 19000 sq. ft.

Excess available for book storage and other purpose ------------------------- 3000 sq. ft.

Tsing Hua High School

DORMITORY GROUP, Continued

DORMITORIES

Each Dormitory is a full three-story building; the North-East, Centre- East, Centre-West and South-East Dormitories being composed entirely of large rooms (4 students in each room;) allowance is also made for two (2) bachelor instructors on each floor, and each floor contain a central Lavatory.

The North-East Dormitory is composed entirely of double rooms (2 students in a room;) allowance is also made for two (2) bachelor instructors on each floor; and each floor would contain a central lavatory.

North-East Dormitory --- 175 Students

South-East Dormitory --- 175 Students

Centre-West Dormitory --- 200 Students

Centre-East Dormitory -- 200 Students

North-East Dormitory -- 50 Students

800 Students

Tsing Hua High School

DORMITORY GROUP, Continued

Taking the two centre Dormitories as typical of their kinds, the Dormitories will each contains:

50 Rooms accommodating 4 students in each / 300 sq. ft. per room ------ 15000 sq. ft.

6 Suites for Bachelor Instructors / 300 sq. ft. -------------------------------- 1800 sq. ft.

Toilet and Lavatory accommodation --- 1200 sq. ft.

18000 sq. ft.

Stairs, Corridors, etc. --- 3600 sq. ft.

Total --- 21600 sq. ft.

The North-West Dormitory will contain:

25 Rooms (2 students in a room / 320 sq. ft. each) --------------------------- 8000 sq. ft.

6 Teachers' Suites / 300 sq. ft. each --- 1800 sq. ft.

Toilet and Lavatory accommodation -- 600 sq. ft.

10400 sq. ft.

Stairs, Corridors, etc. --- 2600 sq. ft.

Total --- 13000 sq. ft.

Tsing Hua High School

DORMITORY GROUP

DINING HALL

800 students / 19 sq. ft. -- 15400 sq. ft.

Owing to the limited space available, it has been found necessary to make this building one large single Hall with a central Kitchen and Pantry, capable of serving 1000 students. The large Hall is a single high story, with servant's quarters over the Kitchen.

Seating accommodation for 800 students at 14 sq. ft. per student --------- 11200 sq. ft.

Kitchen and Pantry accommodation for 800 students at 2 sq. ft. per student ---- 1600 sq. ft.

12800 sq. ft.

Halls, Corridors, etc. --- 2600 sq. ft.

Total -- 15400 sq. ft.

Note:

The unit used for seating accommodation, compares closely with the unit used at the College of New Rochelle; and also in the Restaurant of the Vanderbilt Hotel, both of New York City.

Tsing Hua Middle School

CLASSROOM GROUP

THREE CLASSROOM BUILDINGS

Total accommodation for 1000 students (44 sq. ft. per student) ------------ 44100 sq. ft.

Three Buildings containing 42 Classrooms, 24 students in each (23 sq. ft. per student)

--- 23000 sq. ft.

Assembly Halls in two of the Buildings, seating 300 (10 sq. ft. Per student) ---- 6000 sq. ft.

Rooms for 42 teachers (14 rooms, 3 in a room / 300 sq. ft. each) --------- 2000 sq. ft.

Headmaster's Room --- 450 sq. ft.

Toilet, Lavatory & Coat Room accommodation ----------------------------- 3000 sq. ft.

34450 sq. ft.

Stairs, Corridors, etc. -- 9650 sq. ft.

Total -- 44100 sq. ft.

Note:

This Classroom unit compares closely with the unit used in Classrooms, Loomis Institute, Windsor, Conn.

Total Floor area of three Arts Buildings -------------------------------------- 51000 sq. ft.

Floor Are required by this Department -------------------------------------- 44100 sq. ft.

Excess available for book storage and other purposes ----------------------- 6900 sq. ft.

Tsing Hua Middle School

DORMITORY GROUP. Continued

DORMITORIES

200 students in each (108 sq. ft. per student) ------------------------------- 21600 sq. ft.

Each Dormitory is a full three-story building, being composed of 50 rooms (each room accommodating 4 students;) allowance is also made for two bachelor instructors on each floor; and each floor would contain one General Lavatory.

The five (5) Dormitories shown in color on the General Plan would accommodate 1200.

Each Dormitory contains:

50 Rooms (accommodating 4 students in each / 300 sq. ft. per room) ------- 15000 sq. ft.

6 Suites for Bachelor Instructors / 300 sq. ft. --------------------------------- 1800 sq. ft.

Toilet and Lavatory accommodation --- 1200 sq. ft.

                                                                                18000 sq. ft.

Stairs, Corridors, etc. --- 3600 sq. ft.

Total --- 21600 sq. ft.

Tsing Hua Middle School

DORMITORY GROUP

DINING HALL. Seating

1000 / 19 sq. ft. --- 19200 sq. ft.

Owing to the limited space available, it has been found necessary to make this building one large single Hall with a central Kitchen and Pantry, capable of serving 1000 students. The large Hall is a single high story, with servant's quarters over the Kitchen.

Seating accommodation for 1000 students (14 sq. ft. per student) --------- 14000 sq. ft.

Kitchen and Pantry accommodation for 1000 (2 sq. ft. per student) -------- 2000 sq. ft.

                                                                                16000 sq. ft.

Halls, Corridors, etc. -- 3200 sq. ft.

Total --- 19200 sq. ft.

Note:

The unit used for seating accommodation compares closely with the unit used at the College of New Rochelle; and also in the Restaurant of the Vanderbilt Hotel, both of New York City.

TSING HUA UNIVERSITY

Total number of students provided for in the foregoing analysis as follows:

ACADEMIC GROUP

College of Arts and Letters -- 500 Students

School of Chinese Classics -- 200 Students

School of Education -- 400 Students

School of Law -- 300 Students

School of Journalism -- 200 Students

School of Architecture -- 50 Students

School of Music -- 50 Students

1700 Students

SCIENCE GROUP

School of Engineering -- 600 Students

School of Agriculture & Forestry --- 300 Students

900 Students

MEDICAL GROUP

School of Medicine
School of Dentistry -- 200 Students

200 Students

Middle School Group -- 1000 Students

High School Group --- 800 Students

1800 Students

Total --- 4600 Students

清华大学

学院建筑组群

文学与艺术学院

500名学生（41平方英尺／学生）----------------------------- 20500 平方英尺

- 英国文学系
- 历史系
- 经济系
- 语言系
- 政治与社会科学系

500名学生所需教室——每班30～40人（15平方英尺／学生）--- 7500 平方英尺

一间250座的讲堂（10平方英尺／学生）------------------------ 2500 平方英尺

10000册藏书量的图书室 ---------------------------------- 1500 平方英尺

8位教授（8间房／每间175平方英尺）------------------------ 1400 平方英尺

20位教师（10间房／每间160平方英尺）------------------------ 1600 平方英尺

清洁管理人员（4间房／每间150平方英尺）---------------------- 600 平方英尺

一些杂用小房间 -------------------------------------- 1000 平方英尺

厕所、盥洗室及衣帽间 ----------------------------------- 1000 平方英尺

17100 平方英尺

楼梯、走廊等 --------------------------------------- 3400 平方英尺

该院总计 --- 20500 平方英尺

注：

教室单元与纽约大学的教室单元相仿。

中国文学系——200名学生（30平方英尺／学生）----------------- 6000 平方英尺

200名学生所需教室——每班30～40人（15平方英尺／学生）--- 3000 平方英尺

4位教授（4间房／每间170平方英尺）------------------------ 680 平方英尺

4位讲师（2间房／每间170平方英尺）------------------------ 340 平方英尺

图书室与自习室，5000册藏书量 ---------------------------- 500 平方英尺

厕所、盥洗室及衣帽间 ----------------------------------- 500 平方英尺

5020 平方英尺

楼梯、走廊等 --- 980 平方英尺

该系总计 -- 6000 平方英尺

两文学院系楼总面积 --- 43000 平方英尺

文学与艺术学院所需面积 ------------------------------------- 20500 平方英尺

中国文学系所需面积 --- 6000 平方英尺

26500 平方英尺

剩余面积 -- 16500 平方英尺

300名其他院系学生前来上课所需面积（41平方英尺／学生）--- 12300 平方英尺

剩余可用于藏书以及其他用途的面积 -------------------------- 4200 平方英尺

教育学院

400名学生（45平方英尺／学生）----------------------------- 18000 平方英尺

300名学生所需教室——每班24～40人（15平方英尺／学生）---- 4500 平方英尺

一间大型讲堂，200座（10平方英尺／学生）------------------ 2000 平方英尺

幼儿园 -- 900 平方英尺

模拟学校教室 -- 500 平方英尺

图书室和自习室 -- 3000 平方英尺

其他小房间 -- 800 平方英尺

6位教授（6间房／每间170平方英尺）------------------------ 1020 平方英尺

20位教师（10间房／每间170平方英尺）---------------------- 1700 平方英尺

清洁管理人员（2间房／每间150平方英尺）------------------- 300 平方英尺

厕所、盥洗室及衣帽间 --------------------------------------- 900 平方英尺

15620 平方英尺

楼梯、走廊等 --- 2380 平方英尺

该系总计 --- 18000 平方英尺

注：

　　受当地条件所限，很难将其与相近功能的西方国家大学建筑做比较。但上述功能安排大致与纽约的哥伦比亚大学师范学院相似。

　　预计可能有100名或更多该系学生将去文学与艺术学院或其他院系上课，即需要保证解决剩下300名学生在该建筑中的活动需求。

该建筑总面积 --- 19500 平方英尺

教育学院所需面积 --- 18000 平方英尺

剩余可用于藏书以及其他用途的面积 ---------------------------- 1500 平方英尺

法学院与新闻学院

这两所学院预计会设计在同一座建筑中。如果需要扩建，它们可以分别获得规划中的一幢校舍建筑。

法学院——300名学生（41平方英尺／学生）------------------ 12300 平方英尺

225名学生所需教室——每班24至40人（15平方英尺／学生）---- 3475 平方英尺

200座讲堂（10平方英尺／学生）------------------------------ 2000 平方英尺

辩论室 -- 900 平方英尺

图书室与自习室，9000册藏书量 ------------------------------ 1000 平方英尺

一些杂用小房间 -- 750 平方英尺

3位教授（3间房／每间170平方英尺）------------------------- 510 平方英尺

8位教师（4间房／每间170平方英尺）------------------------- 680 平方英尺

2位清洁管理员（2间房／每间150平方英尺）-------------------- 300 平方英尺

厕所、盥洗室及衣帽间 -------------------------------------- 600 平方英尺

　　　　　　　　　　　　　　　　　　　　　　　　　　　　　10215 平方英尺

楼梯、走廊等 -- 2085 平方英尺

　　　　　　　　　　　　　　　　　　　　　　　　　　　　　12300 平方英尺

注：

该部分设计所用指标与宾夕法尼亚大学法学院相似。

预计将有75名或更多该院系学生前往文学与艺术学院或其他院系上课，即需要保证解决剩下225名学生在该建筑中的活动需求。

新闻学院——200名学生，30平方英尺／学生 ------------------ 6100 平方英尺

150名学生所需教室，15平方英尺／学生 ---------------------- 2250 平方英尺

图书馆及自习室，14000册藏书量 ---------------------------- 2500 平方英尺

2位教授（2间房／每间170平方英尺）------------------------- 340 平方英尺

4位教师（2间房／每间170 平方英尺）------------------------ 340 平方英尺

厕所、盥洗室及衣帽间 -------------------------------------- 500 平方英尺

　　　　　　　　　　　　　　　　　　　　　　　　　　　　　5930 平方英尺

楼梯、走廊等 --- 1170 平方英尺

该院总计 -- 7100 平方英尺

预计将有50名或更多该院系学生前往文学与艺术学院或其他院系上课，即需要保证解决剩下150名学生在该建筑中的活动需求。

该建筑总面积 -- 19500 平方英尺

法学院所需面积 -- 12300 平方英尺

新闻学院所需面积 -- 7100 平方英尺

19400 平方英尺

剩余可用于藏书的面积 -- 100 平方英尺

音乐系

50名学生（160平方英尺／学生）-------------------------------- 8200 平方英尺

演奏厅（225座，8平方英尺／人），有管风琴，舞台40英尺宽、30英尺深

-- 3000 平方英尺

15间个人练习室／每间100平方英尺 ---------------------------- 1500 平方英尺

3间25人教室，15平方英尺／学生 ------------------------------ 1125 平方英尺

合唱俱乐部室 -- 500 平方英尺

指导教师室 -- 200 平方英尺

指导老师助理室 -- 170 平方英尺

清洁员室 -- 150 平方英尺

厕所、盥洗室及衣帽间 -- 200 平方英尺

6845 平方英尺

楼梯、走廊等 -- 1355 平方英尺

8200 平方英尺

该建筑总面积 -- 9000 平方英尺

该系所需面积 -- 8200 平方英尺

剩余可用于藏书以及其他用途的面积 ---------------------------- 800 平方英尺

艺术与建筑学院

50名学生（172平方英尺／学生）-------------------------------- 8600 平方英尺

制图室 -- 2400 平方英尺

铸造工作室 -- 1200 平方英尺

50座讲堂（10平方英尺／学生）----------------------------------- 500 平方英尺

2间开放的小手工工作室（为其他院系同学准备），每间800平方英尺

---1600 平方英尺

会议与自习室，6000册藏书量 ------------------------------ 500 平方英尺

主任室 --- 170 平方英尺

2位教师（1间房）--- 170 平方英尺

清洁管理员（2间房／每间150平方英尺）----------------------- 300 平方英尺

厕所及寄存间 --- 300 平方英尺

7140 平方英尺

楼梯、走廊等 --- 1460 平方英尺

8600 平方英尺

这类学院的条件差异很大，丹纳先生[①]根据他任耶鲁大学建筑学院系主任的经验提供了上述数值。

该建筑总面积 --- 9000 平方英尺

该院系所需面积 -- 8600 平方英尺

剩余可用于储藏的面积 --- 400 平方英尺

工学院

600名学生（63平方英尺／学生）------------------------------- 38000 平方英尺

土木工程系 --- 200名学生

卫生工程系 --- 50名学生

机械工程系 --- 150名学生

电子工程系 --- 100名学生

总计 --- 600名学生

100名学生的制图室（50平方英尺／学生）--------------------- 5000 平方英尺

200座讲堂（10平方英尺／学生）------------------------------- 2000 平方英尺

100座讲堂／10平方英尺 -- 1000 平方英尺

150名学生的教室，每班25～40人（15平方英尺／学生）------- 2250 平方英尺

175名学生的综合工程实验室，每间20～25人（40平方英尺／学生）

-- 3000 平方英尺

仪器室及其他相邻房间 --- 1500 平方英尺

2间小型个人工作实验室 -- 600 平方英尺

①
编者注：指墨菲事务所的合伙人Henry Richard Dana.

卫生工程实验室，供25名学生使用（40平方英尺／学生）----- 1000 平方英尺

仪器室及其他相邻房间 -- 500 平方英尺

16850 平方英尺

工学院（续）

前述合计 -- 16850 平方英尺

小型个人工作实验室 -- 300 平方英尺

电子实验室，25名学生使用（40平方英尺／学生）------------ 1000 平方英尺

仪器室及其他相邻房间 -- 500 平方英尺

小型个人工作实验室 -- 300 平方英尺

矿业实验室，40名学生使用（75平方英尺／学生）------------ 2800 平方英尺

博物馆与会议室 -- 2500 平方英尺

一些杂用小房间 -- 1500 平方英尺

6位教授（6间房／每间150平方英尺）-------------------------- 900 平方英尺

6间教授研究实验室，每间170平方英尺 ------------------------ 1020 平方英尺

20名教师（6间房／每间150平方英尺）-------------------------- 1700 平方英尺

6名清洁员（6间房／每间150平方英尺）-------------------------- 900 平方英尺

厕所、盥洗室及衣帽间 -- 1000 平方英尺

31270 平方英尺

楼梯、走廊等 -- 6730 平方英尺

38000 平方英尺

注：

该建筑设计所采用指标与哥伦比亚大学工程楼相似。

预计将有100名或更多该院系学生前往物理馆和化学馆或其他院系上课，或前往工场、发电厂等，即需要保证解决剩下500名学生在该建筑中的活动需求。

工程楼总面积 -- 44500 平方英尺

工程学院所需面积 -- 38000 平方英尺

剩余可用于仪器储藏的面积 ------------------------------------ 6500 平方英尺

农业与林业学院

300名学生（41平方英尺／学生）------------------------------ 12300 平方英尺

农业学 -- 200名学生

林业学 -- 100名学生

总计 -- 300名学生

农业与林业实验室

75名学生使用（40平方英尺／学生）--------------------------- 3000 平方英尺

仪器室及其他相邻房间 --- 1000 平方英尺

供100名学生使用的讲堂（10平方英尺／学生）----------------- 1000 平方英尺

80名学生使用的教室（15平方英尺／学生）-------------------- 1200 平方英尺

博物馆和展览室 -- 2000 平方英尺

3名教授（2间房／每间170平方英尺）-------------------------- 340 平方英尺

2间教授研究实验室／每间150平方英尺 ------------------------- 300 平方英尺

5名教师（3间房／每间170平方英尺）-------------------------- 510 平方英尺

清洁管理员（2间房／每间150平方英尺）----------------------- 300 平方英尺

厕所、盥洗室及衣帽间 --- 600 平方英尺

10250 平方英尺

楼梯、走廊等 -- 2050 平方英尺

12300 平方英尺

预计将有100名或更多该院系学生前往物理楼、化学楼或其他学院学习，即需要保证解决剩下200名学生在该建筑中的活动需求。

该建筑总面积 -- 14000 平方英尺

农业学院所需面积 --- 12300 平方英尺

剩余可用于储藏的面积 --- 1700 平方英尺

一个实验站，包括以下部分：

· 乳制品生产场

· 马厩

· 家禽饲养场

· 猪圈

· 犬舍

· 暖房

以及其他与农业有关的杂项建筑物，将会占用大量土地（至少10英亩），标注在总规划图上以待购买。

物理和植物学实验室 -- 18150 平方英尺

物理实验室，75名学生使用（40平方英尺／学生）------------ 3000 平方英尺

仪器室、储藏室以及其他相邻房间 -------------------------------- 1500 平方英尺

植物学实验室，50名学生使用（50平方英尺／学生）---------- 2500 平方英尺

仪器室、储藏室以及其他相邻房间 -------------------------------- 1000 平方英尺

200名学生使用的讲堂（10平方英尺／学生）-------------------- 2000 平方英尺

2间教室，50名学生使用（15平方英尺／学生）---------------- 1500 平方英尺

4名教授（4间房／每间170平方英尺）--------------------------- 680 平方英尺

2间教授研究实验室／每间160平方英尺 ------------------------ 320 平方英尺

10名教师（10间房／每间170平方英尺）------------------------ 1700 平方英尺

清洁管理员（3间房／每间150平方英尺）------------------------ 450 平方英尺

厕所、盥洗室及衣帽间 -- 500 平方英尺

15150 平方英尺

楼梯、走廊等 -- 3000 平方英尺

18150 平方英尺

注：

实验室单元设计与哥伦比亚大学费耶维舍楼相似。

由于该建筑将完全由其他院系学生使用，如工程学院、农业学院等，因此很难计算单元基本数据。如按照上述数据，可一次承载400余人在建筑中活动。

该建筑总面积 -- 19500 平方英尺

该部门所需面积 -- 18150 平方英尺

剩余可用于储藏或其他功能的面积 ------------------------------ 1400 平方英尺

化学馆 -- 30300 平方英尺

将由工学院、医学院、农业学院及建筑学院学生使用。

综合化学实验室，包括化学定量、化学定性、冶金、化学分析等其他实验室，供200名学生使用，50平方英尺／人 ------------------------------ 10000 平方英尺

仪器室以及其他相邻房间 -- 4500 平方英尺

小型个人工作实验室，每间200平方英尺 -------------------- 1600 平方英尺

大型讲堂，250人使用（10平方英尺／人）-------------------- 2500 平方英尺

小型讲堂，100人使用（10平方英尺／人）-------------------- 1000 平方英尺

150名学生使用的教室（15平方英尺／学生）---------------- 2250 平方英尺

6名教授（每间170平方英尺）------------------------------------ 1020 平方英尺

2间教授研究实验室（每间160平方英尺）------------------------ 320 平方英尺

12名教师（6间房，每间170平方英尺）------------------------ 1020 平方英尺

清洁管理员（3间房，每间150平方英尺）------------------------ 450 平方英尺

厕所、盥洗室及衣帽间 -- 600 平方英尺

25260 平方英尺

楼梯、走廊等 -- 5040 平方英尺

30300 平方英尺

注：

实验室设计所用指标与哥伦比亚大学化学实验室相似。

该建筑总面积 -- 32000 平方英尺

上述所需面积 -- 30300 平方英尺

剩余可用于储藏或其他功能的面积 ---------------------------- 1700 平方英尺

医学建筑组团

医学院及牙科学院

200名学生 -- 33000平方英尺

该学院分为两座建筑，一座全部为教室与讲堂，另一座为医学和生物学实验室。

由于很难计算医学院这种需求特殊的学科所需各功能面积，我们没有通过平面的计算得出结果，而选择沿用宾夕法尼亚州西部大学（the Western University of Pennsylvania）新竣工的医学院楼的规制。这是一座拥有200名学生，总面积32000平方英尺，即人均160平方英尺的建筑。

宿舍建筑组团

宿舍楼

120名学生一栋楼（160平方英尺／学生）---------------------- 19200平方英尺

宿舍楼都是三层建筑。一、二层房间是由两间卧室一间书房组成的套房（每间卧室可住两名学生）。总共能住120名学生。每层有一间为获得津贴的学士教师准备的房间。每层还有一个公用卫生间。

总图上用颜色标注出来的12栋宿舍楼总共可容纳1440名学生。一层为合作社商店的西南宿舍楼可住90名学生。因此，在宿舍区的中央空地以及地形图中"待购买"的地块上，无须再建其他建筑就已经可以容纳1530名学生。为了使可容纳学生

总数达到2800人，还需要在"待购买"的土地上建设宿舍楼。这样一来，宿舍建筑组团便可以在阔大、开敞的校园中保留其优美的建筑形态。

宿舍楼（续）

每栋宿舍楼包括：

30间套房（每间2名学生）每间320平方英尺 -------------------- 9600平方英尺

30间宿舍（每间2名学生）每间150平方英尺 -------------------- 4500 平方英尺

3间学士教师套房（每层一间）每间300平方英尺 ---------------- 900 平方英尺

厕所和盥洗室 --- 1000 平方英尺

16000 平方英尺

楼梯、走廊等 --- 3300 平方英尺

19300 平方英尺

每栋宿舍楼面积 --- 21000 平方英尺

上述所需面积 --- 19300 平方英尺

剩余可用于储藏或其他功能的面积 ----------------------------- 1700 平方英尺

食堂

每栋建筑容纳1200名学生（19平方英尺／学生）-------------- 23500 平方英尺

为了避免出现过大的单个饭厅，并且保留扩建适当大小的饭厅的可能性，食堂最终选择了可容纳600名学生的尺寸。每一食堂建筑布置了两个这种能容纳600人的饭厅、一个中央厨房和储藏室，总共可容纳1200名学生。饭厅为单层空间，服务人员用房在厨房旁边。

1200名学生的座位（14平方英尺／学生）--------------------- 16800 平方英尺

可供应1200人就餐的厨房和储藏室（2平方英尺／人）--------- 2400 平方英尺

19200 平方英尺

门厅、走廊等 --- 4300 平方英尺

总计 --- 23500 平方英尺

注：

用餐区单元设计与纽约新罗谢尔学院及纽约范德比尔特酒店的餐厅相似。

医务所

66个床位——275平方英尺／病人 ----------------------------- 18300 平方英尺

60个床位（150平方英尺／床位，每位病人拥有独立房间）---- 9000 平方英尺

急诊病房（6个床位，80平方英尺／床位）----------------------- 480 平方英尺

营养厨房 --- 500 平方英尺

药房 --- 300 平方英尺

2间医生套房（每间300平方英尺）----------------------------- 600 平方英尺

6间护士室（每层两间，每间150平方英尺）-------------------- 900 平方英尺

一层的康复活动室 -- 3500 平方英尺

15280 平方英尺

楼梯、走廊等 --- 3020 平方英尺

总计 --- 18300 平方英尺

注：

病床部分的设计指标与纽约市法国医院（the French Hospital）相似。

大礼堂 4000 座

主层 --- 3000 座

挑台 --- 1000 座

合计 --- 4000 座

主层3000座区域及过道（8平方英尺／人）---------------------- 24000 平方英尺

楼梯、走廊 --- 6000 平方英尺

舞台，70英尺宽×40英尺深 ----------------------------------- 2800 平方英尺

后台 --- 800 平方英尺

厕所 --- 1600 平方英尺

合计 --- 35200 平方英尺

注：

座席部分的设计指标与密歇根大学希尔纪念礼堂相似。

图书馆

225000册藏书量 -- 10800 平方英尺

首层的行政办公室包括：

- 校长办公室
- 教务长办公室
- 注册中心
- 秘书办公室
- 经理办公室
- 其他项办公室

主层：

可容纳250名学生的阅览室（22平方英尺／学生），内含可放置25000册图书
的书架 -- 5500 平方英尺

200000册藏书量的书库（4层）（每1000册55平方英尺）------ 2750 平方英尺

图书管理员办公室 -- 250 平方英尺

图书管理员助理办公室 -- 150 平方英尺

8650 平方英尺

门厅、楼梯和走廊 -- 2150 平方英尺

合计 -- 10800 平方英尺

注：

这座建筑中各功能单元的设计指标与纽约大学图书馆（已在1/16英寸的图纸中
呈现）相似，数据由纽约图书馆管理局（The Library Bureau）给出。

校友纪念堂   11525 平方英尺

这座建筑将由清华校友出资建设，作为送给学校的一份礼物。

- 位于主层的1000座礼堂
- 位于二层的俱乐部及教职工室
- 位于地下室的厕所及储藏室

1000座坐席（8平方英尺／座）-- 8000 平方英尺

楼梯、走廊等 -- 2000 平方英尺

舞台，50英尺宽×25英尺深 -------------------------------- 1025 平方英尺

后台等 -- 500 平方英尺

总计 --- 11525 平方英尺

注：

坐席部分设计指标与密歇根大学希尔纪念礼堂相似。

天文台   1400 平方英尺

天文台（及穹顶）-------------------------------------- 600 平方英尺

仪器室 -- 300 平方英尺

记录室及图书馆 --- 200 平方英尺

指导员室 -- 170 平方英尺

                                             1270 平方英尺

门厅、走廊等 --- 130 平方英尺

总计 --- 1400 平方英尺

注：

由于该建筑将被其他院系学生使用，因此无法确切估计使用者将会有多少人。该建筑将采用与纽约哥伦比亚大学天文台相似的规制。

体育馆

1500名学生 -- 20850 平方英尺

主层：

体育馆宽90英尺, 长125英尺, 有一条环形跑道, 20圈合1英里 ---- 11250 平方英尺

更衣室位于二层, 在建筑一翼, 可容纳1500人 -------------------- 4800 平方英尺

更衣室楼下的一层部分为淋浴室、教职工更衣室、手球室等。

泳池及附属用房在建筑另一翼 ---------------------------- 4800 平方英尺

                                             20850 平方英尺

注：

该建筑各功能单元设计与纽约大学体育馆相似（该体育馆按照纽约市的A.G.斯波尔丁先生（Mr. A.G. Spalding）给出的范图设计，已在1/16英寸的图纸中呈现）。

清华学校高等科

教室建筑群

教学楼
- 西北教学楼：400人
- 西南教学楼：400人
- 合计：800人

西北教学楼——400名学生（34平方英尺／学生）------------- 13600 平方英尺

16间教室（每间24名学生，23平方英尺／学生）---------------- 8812 平方英尺

16位教师（5间房，3人一间，每间300平方英尺）-------------- 1500 平方英尺

厕所和盥洗室 -- 1000 平方英尺

11312 平方英尺

楼梯、走廊等 -- 2288 平方英尺

合计 -- 13600 平方英尺

该建筑总面积 -- 15215 平方英尺

上述所需面积 -- 13600 平方英尺

剩余可用于储藏或其他功能的面积 ------------------------------ 1615 平方英尺

西南教学楼

400名学生（49平方英尺／学生）------------------------------- 19000 平方英尺

16间教室（每间24名学生，28平方英尺／学生）---------------- 9832 平方英尺

300座会堂（10平方英尺／学生）------------------------------- 3000 平方英尺

16名教师（5间房，3人一间，每间300平方英尺）-------------- 1500 平方英尺

厕所、盥洗室及衣帽间 --- 1000 平方英尺

15332 平方英尺

楼梯、走廊等 -- 3668 平方英尺

19000 平方英尺

该建筑总面积 -- 22000 平方英尺

上述所需面积 -- 19000 平方英尺

剩余可用于藏书或其他功能的面积 ------------------------------ 3000 平方英尺

宿舍建筑群（续）

宿舍楼

宿舍楼均为三层建筑，包括东北楼、正东楼、正西楼和东南楼。上述楼内均为可住4名学生的大房间。同时，每层为2名学士教师准备津贴宿舍，另设一个公共卫生间。

此外，东北楼内还有可住2名学生的两室房间。每层有两间为单身教师准备的单人房间（作为其津贴补助），每层都有一个公共卫生间。

东北楼 --- 175人
东南楼 ---175人
正西楼 ---200人
正东楼 --- 200人
东北楼 --- 50人
　　　　　　　　　　　　　　　　　　　　　　　　　　　　800人

宿舍建筑群（续）

正西楼和正东楼包括：

50间4人间（每间300平方英尺）----------------------------- 15000 平方英尺
6间学士教师套房（每间300平方英尺）-------------------------- 1800 平方英尺
厕所和盥洗室 --- 1200 平方英尺
　　　　　　　　　　　　　　　　　　　　　　　　18000 平方英尺

楼梯、走廊等 --- 3600 平方英尺
总计 --- 21600 平方英尺

西北楼包括：

25间2人间（每间320平方英尺）----------------------------- 8000 平方英尺
6间学士教师套房（每间300平方英尺）-------------------------- 1800 平方英尺
厕所和盥洗室 --- 600 平方英尺
　　　　　　　　　　　　　　　　　　　　　　　　10400 平方英尺

楼梯、走廊等 --- 2600 平方英尺
总计 --- 13000 平方英尺

宿舍建筑群

食堂

800名学生（19平方英尺／学生）------------------------------ 15400 平方英尺

由于面积限制，该建筑内包括一个可容纳1000名学生大型独立饭厅、一个中央厨房和储藏室。此大厅为层高较高的单层建筑，服务人员用房在厨房旁边。

800名学生的座位，14平方英尺／学生 ------------------------ 11200 平方英尺

可供应800人就餐的厨房和储藏室，2平方英尺／人 ----------- 1600 平方英尺

12800 平方英尺

门厅、走廊等 -- 2600 平方英尺

总计 -- 15400 平方英尺

注：

用餐区单元设计与纽约新罗谢尔学院及纽约范德比尔特酒店的餐厅相似。

清华学校中等科

教室建筑群

三座教学楼

1000名学生（44平方英尺／学生）----------------------------- 44100 平方英尺

三座建筑共有42间教室，每间可容纳24名学生（23平方英尺／学生）

--- 23000 平方英尺

两座建筑中各有一300座会堂（10平方英尺／学生）----------- 6000 平方英尺

42位教师的房间（14间房，每间3人，每间300平方英尺）----- 2000 平方英尺

校长室 -- 450 平方英尺

厕所、盥洗室和衣帽间 ------------------------------------- 3000 平方英尺

34450 平方英尺

楼梯、走廊等 -- 9650 平方英尺

总计 -- 44100 平方英尺

注：

教室单元的设计与康涅狄格州温莎市的卢弥斯学校相似。

以上建筑总面积 ----------------------------------- 51000 平方英尺

上述部门所需面积 ----------------------------------- 44100 平方英尺

剩余可用于藏书或其他功能的面积 ----------------------------- 6900 平方英尺

宿舍建筑群（续）

宿舍楼

每栋200名学生（108平方英尺／学生）----------------------- 21600 平方英尺

每栋宿舍楼都为3层建筑，内有50间4人间。每层为2名学士教师准备津贴宿舍。每层有一个公共卫生间。

设计总图中用颜色标出的五个宿舍楼总共可容纳1200名学生。

每栋宿舍楼包括：

50间4人间（每间300平方英尺）--------------------------- 15000 平方英尺

6间学士教师套房（每间300平方英尺）---------------------- 1800 平方英尺

厕所和盥洗室 ----------------------------------- 1200 平方英尺

18000 平方英尺

楼梯、走廊等 ----------------------------------- 3600 平方英尺

总计 --- 21600 平方英尺

宿舍建筑群

食堂

1000座（19平方英尺／人）------------------------------ 19200 平方英尺

由于面积限制，该建筑内包括一个可容纳1000名学生的大型独立饭厅、一个中央厨房和储藏室。此大厅为层高较高的单层建筑，服务人员用房在厨房旁边。

1000名学生的座位（14平方英尺／学生）---------------------- 14000 平方英尺

可供应1000人就餐的厨房和储藏室（2平方英尺／人）--------- 2000 平方英尺

16000 平方英尺

门厅、走廊等 ----------------------------------- 3200 平方英尺

总计 --- 19200 平方英尺

注：

用餐区单元设计与纽约新罗谢尔学院及纽约范德比尔特酒店的餐厅相似。

清华大学

上述记录中学生人数合计如下：

文学院系

文学与艺术学院 -- 500人

中国文学系 -- 200人

教育学院 -- 400人

法学院 -- 300人

新闻学院 -- 200人

建筑学院 --- 50人

音乐系 --- 50人

共计：1700人

科学院系

工学院 -- 600人

农业与林业学院 -- 300人

共计：900人

医学院系

医学院、牙科学院

共计：200人

中等科 -- 1000人

高等科 --- 800人

共计：1800人

以上合计：4600人

（梁曼辰转录、翻译）

2.3 Jeffery W. Cody. Building in China: Henry K. Murphy's "Adaptive Architecture," 1914—1935. Hong Kong: The Chinese University Press, 2001. 关于墨菲与清华建设部分节译之一（pp.44-51）

Beijing, June 1914

As Murphy's train chugged to Beijing on June 6, he realized that major changes would have to be made at both the Tokyo and Changsha sites. He quickly wrote to both his partner Dana and to his Yali contact in New Haven, F. Wells Williams, Chairman of Yale-in-China's Executive Committee, so both men could make plans to confer with Murphy upon his return to New York at the end of July. As invigorated as Murphy was on the way to Beijing, he was probably unable to imagine the good fortune about to come his way soon after he arrived in the Chinese capital.

Through an unknown contact, Murphy was introduced to President Y. T. Tsur and Dean Zhao, of Qinghua College, who engaged the dynamic young American as their architect for "new buildings to be built immediately and to make the scheme for future architectural developments." This unforeseen accomplishment came about only a few days after Murphy's first exhilarating experience in the Forbidden City, and he felt "naturally, much elated." Based upon what Tsur said he wanted Murphy to design, the commission represented more work than all the rest of the architects' job in Asia combined, and miraculously the engagement was made "after only a few hours' talk, without [Tsur] ever having seen me before, or any of our work!" It is unclear why Tsur and Murphy had such an immediate rapport. It is possible that Murphy & Dana's Yale affiliation had once again been providential; Tsur had graduated from Yale University in 1909.

The American seemed stunned by the break, proudly writing home that it was "something of an achievement," particularly since at least two other American architects had been competing for Tsur's favor. The first was William Merrell Vories, a University of Colorado graduate, who began teaching English in Japan in 1905, and then opened an architectural practice in Kyoto two years later, supervising the construction of the YMCA there, as well as the Omi Mission. Although Vories did not succeed in securing the Qinghua contract, he became very successful in Japan. William K. Fellows was the other American architect passed over when Tsur gave Murphy the nod. He was a partner in the Chicago firm of Perkins, Fellows & Hamilton, which specialized in architecture for schools, and which had designed parts of Nanjing University and Shandong Christian University in China.

On June 13, 14 and 15 Murphy conferred at length with Tsur about how to proceed with the Qinghua expansion. Tsur was enthusiastic and Murphy was encouraging, but the knowledgeable young Chinese administrator cautioned Murphy that he could give no definite assurances about what might happen at Qinghua. Nonetheless, prospects were promising. In 1914 Qinghua was not yet technically a university; rather, it was both a middle school and high school, founded in 1909 on the grounds of a park east of the large Yuan Ming Yuan garden, and located approximately seven miles northwest of the Forbidden City. What became known as the Qinghua Park also contained a yamen, an official compound that served largely a judicial and administrative function during the Qing Dynasty. The Qinghua yamen was probably constructed between 1662 and 1722, during the early Qing, when the park was also landscaped with a pond and a series of artificial hills which simulated mountain ranges. Emil Fischer, the Austrian builder who probably designed the first educational buildings at Qinghua in 1909, used a variety of foreign styles. Murphy described them as follows:

The blunted gables of the roofs in the High School Dormitories and the Faculty Houses are English; the windows are casements of the French type; the Middle School gables have a strongly Spanish character, while the large High School Building is of a nondescript style of architecture, with a red French tile roof, treated at the comers with mansards.

The reason this park-school complex became the focus of renewed educational attention in 1914 was due to one of the financial results of the Boxer Uprising in 1900. To give reparations for foreign property damaged during the rebellion, an indemnity fund was established in 1908 by the Chinese Government for each pertinent foreign nationality. The United States Government, in a seemingly magnanimous but also politically expedient gesture, arranged to remit part of its indemnity if China agreed to certain conditions, among them to establish a National University for the training of young Chinese scholars, many of whom would eventually be permitted to study at American universities under the auspices of scholarships financed by the Boxer Indemnity Fund. The Chinese Government chose the Qinghua site for this university. The American connection to Qinghua also helps explain why Tsur was probably only considering American architects such as Vories, Fellows and Murphy for the Qinghua commission. The new university was to follow "in general the plan of the American universities, rather than the English plan of separate small Colleges." It is likely that Tsur was given this instruction by the Government, although no proof of this exists. The Government certainly controlled the purse strings, but in the uncertain post-dynastic state in which China found itself in 1914, it was not always clear what "the

Government" was and it was less predicable still to foretell how the purse strings might be tugged.

But Tsur retained considerable decision making power, as was demonstrated during his conferences with Murphy. What did they discuss? Certainly one of the first concerns was the monetary one: how much was Murphy's fee, and what was he supposed to do to get it? In exchange for a preliminary retainer of $500 (Gold) that Murphy could collect in New York upon his return, and in addition to double the amount of any bills incurred by Murphy & Dana for the work, the American promised to deliver the following:

(1) Plan of entire present group of buildings, showing the proposed location of the following new buildings:

   a) Auditorium

   b) Library

   c) Science Building

   d) Gymnasium

(2) Plans and Perspective [for each of the above]

(3) Group-plan of future University buildings.

The men then conferred about what Tsur envisioned for the university's future and they discussed what Murphy needed in order to furnish these drawings. Tsur willingly provided the requisite data. Given the unpredictability about enrollment and funding, Tsur optimistically told Murphy to allow in his designs for expansion within each department.

Three issues remained: the question of style, the logistics of managing construction, and the schedule or timetable of building. Style was of chief concern to Murphy, partially because of the fundamental question of how buildings designed by him would appear, but also because of the critical relationship between style and actual construction. If Chinese style were used, then the logistics of building would be different than if Murphy were putting up another American Colonial Revival style, academic structure. And these logistics would then affect the schedule. Murphy and Tsur concurred that there were two principal reasons for choosing a Chinese style in the new buildings. First, the "really beautiful yamen" and the landscape features reminiscent of traditional Chinese gardens suggested that the future campus take its stylistic cuts from the historical precedent which had already been set. Also, the educational value of a "group of modern buildings in the Chinese style" was considerable.

But there were also important reasons for rejecting Chinese style at the new Qinghua. One was that none of Fischer's buildings on the site were Chinese in style, even if the yamen was, and therefore if precedent were being followed, a designer might reasonably opt for one of the other styles evident on the site. Another logical reason for not being guided by Chinese style, Tsur and Murphy frankly suggested, was that "the Chinese style, if at all well carried out, imposes many restrictions and limitations, from the utilitarian point of view, on the design of buildings intended for classroom and dormitory purposes." Finally, Tsur assumed that construction of a modem version of a Chinese building would be less economical than one which used imported construction techniques and an imported style. As the two men weighed the pros and cons of modeling the new campus on Chinese or non-Chinese style, they realized that a compromise might be the most accommodating solution. They concluded that a "fine, purely Chinese entrance gate might be made a feature of the new university campus," that gray brick which was similar to older campus structures would be used, and that the new buildings would be kept "as low as would be consistent with economy of construction."

Construction economy was also related to site management and the schedule. Recalling what he had learned in Tokyo and Changsha about the importance of an American architectural superintendent, Murphy "strongly advised [engaging] a man to go out from New York, in the fall of 1914, to supervise the letting of contracts and the actual construction of the buildings Unless this precaution is taken, through ignorance of modern, scientific, American building methods, great financial losses [will result]." Tsur agreed, and he promised that the university could house this person rent-free and even pay his steamer fare, but he explained that it might be difficult to secure official permission for a foreigner to supervise construction at a site ostensibly controlled by the Chinese Government. A month later he had "spoken to the Government about it, but 1 have not yet asked the approval of thc Minister in charge. 1 am going to see him next week." In China, such matters took time.

But Murphy and Tsur realized that time was of the essence in coordinating the Qinghua expansion efficiently. Fortunately, Tsur had already arranged to travel to the United States from late August until mid-October, 1914, but this did not give Murphy much time to reflect, consult and draw. He would have to start sketching as he crossed the Siberian steppes, and he would need help from his partner in New York. Murphy directed Dana to "put out an alarm for the best draughtsman we can get, a really good man" who would work on both the Qinghua and St. Paul's drawings, with assistance from two of the firm's other draftsmen, and with direction from Murphy, who would arrive in New York about July 17, after visiting Moscow, Berlin, Frankfurt, Cologne, Paris, London and Dublin.

By taking the Trans-Siberian railway westward through Russia and Europe and then

New York, instead of catching a steamer eastward to Seattle or San Francisco, Murphy was making his 1914 trip an around-the-world journey. This certainly would be the "time for a change" he had hoped for, especially if one recalls that the Austrian Archduke Franz Ferdinand was assassinated in Sarajevo on June 28, as Murphy was crossing the steppes toward Europe. Furthermore, in July, when Murphy traveled to Dublin from London he would have certainly been aware of the bitter political divisions within the United Kingdom over the question of Irish home rule.

Beyond the outbreak of the war and beside the geographical circumnavigation of the globe, Murphy's brief stops between Beijing and New York suggest that he was widening his architectural horizons even more broadly than his exposure to Japanese and Chinese architecture already imply. If one considers his itinerary more carefully, it is curious to note that he made three stops within Germany, a country he had not visited on his first European trip in 1906, before arriving in Paris and London.

Although Murphy did not write extensively about the German portion of his trip, except calling his journey in general a "great success," it is noteworthy to recall what significant architectural sites he would have likely visited during his brief stops in Berlin, Frankfurt and Cologne. Given Murphy's Beaux-Arts background and his use of classical elements in a revivalist sense for American middle-class residences, he was probably drawn to Schinkel's major monuments in Berlin such as the Altes Museum and the Schauspielhaus. It is also possible that he would have seen Peter Behrens's relatively new AEG Turbine Factory or his AEG Small Motors Factory, both completed in 1910, and given Murphy's interest in British country houses and the reworking of English vernacular styles in new construction, it is possible that Murphy also saw Hermann Muthesius's Freudenberg house in Berlin, completed in 1908, five years after Muthesius's publication of Das Englische Haus.

Although Murphy never indicated that he was swayed by the work of Behrens, Muthesius, Gropius or other Germans associated with the dramatic architectural shifts of the early twentieth century, the fact that he traveled to Cologne in 1914 suggests that he was both aware of, and curious about gaining exposure to novel kinds of architectural expression, especially that associated with the Deutscher Werkbund Exhibition in Cologne, which he could not have missed seeing. There Murphy would have gazed at Walter Gropius's Exhibition Hall and Bruno Taut's Glass Pavilion.

However, there were no impacts from having seen this new German architecture in the sketches for Qinghua University that Murphy was carrying in his briefcase. Nonetheless, the chance to see firsthand some of the key sites associated with creative architectural expression just as World War I was erupting was another measure of just how significant Murphy's 1914 trip was to his career. That career proceeded apace once Murphy docked in New York in mid-July, when he further prepared his Qinghua drawings and when he convinced Richard Dana that their firm's success depended not just on Connecticut, New York and domestic commissions, but on Asian work as well.

In early August, just as Germany, Austria, Russia, Serbia, France, Belgium and Britain were declaring war, Tsur wanted Murphy to send sketches of the Science Building to two American professors who taught at Qinghua, but who were in the U.S. on vacation, so they could make suggestions. He then wanted all drawings to be sent to San Francisco by early September, so that as Tsur crossed the country by train, he could review the plans and be ready to discuss them with his architects on Madison Avenue by September 21. This was approximately one week prior to when Dr. Reifsnider, of St. Paul's, would arrive from Tokyo to discuss his project's final drawings. It is not surprising, then, that Murphy cautioned Dana, "we shall have to look out not to get behind."

When Tsur met with Murphy & Dana in late September, they amended, discussed and agreed on final plans. Tsur also met with the American superintendent and they cruised back to China together in mid-October, arriving almost a month later. The American superintendent, like Murphy and Dana, would have to hustle not to trail behind.

[He] will devote the next four months to a careful study, on the ground, of all the conditions affecting the letting of contracts, methods of construction, materials, etc., and will interview as many as possible of the best engineers and architects in Peking and other cities in China, in order that the Tsing Hua work may be done in the best way at lowest total cost consistent with good work. He will also collect ... data to enable him later to amplify the outline specifications to be prepared in New York.

On February 15, 1915 Tsur wanted those specifications sent to China so that a month later the American superintendent could send out full plans for competitive bids by the capital's best contractors. Hopefully, by mid-April actual construction could begin and by September 1916 (one and one-half years later) the new buildings could be ready for occupancy.

This scenario implied that Murphy & Dana would be committed to work in northern China for at least two years. In reality, the commitment was much longer; it lasted for the firm until 1921, when the partnership disbanded, but it lasted for Murphy himself until 1935. In 1914 an elated Henry Murphy wrote to his partner about "the great success" of his around-the-world journey. He was so encouraged by how well his meetings had gone that he asked Dana to rent an additional small office, "as high up as possible," at 331 Madison

Avenue. Murphy called this room his "Headquarters for the Oriental Department." During the next four years, before his second trip to Asia in 1918, Murphy worked frequently in this office, overseeing the details that fleshed out the skeleton of the work he had helped piece together during his 1914 journey.

Before Murphy had left New York for East Asia, Tokyo and Changsha were mere names that appeared in the corners of the architect's drawings. In the summer of 1914, however, those places were vivid settings awaiting the architect's future imprint in the form of university campuses in a missionary environment. Situated in different countries where the languages, architecture, building practices, and problems were distinct, Murphy had at least known about these two projects before his trip. But at Qinghua the improbable had become manifest. By being the right person on the scene at the right time, and by virtue of his skill in negotiating amiably with the Qinghua principal, Murphy had doubled the stakes of his Asianwork. He was learning how to hustle in the East.

These projects challenged the cosmopolitan Connecticut Yankee. They forced him to expand his architectural horizons, made it sensible for him to expand his New York office staff and rent more space, and helped make Murphy a specialist about architecture in Asia. In 1914, at age 37, Murphy's life was taking a turn, veering him away from the Atlantic seaboard and edging him across the distant Pacific. He was finding that working with and for the Protestants was intriguing, challenging and potentially profitable. As he rose to the missionaries' challenge, he was slowly becoming a sinophile, although he never did learn to speak, read or write Mandarin. Murphy was attracted immediately to Chinese architectural forms, especially those dating from the Ming and Qing dynasties in Beijing's Forbidden City. The missionary architect was also becoming an architect with his own mission, to practice architecture abroad.

How did this happen? As his projects developed his involvement intensified, compelling him in the summer of 1918 to choose a new base of operations and to become another foreign architect with an office in China, but one who was distinctive. He had demonstrated considerable personal skills as he first dealt with clients in China: expediency, deference, a down-to-earth attitude, an attention to detail, confidence, energy, resourcefulness and variability. But he also faced pitfalls. Murphy encountered new challenges in Asia as he established a China-based practice, one of the most pressing being how to carve out his professional activities in the context of competition. （pp.44-51）

1914年6月，北京

6月6日，当墨菲乘坐的火车隆隆驶向北京时，他意识到在东京和长沙两地的项目都必须做出重大的改变。他很快写信给他的合伙人丹纳以及在纽黑文的雅礼会执行委员会主席F. 威尔士·威廉姆斯（F. Wells Williams），以便在自己7月底返回纽约前，两人都能安排好与他进行磋商的计划。在去北京的路上，精神振奋的墨菲可能不知道，在他抵达中国的首都后，好运即将降临。

在一次接触中，墨菲被介绍给清华大学的周诒春和赵国材，他们想聘请这位充满活力的年轻美国建筑师来负责"即将建造的新建筑，并为未来的建设制订规划"。这一惊喜就发生在墨菲初次造访故宫的几天后，兴奋还未散去，他又收获了"兴高采烈"。按照周诒春的说法，他想让墨菲设计的这个项目比在亚洲的其他项目加起来的规模还要大，出乎意料的是，这次委托"在仅仅几个小时的谈话后就达成了，（周）之前从未见过我，也没有看过我们的作品"。目前还不清楚为什么周诒春和墨菲会如此一拍即合。有可能是因为茂旦洋行和耶鲁的关系又一次促成了这冥冥中的缘分；周诒春是1909年从耶鲁大学毕业的。

这位美国人似乎被这种突然惊呆了，他自豪地写道，这是"某种成就"，尤其是还有至少另外两位美国建筑师一直在向周争取这个项目。一位是威廉·梅里尔·沃利兹（William Merrell Vories），他毕业于科罗拉多大学，1905年起在日本教英语，两年后在京都开设了一家建筑公司，负责监管那里的基督教青年会建设。虽然他没有成功签下清华的合同，但他在日本取得了很大的成就。另一位是威廉·K. 费洛斯（William K. Fellows），他是芝加哥 Perkins, Fellows & Hamilton 事务所的合伙人，该事务所专门从事校园建筑设计，此前设计的作品包括金陵大学和齐鲁大学。

6月13日、14日和15日，墨菲与周诒春详细讨论了如何进行清华扩建的事宜。周很热情，墨菲也很兴奋。但这位知识渊博的年轻校长也向墨菲坦言，他无法对清华项目最后的结果作出保证。当然，前景还是很有希望的。1914年的清华严格来讲还不是一所大学，而是一所于1909年在圆明园以东、故宫西北约7英里处的一座园林里创办的预科学校。这个被称为"清华园"的地方还留存了一个"衙门"，是清朝的司法和行政官邸。清华园的"衙门"可能建于清初的1662年至1722年间，当时园子内还有一个池塘和一些假山。清华最早的一批校舍可能是奥地利建筑商斐士于1909年设计的，他采用了多种外来风格。墨菲是这样描述它们的：

> "高等科宿舍和教工宿舍屋顶的山墙是英式的；窗户是法式的平开窗；初等科建筑的山墙有着强烈的西班牙特色，而庞大的高等科建筑则是一种不伦不类的建筑风格，有着红色的法式瓦屋顶，在拐角处用孟莎式折面屋顶处理。"

1900年义和团运动结束后的庚子赔款等事件，使这座位于园林之中的学校在

1914年成为教育界关注的焦点。1908年，中国政府开始向在运动中受损的各国进行（庚子）赔款。美国政府以一种看似宽宏大量的姿态、实则出于政治考量的目的，采取某些特殊方式免除部分赔款，其中包括建立一所培养年轻中国学者的国立大学，其中许多人将通过庚子赔款的资助赴美国留学。中国政府将清华园选为这所大学的校址。美国与清华的这种联系也有助于解释，为什么在清华项目中，周诒春或许只能考虑像沃利兹、费洛斯和墨菲这样的美国建筑师。这所新学校应遵循"美国大学的总体设计规划，而不是英国大学那种分立的小型学院布局"。尽管没有证据证明，但很有可能是政府给了周诒春这一指示，毕竟钱袋子毫无疑问是掌控在政府手中。但是1914年的中国已然是一个王朝破灭后风雨飘摇的国家，很难说清楚"政府"究竟是什么，更不容易预测这手中的钱袋子将因何而敞开。

不过，正如周诒春与墨菲的会谈中体现的那样，周仍保留了相当大的决策权。他们讨论了什么？显然首要关心的是财政问题：墨菲的薪酬是多少？他的工作任务是什么？校方将在墨菲回到纽约后支付其500美元（黄金）的定金，并承担茂旦洋行在项目中的花销，同时，墨菲须承担以下工作：

（1）现有建筑的总平面图以及下列建筑的拟定位置

a）大礼堂

b）图书馆

c）科学馆

d）体育馆

（2）平面图与透视图（上述每一个建筑）

（3）未来大学部建筑的整体规划

周欣然地提供了相关的数据。尽管招生人数和资金存在着不确定性，但周还是乐观地让墨菲对设计中的各部分进行扩大。

此外，还存在三个问题：风格问题、建设中的后勤管理问题以及建设工期的安排问题。风格是墨菲最关心的问题，一是因为这关乎他的设计以何面貌面世，二是因为这关乎如何从风格设计到实际建造。如果采用中式风格，那么建筑的后勤组织工作将不同于墨菲提出的另一种美国殖民复兴风格，即学院风格。这些后勤组织工作会影响到工期。墨菲和周诒春一致认为新建筑应采用中式风格，这主要有两个原因：首先，那座"美极了的衙门"和令人想起中国传统园林的景观特征已然暗示着未来的校园应从已经建立的历史先例中汲取风格；其次，"中式风格的现代建筑群"在教育意义上的价值也不容忽视。

但摒弃中式风格也有其重要理由。一是除了衙门，原先斐士的设计中没有一个是中式风格的；因此如果遵循先例，或许设计师选择基地中更为突出的其他风格更为合理。另一个原因是，如周诒春和墨菲所坦言的，"如果要完全顺应中式风格，从实用的角度来看，它会对教室和宿舍建筑的设计产生许多限制和局限"，最终周诒春认为，采用国外建筑技术和风格比采用中式风格的现代建筑更经济。在权衡新校园采用中式或非中式风格的利弊时，他们认为，最合适的解决方案是折中。他们最终得出的结论是"一座精美的、纯中式的大门可能成为新大学校园的一大特色"，大门将使用与旧校园建筑相似的灰色砖块，并"尽可能低矮，以符合建设的经济性"。

建设的经济性也与现场管理和工期有关。墨菲回忆起自己在东京和长沙所意识到的聘请一位美国驻场建筑师的重要性，他"在1914年秋天强烈建议从纽约（聘请）过来一个人，从事合同外包和建筑施工的监管工作。这种防范措施是必要的，否则会由于缺乏对现代的、科学的美式建筑做法的了解，而造成巨大的经济损失"。周诒春同意了，并免费为这个人提供宿舍，乃至报销他的船票，但他补充说，让外国人在中国的工地上监理施工进展很难获得官方许可。一个月后，他说自己"已与政府谈过此事，但还未向负责此事的部长请求许可。我会在下周去见他"。在中国，这样的事需要费些时间。

墨菲和周诒春意识到，协调清华扩建的事宜非常紧迫。幸运的是，周已经安排了从1914年8月下旬到10月中旬的美国之行，然而这并没有给墨菲留出足够的构思、查阅和绘图的时间。在穿越西伯利亚大草原时，他不得不开始画草图了，此时的他急需来自他的纽约合伙人的帮助。墨菲告诉丹纳："万分火急！急需找到一个我们所能找到的最好的绘图员，一定要非常非常出色！"此人将在公司其他两名绘图员的协助和墨菲的指导下，同时完成清华和圣保罗两个项目的图纸。而墨菲将在访问了莫斯科、柏林、法兰克福、科隆、巴黎、伦敦和都柏林之后，于7月17日左右抵达纽约。

墨菲沿西伯利亚铁路向西穿过俄罗斯和欧洲，然后到达纽约，而没有选择乘轮船向东前往西雅图或旧金山，这让他的1914年之旅成为一次环球旅行。这无疑是他所希求的"迎来改变的时刻"，尤其是6月28日奥地利大公弗朗茨·斐迪南在萨拉热窝遇刺时，墨菲正穿越大草原向欧洲进发。此外，7月从伦敦前往都柏林时，墨菲肯定也已发觉了英国内部有关爱尔兰自治问题的激烈政治分歧。

除了战争的爆发和环球航行外，墨菲在北京至纽约途中的多次短暂停留表明，他正在积极拓宽他的建筑视野，且远比他在接触日本和中国建筑时所开放的眼界更加宽广。如果仔细研究一下他的行程，我们会注意到一个有趣的事实，他在1906年第一次欧洲之行中没有去过的德国境内停留了三次，然后才抵达巴黎和伦敦。

尽管墨菲并没有大书特书他此行的德国部分，只是称他的旅程总体上是一次"巨大的成功"，但值得注意一下他在柏林、法兰克福和科隆短暂停留期间可能会参观哪些重要的建筑。考虑到墨菲所受的"布扎"式训练背景和他在美国中产阶级住宅中对古典元素的复兴主义式运用，他很可能会被申克尔在柏林的一些经典大作所吸引，比如柏林老博物馆（Altes Museum）和御林广场剧院

（Schauspielhaus）；他也有可能会去参观彼得·贝伦斯新近设计的AEG透平机车间或AEG小型发动机车间，二者都于1910年完工。同样，鉴于墨菲对英国乡间别墅的兴趣和他在新建筑中对英式本土风格的再造，他也有可能参观了赫尔曼·穆特修斯在柏林于1908年建造的费登伯格楼，该建筑竣工于穆特修斯出版《英国建筑》的五年之后。

尽管墨菲从未指出他受过贝伦斯、穆特修斯、格罗皮乌斯或其他推动了20世纪初建筑革命性发展的德国建筑师影响，但他1914年前往科隆的事实表明，他在意且好奇那些新颖的建筑表现形式，他更不会错过在科隆举行的德意志制造联盟展览。在那里，墨菲也许会在瓦尔特·格罗皮乌斯的展览厅和布鲁诺·陶特的玻璃展馆里目不转睛。

然而，从墨菲公文包里那些为清华学校绘制的草图上，并没有看到这种来自德国新建筑的影响。尽管如此，在第一次世界大战爆发之际，有机会亲眼见到一些代表着创造性建筑形式的经典作品，足以证明1914年的旅行对墨菲的职业生涯有多么重要的影响。7月中旬，墨菲抵达纽约，完善了清华的设计，并告诉合伙人丹纳，他们公司要想走向成功，不应仅局限于康涅狄格州、纽约州和国内的机遇，还要到亚洲去开拓业务。

8月初，正当德国、奥地利、俄罗斯、塞尔维亚、法国、比利时和英国宣布开战之际，周诒春希望墨菲将科学馆的草图发给两位在清华任教但正回国度假的美国教授，好让他们提出建议。然后，他希望所有图纸在9月初都能送到旧金山，这样当周乘火车穿越美国时，可以好好地审视一下这些方案，以做好9月21日与在麦迪逊大道与其建筑师们讨论方案的准备。这大约是在负责圣保罗项目的赖夫斯奈德博士从东京赶来讨论项目终稿的前一周。于是毫不意外地，墨菲提醒丹纳："我们必须跟上进度。"

9月底，周诒春到茂旦洋行会商，他们修改、讨论并商定了最终计划。周还会见了那位美国监理，他们于10月中旬一起乘船返回中国，差不多一个月后抵达。这位美国监理，和墨菲与丹纳一样，都在与时间赛跑。

11月10日左右，代表茂旦洋行的监理将抵达中国，并将在接下来的4个月中对场地情况、合同设立条件、建设手段、建筑材料等方面进行深入研究。同时，为了用最低的开销将清华的工程做到最好，他将会面试大量来自北京和中国其他城市的优秀工程师和建筑师。于此期间，他还会收集更多的资料，以便在回到纽约之后充实简要标准规范。

1915年2月15日，周诒春希望这些标准规范能尽快发到中国来，以便一个月后美国监理可以向首都最好的施工队发出完整的投标方案书。如果一切顺利的话，希望在4月中旬能开始实际施工，到1916年9月（也就是一年半后）新建筑可以投入使用。

　　这意味着茂旦洋行将在这个中国北方的项目上至少耕耘两年；而实际耗费的时间要长得多；它一直持续到1921年茂旦解散，此后由墨菲自己接手，又持续到1935年。1914年，兴高采烈的墨菲写信给他的合伙人，讲述他环球旅行的"伟大成功"。他因洽谈的顺利备受鼓舞，于是让丹纳在麦迪逊大道331号"尽可能高的地方"再租一间小办公室。墨菲称这个房间为他的"东方事务总部"。直到1918年再次前往亚洲，4年间，墨菲经常在这个办公室工作，为他在1914年旅程中搭建出的工作框架补充细节。

　　在墨菲离开纽约前往东亚之前，东京和长沙只是出现在这位建筑师图纸角落里的小小地名。然而，在1914年夏天，这些地方变成了一个个喧嚣的工地，等待着建筑师在一派传教士的主导下，将它们变成大学校园并长存于世。由于不同国家在语言、建筑、实际项目和挑战方面大相径庭，墨菲在出访前已对这两个项目进行了一些了解。但在清华，意想不到的事情发生了。墨菲成了那个在正确的时间、正确的地点出现的正确的人，并凭借他在与清华校长洽谈中表现出的愿意就任何议题商谈的和蔼态度，获得了这一亚洲的重要工程。他逐渐开始在东方拓展其业务。

　　这些项目给这个来自康涅狄格州、见多识广的美国人带来了挑战。它们迫使墨菲拓展自己的建筑视野，并顺理成章地扩大纽约公司的人员规模、租用更多的空间，从而使自己成为一位亚洲建筑的专家。1914年，37岁的墨菲的人生发生了转变，他离开了大西洋海岸，在遥远的太平洋上艰难前行。他发现，与传教士们一起工作是一件有趣而富有挑战性的事情，而且可能有利可图。尽管他从未学着说过、读过或写过中文，但他在传教士们面前渐渐地成了一名亲华派。墨菲很快被中国的建筑形式所吸引，尤其是北京故宫明清时期的建筑样式。这位教会建筑师正在成为一名拥有自己使命的建筑师——在国外从事建筑实践。

　　这一切是如何发生的呢？随着项目的进展和参与度的不断提高，墨菲在1918年夏天成立了一个新事务所，从而成了又一位在华开设事务所的外国建筑师。但他是与众不同的。他在第一次与中国客户打交道时就展现了相当的个人能力：善用权宜之计、尊重他人、脚踏实地、关注细节、充满自信、朝气蓬勃、足智多谋又灵活多变。但墨菲也面临着来自亚洲的挑战，那就是在中国开展业务，需要在激烈的竞争环境中开创出自己的一番事业。（原书pp.44-51）

2.4 Jeffery W. Cody. Building in China: Henry K. Murphy's "Adaptive Architecture," 1914—1935. Hong Kong: The Chinese University Press, 2001. 关于墨菲与清华建设部分节译之二（pp.67-70）

Murphy in Beijing, May 1918

After delivering an address to students at Chosen, Murphy and Lane proceeded to Beijing. As Lane showed his superior the status of the work at Qinghua University and on Legation Street, where the IBC branch was to be built, Murphy was both pleased and troubled. Regarding Qinghua, the work was "thrillingly satisfactory", but the bank was "thrillingly mixed-up, but hopeful." Four Qinghua structures were by then completed: the Library, the Auditorium, the Science Building and the Gymnasium. Lane and his Chinese assistant superintendent, Zhuang Jun, tendered bids from foreign and Chinese contracting companies, shopped for supplies of foreign hardware, bargained for building materials, and oversaw the simultaneous construction of four buildings on the site. They had done their jobs well. "These buildings ... are generally considered to present in many respects the best examples of first class modern construction yet attained in China." Both foreign and Chinese building materials were utilized, which was not particularly unusual. However, what made Qinghua more "first class" and "modern" from the standpoint of some observers was that Murphy's building used contemporary American technologies, such as steel trusses supplied by the U.S.Steel Products Company, even if the construction of the Library and Gymnasium was supervised by a contracting company, Telge & Schroeter, that had more European than American linkage and even if a Chinese firm, Kung Sung Kee, built the Science Building and the Auditorium.

Some Chinese contemporaries saw the buildings as inspiring "paragons" which demonstrated the growth of Chinese learning from an "embryo" state:

[The new building] are an inspiration to the development of the nation. It is only of late that China has begun to open public libraries, but as yet such a paragon as our new library has rarely, if ever, been seen. It is certainly one of the best, if not the best, in China. Perhaps to a greater extent than the public Library has athletics been neglected by the Chinese. The construction of a gymnasium like ours is almost incomparable throughout

the Republic. As to sciences, they are in an equally embryo state in this country. The sight of such a Science Building as we are undertaking to produce is in reality a rare pleasure. Again the art of public speaking is undeveloped among our people. Our Auditorium may well be regarded as Forum Romanum where budding Ciceros will deliver their orations.

By relating the "Italian Renaissance" style of the new buildings' architecture to the Roman Forum, this critic demonstrated that style was matched with both educational and democratic values that reverberated from a western classical tradition. Although Murphy never mentioned precisely which precedents he might have used for the Qinghua Auditorium, given the pedimented portico attached to central dome it is likely that he had in mind the Pantheon in Rome, Palladio's Villa Rotonda, Thomas Jefferson's rotunda at the University of Virginia and McKim, Mead and White's newly completed library at Columbia University in New York. Within a year, in May 1919, Chinese students would stage demonstrations outside the Gate of Heavenly Peace of the Forbidden City, protesting the terms of the Treaty of Versailles negotiated by western leaders, which passed control of Shandong Province from the Germans to the Japanese, setting in motion a cycle of political demonstrations that still reverberates in China. Spaces such as Murphy's auditorium at Qinghua, based upon several western icons with democratic overtones, become settings for political discussions by "budding Ciceros", although there is no proof that Murphy designed them with a political motive in mind. The same Chinese commentator saw Murphy's structures as "full of dignity and repose … [which]will add to the charms of the Tsing Hua [Qinghua] campus, facilitate the education of this college, and represent to our children, and to our children's children the untiring efforts of labor, the unaltering value of self-sacrifice, and the culminating power of discreet and sagacious management."

At the Qinghua Commencement Exercises in June 1918, Dr. Paul S. Reinsch, the American Government's chief representative in China, voiced similar ideals regarding what he termed "efficiency, progress and political considerations."

[It would be helpful] to bridge the fearsome passage from the old literary culture to that regime of efficiency which modern conditions require. The venerable philosophic and literary tradition of China is looked upon by Americans with the greatest respect and the hope that the essential virtues of this national inheritance and also its artistic emanations will be preserved; but it is also believed that this can be done only if the guiding impulse of education is practical in the sense of the word that it must concentrate its efforts upon greater perfection in action … This can be done only through modern processes of education … If such an ideal is successful in China, I am sure the American people will be most happy, not only because of political considerations but because the existence of a

progressive, free and efficient national life in China is of itself a safeguard of all the best that America hopes for the world.

Murphy's buildings, therefore, were "thrilling satisfactory" not only because logistically their construction was progressing smoothly, but also because they measured up to the ideals articulated above. His "dignified" spaces were conceived by at least some contemporaries for "free" experimentation by China's youth.

While sorting out some of the technical details of his building sites, Murphy became better informed about two other projects during his short stay at Beijing's Grand Hotel des Wagons-Lits. The first was a potential new commission while the second, the Peking Union Medical Collge (hereafter PUMC), funded by John D. Rockfeller, Jr., was being erected under a gathering cloud of suspicion hovering over the architect, Harry Hussey. Murphy's opinions about these two matters had direct bearing upon the building of his office in Shanghai and it is thus useful to scrutinize them in some detail.

Murphy learned from Lane that a "big Post Office building" was to be built in Beijing, the program for which was to be announced on 15 July 1918. Preliminary sketches from the Government indicated a preference for a long colonnaded facade, similar to "the New York Post Office off Pennsylvania Station", and because this was a Government project, a competition had to be conducted to choose the architect. The Postmaster, however, was unskilled in "square" architectural competitions because most architectural work in the capital that involved foreigners was assigned on the basis of word-of-mouth. After Murphy helped "educate" the Postmaster about how to conduct the competition, he confidently wrote to Dana that with Forsyth in China to help Lane, "we ought to be able to win it." However, Murphy felt his practice was handicapped without some of the architectural reference works that graced his Madison Avenue office.

I thought of cabling you [Dana] to send out some good books to help in the competition; but decided there would not be time on this job. There will be others coming along, however; so please send ... American Competitions; McKim, Mead & White Monograph (select parts applicable to public work); any other books you think of, for our purpose; and 20 sheets of ... paper best adapted for competitions and show perspectives. $100 invested as above might easily bring in several thousands in profits on large jobs. [The books] will be used not only in making drawings to get work, but also in detailing work we get.

Here Murphy demonstrated his respect for, and reliance upon McKim, Mead & White, he shows how optimistic he was about future work (perhaps in part because he was "educating" potential clients) and suggests why reference works would be so useful in accomplishing that work. In the best tradition of his architectural training, Murphy was reverting to past precedent in order to solve contemporary architectural problems. His two Beijing projects were already linked to McKim, Mead & White. Not only was the Qinghua Auditorium patterned after that prominent firm's work at Columbia University, but also the IBC branch was likewise stylistically tied to renovations made by the firm at the bank's main office at 55 Wall Street in New York. Furthermore, Murphy & Dana's New York general manager, Norman Nims, had worked for McKim, Mead & White for six years. (pp.67-70)

--- * * * ---

1918年5月，墨菲在北京

在朝鲜向学生们发表了演讲后，墨菲和雷恩一同继续前往北京。雷恩向自己的上司汇报了清华学校和将要兴建的、位于北京东交民巷的国际银行股份有限公司（IBC）分行的工作进展，墨菲喜忧参半。喜的是他在清华的作品"令人震惊地满意"，忧的是银行"同样令人震惊，风格混杂，但还有（挽救的）希望"。那时清华已有四栋建筑落成：图书馆、大礼堂、科学馆和体育馆。雷恩和他的中国助手庄俊通过向一众中国和外国承包商招标，采购外国硬件设施，为建材讨价还价，并监督四幢建筑的建造。他们工作很出色。"这些建筑……在许多方面都被认为是中国迄今为止一流的现代化建筑的最佳范例。"其中同时运用了中国和外国的建材，这一点其实并不罕见。然而，从一些观察家的视角看来，使清华更"一流"和"现代"的正是墨菲的建筑采用了当代的美国技术，如美国钢铁产品公司提供的钢桁架，即便图书馆和体育馆是由一家与欧洲联系比与美国联系更紧密的承包商Telge & Schroeter负责施工的，而科学馆和大礼堂则是由一家名为公顺记的中国公司建造的。

一些同时代的中国人把这些建筑看作鼓舞人心的"典范"，认为它们代表着中国学术教育的"萌芽"：

[新建筑]是国家发展的灵感。近来中国开始开放公共图书馆，但目前为止，像我们新图书馆这样的典范还很少出现。至少在中国，这是最好的图书馆。也许，比起公共图书馆，体育运动没那么被中国人所重视。像我们这样的体育馆在整个民国几乎是无与伦比的。至于科学，在这个国家同样处于萌芽状态。看到我们正在建设的科学馆着实令人喜悦。同样，在中国的民众中，演讲艺术还没有发展起来。我们的大礼堂也许可以被看作那些初露头角的西塞罗们未来发表他们演讲的古罗马广场。

通过把新建筑的意大利文艺复兴风格与古罗马广场联系起来，这位评论家想说明这种风格与西方古典传统的教育和民主价值观是相符的。虽然墨菲从来没有明确提到过他在清华大礼堂中参考了哪一个先例，但基于大礼堂那附在中央穹顶旁的山墙和柱廊判断，他有可能是借鉴了罗马的万神庙、帕拉迪奥的圆厅别墅、托马斯·杰斐逊的弗吉尼亚大学圆厅图书馆，以及McKim, Mead and White公司在纽约新盖的哥伦比亚大学图书馆。不到一年后的1919年6月，中国学生将在紫禁城的天安门外组织示威游行，抗议西方列强商定的《凡尔赛条约》中将中国山东的控制权从日本移交给德国的条款，并在中国引发了一场旷日持久的政治运动。像墨菲的清华大礼堂这样带有些许西方民主象征的地点，就成了"初出茅庐的西塞罗们"进行政治讨论的场所，尽管并没有证据表明墨菲设计这些空间时带有政治动机。同一位中国评论家认为墨菲的设计"充满了高尚和宁静……这将提升清华校园的魅力，促进这所学校的教育，并且向子孙后代展示什么是不懈努力、勇于献身、敏锐自律的崇高力量"。

在1918年6月清华的毕业典礼上，美国政府在中国的首席代表芮恩施博士在他关于"效率、进步和政治考量"的演讲中表达了类似的观点：

[这将有助于]跨越从旧文学、旧文化到现代化所要求的效率制度之间的巨大鸿沟。美国人对中国古老的哲学和文学传统极为尊重，并希望这一民族遗产的美德精华和艺术魅力能够得到保留；但也相信只有在教育的指导思想切实可行的情况下，也即必须把精力集中在更完善的措施上，才能做到这一点……这只能通过现代化的教育手段来实现……如果这样一个理想能在中国获得成功，我相信美国人民将为此欢欣鼓舞，这不仅是出于政治考虑，也是因为这样一个进步、自由而高效的中国国民生活，其存在本身就是美国对世界所希冀的一切美好的保障。

因此墨菲的建筑是那样"令人震惊地满意"，不仅仅是因为就物料供应而言施工进程顺畅平稳，更是因为它们符合了上文所述的理念。他"高尚"的空间至少是由当时的一些人为了中国青年进行"自由"尝试而构想出来的。

墨菲在北京的六国饭店短暂停留、处理一些建筑工地的技术细节期间，对另外两个项目进行了更为深入的了解。其一是一个潜在的新项目；其二是由约翰·洛克菲勒二世资助建造的北京协和医学院（以下简称PUMC），其建筑师何士的周遭围绕着颇多谜团。墨菲对这两个项目的看法，直接关系到他在上海的事务所的建立，因此，有必要对这两个项目进行详细的探讨。

墨菲从雷恩那里得知，北京将修建一座"邮政大楼"，并计划于1918年7月15日公布。政府的初步设计图显示出他们倾向于采用长柱廊立面，类似于"宾夕法尼亚车站外的纽约邮局"。由于这是政府项目，需要通过比赛来确定建筑师。然而，这位邮政局长对于如何"坐镇"一场建筑比赛并不熟悉，因为之前首都大多数涉及外国人的建筑工作都是根据口碑分派的。在对邮政局长进行了如何举办比赛的"指导"后，墨菲自信地写信给丹纳说，有福赛斯在中国帮助雷恩，"我们应该能够赢得比赛"。然而，墨菲感觉，少了那些远在麦迪逊大道办公室用以装点门面的建筑参考书籍的帮助，他的作品仍有不足。

我本想发电报给你[丹纳]，让你带一些好书过来协助比赛，但最后还是觉得这样做时间不够。不过，还是会有其他人来帮忙；所以请寄来……《美国竞赛》McKim, Mead & White（选择其中关于公共建筑的部分）和其他任何你认为可能帮助我们赢得比赛的书籍；还有20张……最适合用来比赛和展示透视图的纸。以上大概100美元的投资也许很容易就能在一些大项目中带来数千美元的利润。[这些书籍]不仅仅是为了用来绘制图纸拿下这次的项目，还有助于我们之后细化方案。

墨菲在此处表达了他对McKim, Mead & White公司的尊重和信赖，也展示了他对未来工作的乐观态度（可能部分是因为他"指导"了潜在的客户），并指出了为什么参考资料在完成这次比赛的事情上会如此有用。按照所受建筑训练的优良传统，墨菲将求助于先例来解决当前的建筑问题。他在北京的两个项目已经与McKim, Mead & White公司有所联系。不仅清华大礼堂仿照了这家著名公司在哥伦比亚大学的作品，IBC的分行在风格上同样与由该公司改造时，位于纽约华尔街55号的IBC总部相似。此外，茂旦洋行在纽约的总经理正是曾在McKim, Mead & White公司工作过六年的诺曼·尼姆斯。（原书pp. 67-70）

（刘翘楚转录、翻译，邓可校对）

墨菲档案中墨菲与丹纳、周诒春等人往来书信

3.1 周诒春致墨菲信（1914年7月17日）首页影印与全文转录及翻译

President's Office

校學華清京北
TSING HUA COLLEGE
PEKING

RECEIVED
AUG 24 1914

校長室

Peking, July 17th 1914.

Mr H.K.Murphy,

Murphy and Dana Architects,

331 Madison Avenue, New York City,

U. S. A.

My dear Mr Murphy:-

I have received your letter of June 21st with two coppies of the report of our interviews here in Peking, for which please accept my hearty thanks. I shall make a few corrections and additions and send the revised copy to you early next week via Siberia.

Regarding the two copies of drawings which you need, I labored under the misapprehension that with the blue prints which I gave you, I thought they were sufficient, and so I did not cause new drawings to be made out. I have telegraphed for my student to come up to make out two more drawings which I shall try to send to you by the same mail as the revised copy of the report of our interviews. I am very sorry for my thoughtlessness.

I have cabled to the International Committee of the Y.M.C.A. New York to advance to you $500 gold as our retaining fee for your services, and hope you have already call-ed for the money.

Since your departure, I have sent to your business address a copy of Mr Whitmore's letter in regard to the Manual Departmant. I hope you have received the same.

I shall probably write you another letter before I leave the country. No letter will accompany the revised report, I think.

My wife will come over with me to America and we are looking forward to this trip with a great deal of pleasure.

Peking, July 17th 1914

Mr. H.K.Murphy,

Murphy and Dana Architects,

331 Madison Avenue, New York City,

U. S. A.

My Dear Mr. Murphy:

I have received you letter of June 21st with two copies of the report of our interview here in Peking, for which please accept my hearty thanks. I shall make a few corrections and additions and send the revised copy to you early next week via Siberia.

Regarding the two copies of blue drawings which you need, I labored under the misapprehension that with the prints which I gave you, I thought they were sufficient, and so I did not cause new drawings to be made out. I have telegraphed for my student to come up to make out two more drawings which I shall try to send to you by the same mail as the revised copy of the report of our interviews. I am very sorry for my thoughtlessness.

I have cabled to the International Committee of the Y.M.C.A. New York to advance to you $ 500 gold as our retaining fee for your services, and hope you have already call-ed for the money.

Since your departure, I have sent to your business address a copy of Mr. Whitmore's letter in regard to the Manual Department. I hope you have received the same.

I shall probably write you another letter before I leave the country. No letter will accompany the revised report, I think.

My wife will come over with me to America and we are looking forward to this trip with a great deal of pleasure.

Regarding the suggested building superintendent, I hope to be able to cable affirmatively on August 1st. I have spoken to the Government about it, but I have not yet asked the approval of the Minister in charge. I am going to see him early next week.

Hoping (you) and Mrs. Murphy are well.

I remain,

Yours very sincerely.

北京，1914年7月17日

亲爱的墨菲先生：

您在6月21日发的那封附了两份我们在北京会晤报告的信，我已经收到了，非常感谢。我会对报告做一些修正和补充，在下周早些时候将修订稿通过西伯利亚铁路寄给您。

关于您需要的两份图纸备份，我之前一直有所误解，认为我给您的蓝图数量已经足够了，所以没有再准备其他图纸。现在我已经发电报让我的学生再绘制两份图纸。如能赶上，我会将新的图纸和我修订的报告一同邮寄给您。对于我的大意深表歉意。

我已经向基督教青年会的纽约国际委员会发过电报，他们将预付您500美元（黄金）作为您的费用，希望您已经收到这笔钱了。

在您离开之后，我向您的工作地址寄了一份惠特莫尔先生关于工艺馆[①]的信，希望您已经收到。

在我离开中国前，我应该还会再写给您一封信。但我想报告的修订稿应该会单独寄出。

我太太将会和我一同前往美国，期待这是一次美好的旅行。

关于驻场建筑师的建议，我希望能在8月1日通过电报联系决定。我已与政府谈过此事，但还未向负责此事的部长请求许可。我会在下周去见他。

希望您和墨菲夫人一切安好。

此致
您诚挚的
周诒春

（梁曼辰译）

---

① 编者注：工艺馆（后改称土木工程馆）
在20世纪20年代初由庄俊设计建成，
详见本书上编第4章。

3.2 周诒春致墨菲信（1917年3月1日）影印与转录及翻译

校學華清京北

TSING HUA COLLEGE

President's Office

PEKING

校長室

MAR 29 1917

March 1, 1917.

Mr. H. K. Murphy,
331 Madison Ave.,
New York City,
U. S. A.

Dear Mr. Murphy:

I have been most culpable for not having written you even once a year. Excuses are lame.

Mr. C. E. Lane, about whom I must have written ere now, has been a most satisfactory man and I cannot say enough to thank you for having helped to secure him for us. Also you have in him a most satisfactory representative in carrying out your ideas.

Since we last met at New York in the fall of 1914, times have greatly changed. The world war has pushed up prices tremendously high, and buildings are costing more than double. Besides, the demand by the public for immediate higher education at Tsing Hua has been very pressing, and new ideas meanwhile have also evolved. All these factors have necessitated more or less radical changes in my educational scheme, at least for the present. Briefly, we have to use the present compound for the College and the academy together. Consequently we have to enlarge the Auditorium as well as the Science Building. As the Library and Gymnasium cost us far too much, I gave Mr. Lane explicit instructions to cut down expense on the two buildings as far as possible, compatible with good substantial construction. I think he has done so and also informed you. Under present conditions, it will be a long time before we can realize the University plan in the new compound.

I hope Mrs. Murphy and Prof. and Mrs. Dana have been well.

I want to thank you for your cooperation in giving our students of architecture practical training.

I hope to visit the States this fall and make a longer stay this time.

Very sincerely yours,

YTT:K

March 1, 1917.

Mr. H. K. Murphy,

331 Madison Ave.,

New York City,

U.S.A

Dear Mr. Murphy:

I have been most culpable for not having written you even once a year. Excuses are lame.

Mr. C. E. Lane, about whom I must have written ere now, has been a most satisfactory man and I cannot say enough to thank you for having helped to secure him for us. Also you have in him a most satisfactory representative in carrying out your ideas.

Since we last met at New York in the fall of 1914, times have greatly changed. The world war has pushed up prices tremendously high, and buildings are costing more than double. Besides, the demand by the public for immediate higher education at Tsing Hua has been very pressing, and new ideas meanwhile have also evolved. All these factors have necessitated more or less radical changes in my educational scheme, at least for the present. Briefly, we have to use the present compound for the College and the academy together. Consequently we have to enlarge the Auditorium as well as the Science much. As the Library and Gymnasium cost us far too much, I gave Mr. Lane explicit instructions to cut down with good substantial construction. I think he has done so and also informed you. Under present conditions, it will be a long time before we can realize the University plan in the new compound.

I hope Mrs. Murphy and Prof. and Mrs. Dana have been well.

I want to thank you for your cooperation in giving our students of architecture practical training.

I hope to visit the States this fall and make a longer stay this time.

Very sincerely yours,

Y. Tsur

YTT: K

1917年3月1日

亲爱的墨菲先生：

非常抱歉我近来和您联系甚少，一年都不曾有一次。我知道什么借口都显得苍白无力，这都是我的错。

我不得不在此声明，C.E.雷恩先生非常令人满意，非常感谢您帮我们聘请他来清华。同时，他也是表达您想法的最好人选。

自从我们于1914年秋天在纽约相见之后，时光飞逝。世界大战使得物价飞涨，建筑的造价甚至涨到了两倍。大众期待清华能提供为社会服务的高等教育，教育观念也不断进步。至少从现在来看，种种因素使我有必要对我的教育计划或多或少地加以修订。简而言之，我们必须把现已建成的建筑同时用于大学预科和中学部的教学。因此，我们需要扩大礼堂和科学楼的规模。由于在图书馆和体育馆上花费过多，我已经明确指示雷恩先生在保证质量的同时尽可能地降低这两处建筑的造价。我认为他已经按此实施了，想必也已告知了您。在目前条件下，距离实现您方案中大学部的规划应该还要不少时间。

希望墨菲夫人以及丹纳教授夫妇一切安好。

非常感谢您对于我校学生建筑实践训练的合作指导。

我希望于这个秋天前往美国拜访，这次停留时间可能较以往能长一些。

您诚挚的
周诒春

（梁曼辰译）

3.3 周诒春致墨菲信（1917年11月21日）[①]

Nov. 21st, 1917

Dear Mr. Murphy:

I want to thank you for your kindness to me at New York. I am very sorry that I did not have more time at my disposal, so that I could call to [?] my respect to Mrs. Murphy and Mrs. Dana.

Re. the possibility of securing a Chinese student in architecture for your office for practical training, so that he may likely(?) start a branch office in China for your firm. I have written to Mr. [?]. You can carry in direct correspondence with him whenever you paid it [?].

I met one Mr. Vogel, a sort of missionary architectural firm at Tokyo, at Columbus. O. He seems to be taking some work at the State University [?]. He told me that his firm was competing for the work at Seoul also.

When I left hastily for D.C. on Nov. 14th, the manager of Yale Club has not yet sent me a bill for my staying there. Will you please remind the manager and ask him to send it to Mr. C.H.Hsie, 2023 Kalorama Road, Washington, S. [?]. I have left [?] to [?][?][?].

I have recently heard from Mr. Lane saying that the buildings are processing well, and the Gym and the Library are about finished.

Hope to see you in China before I come again.

Yours Faithfulness,

Y.T.Tsur

P.S.

I forgot to send you this [?] for life.

[①] 此信为手写，字迹不清处以"[?]"代替。

3.4　墨菲致丹纳信（1914年6月22日）[1]

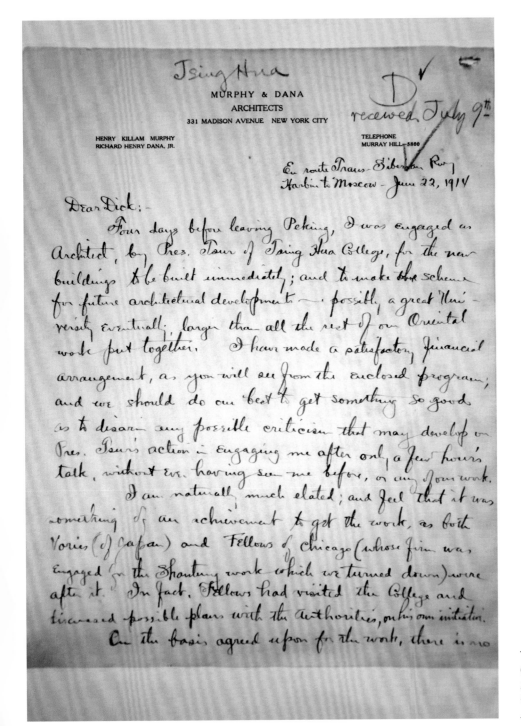

①
此信为手写，字迹不清处以"[?]"
代替。

En route Trans-Siberian Rail

Harbin to Moscow - June 22, 1914

(recorded July 9th)

Dear Dick:

Four days before leaving Peking, I was engaged as Architect, by Pres. Tsur of Tsing Hua College, of the new buildings to be built immediately; and to make the scheme for future architectural developments and possible, a great University eventually, larger than all the rest of our Oriental work put together. I have made a satisfactory financial arrangement, as you will see from the enclosed program; and we should do our best to get something so good as to disavow any possible criticism that may develop on Pres. Tsur's action in engaging me after only a few hours' talk, without ever having seen me before, or any of our work.

I am naturally much elated; and feel that it was something of an achievement to get the work, as both Vories (of Japan) and Fellows of Chicago (whose firm was engaged for the Shantung work which we turned down) were after it. In fact, Fellows had visited the College and discussed possible plans with the authorities, on his own initiations.

On the basis, agreed upon for the work, there is an incentive for us to put any but the best draughtsman we can get, on the work. Of course, I do not know what are the conditions in the office now; but about July 20 we shall need a really good man to go on the Tsing Hua work, steadily until Feb. or March; so I think you'd better send out an alarm for one as soon as you get this; offering $45 to $50 if necessary — as there will be so much to do on St. Paul's and Yale-in-China during July and August that we shall have to look out not to get behind. One thing looks certain is that neither you nor I can be away from N.Y. for more than three or four days at a time until after we get Pres. Tsur off to China about Oct. 15. You see Pres. Reifsuider — who is worth all the others put together, as far as real knowledge of St. Paul's in concurred — will also be in N.Y. in Sept. for final conferences on the completed working drawings for St. Paul's, on which I came bringing back many notes of changes. I judge all my own time this summer will be taken up by Scott House, Tokyo, Tsing Hua and Yale-in-China; and a very large share of your time will also be needed on these jobs. The only way to keep up and give satisfaction will be to have a new first-class man in charge of Tokyo with Teunqle and especially Chapin to help; another new No. 1 man for Tsing Hua, with, say Methuen and Perkin, to help. With a $50 a week in keeping each of these jobs [?] along (and his salary, plus varying amounts, all paid by the clients), we shall [?] about be able to keep up to the schedule. Melyill will of course have to stick to the Scott House, — Donald must absolutely not be kept wasting a single day for anything after I get home; — Somerville will probably have his hands full finishing Loomis and straightly leaving out whatever messes it may get into on the end; and, on the basis of the personnel of the office as I knew it in the dim past this leaves only [?] for the rest of the work with Kellogy as general trouble man. If there is much new work in the office that I haven't heard about, I should very much like to get one of the single empty offices in the building — as high up as possible, — by the month; and wish you would speak for one we could make this extra room Headquarters, for the Oriental Department.

（此后为墨菲向丹纳描述来华处理的雅礼学校事宜，以及墨菲从哈尔滨经俄国游览东欧、法兰克福和巴黎等地的时间安排，从略）

On account of the extension of our stay in Peking, we shall reach Moscow on June 26, (instead of St. Petersburg on June 20 as originally planned). We shall leave Moscow on June 27（旅程略）. We shall probably stop at the office on Fri. July 17, for a short visit; and go up to Killam's Pt. Fri. Aft. If the "[?]" reaches N.Y. late, we shall go up to K. Pt. Sat. [?] In which case, I should very much like a general conference with you late Friday, dining together if possible.

We are both very well, and happy over the great scene of our trip from [?] of [?] both of pleasure and of business. Hope all is well with you.

Regards for Enda.

Henry

3.5 墨菲致丹纳信（1918年6月2日）①

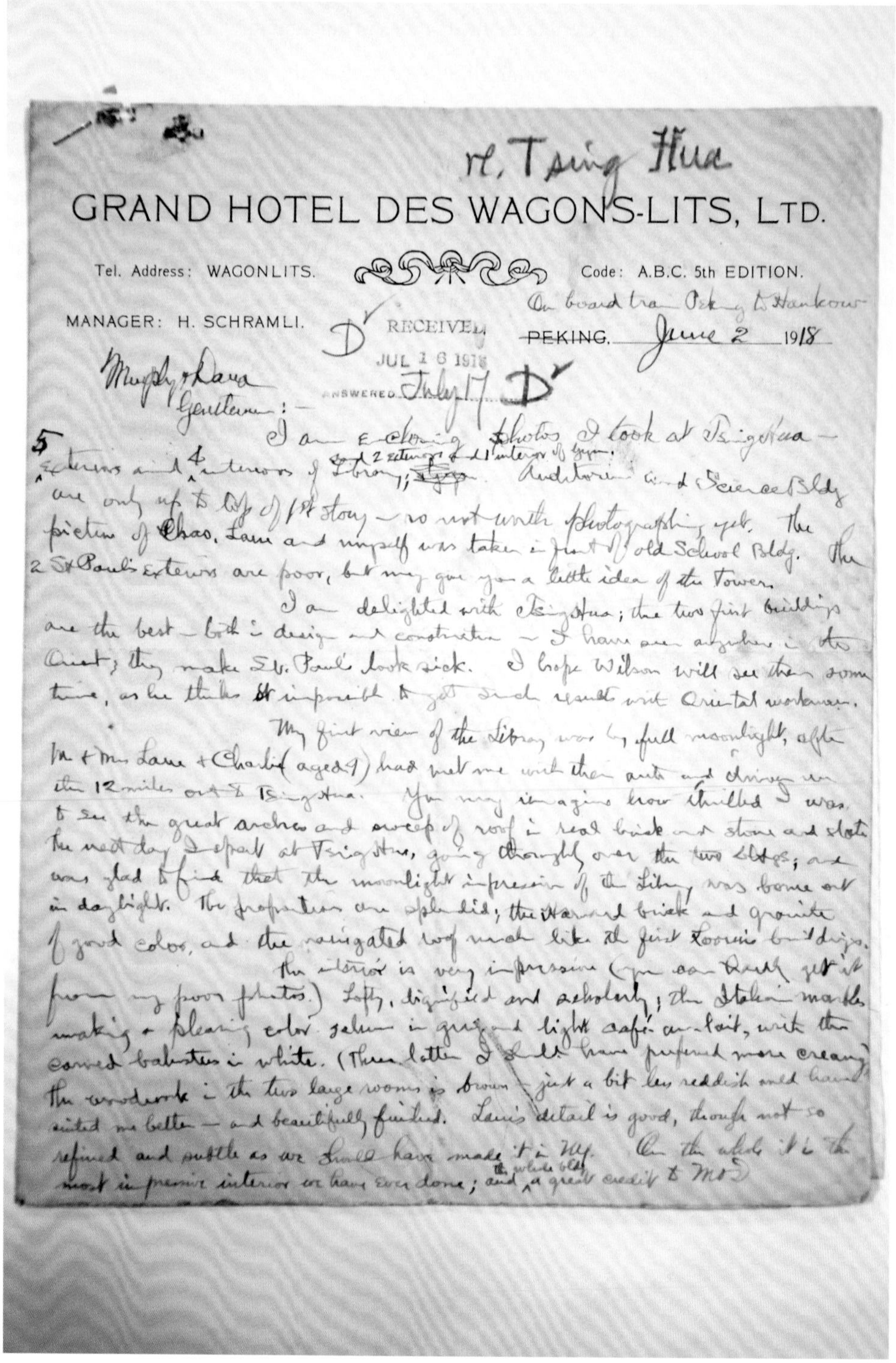

re. Tsing Hua

GRAND HOTEL DES WAGONS-LITS, LTD.

Tel. Address: WAGONLITS.　　Code: A.B.C. 5th EDITION.

MANAGER: H. SCHRAMLI.　　On board train Peking to Hankow

~~PEKING.~~ June 2 1918

RECEIVED JUL 16 1918 ANSWERED July 17 D

Murphy & Dana

Gentlemen:—

I am enclosing photos I took at Tsing Hua — 5 exteriors and 4 interiors of Library; 2 exteriors and 1 interior of Gym. Auditorium and Science Bldg are only up to top of 1st story — so not worth photographing yet. The picture of Chas. Lane and myself was taken in front of old School Bldg. The 2 St Paul's exteriors are poor, but may give you a little idea of the tower.

I am delighted with Tsing Hua; the two first buildings are the best — both in design and construction — I have seen anywhere in the Orient; they make St. Paul's look sick. I hope Wilson will see them some time, as he thinks it impossible to get such results with Oriental workmen.

My first view of the Library was by full moonlight, after Mr & Mrs Lane + Charlie (aged 4) had met me with their auto and driven us the 12 miles out to Tsing Hua. You may imagine how thrilled I was to see the great arches and sweep of roof in real brick and stone and slate. The next day I spent at Tsing Hua, going thoroughly over the two bldgs; and was glad to find that the moonlight impression of the Library was borne out in daylight. The proportions are splendid; the Harvard brick and granite of good color, and the variegated roof much like the first Loomis buildings.

The interior is very impressive (you can hardly get it from my poor photos.) Lofty, dignified and scholarly; the Italian marbles making a pleasing color scheme in gray and light café au lait, with the carved balusters in white. (These latter I should have preferred more creamy.) The woodwork in the two large rooms is brown — just a bit less reddish would have suited me better — and beautifully finished. Lane's detail is good, though not so refined and subtle as we should have made it in N.Y. On the whole it is the most impressive interior we have ever done; and the whole bldg a great credit to M&D.

①
此信为手写，字迹不清处以“[?]”代替。

On board via Peking to Hankow

June 2, 1918

(Received July 16, 1918)

Murphy &. Dana

Gentleman:

I am enclosing photos I took at Tsing Hua,—5 exteriors and 4 interiors of Library, and 2 exteriors and 1 interior of Gym; Auditorium and Science Building are only up to top of 1st story— so not worth photographing yet. The picture of Chao, Lane and myself was taken in front of old school building. The 2 St. Paul's exteriors are poor, but may give you a little idea of the tower.

I am delighted with Tsing Hua; the two fine buildings are the best—both in design and construction—I have seen anywhere in the Orient they make St. Paul's look sick. I hope Wilson will see them sometime, as he thinks it impossible to get sand [?] with Oriental [?].

My quick view of the Library was by full moonlight, after Mr. + Mrs. Lane + Charlie (aged 9) had met[?] me with their auto and driven more than 12 miles out of Tsing Hua. You may imagine how thrilled I was to see the great arches and scope of roof in real brick and stone and slate. The next day I spoke at Tsing Hua, going thoroughly over the two buildings; and was glad to find that the moonlight influence of the Library was borne out in daylight. The proportions are splendid; the Harvard brick and granite of good color, and the [?] roof much like the first Loomis building.

The interior is very impressive (you can hardly get it from my poor photos.) lofty, dignified and [?]; the Italian marbles making a pleasing color [?] in gray and light Café au lait, with the carved balusters in white. (There latter I should have performed [?]). The woodwork in the two large rooms is brown, just a bit less reddish would have suited me better—and beautifully finished. Lane's detail is good, though not so refined and subtle as we should have made it in N.Y. On the whole, it is the most impressive interior we have ever seen; and the whole building a great credit to M. &. D.

The Gym is a special joy to me, as I had more to do with that than with any of the other these buildings. As you will see from the photo, the proportions are good; and the change from the 5 semi-circular windows and straight roof line (as on our final scheme, show on perspective) to the 3 large semi-circular and broke roof line, was a decided success. The granite colonnaded Porch is splendid—gives just the country club effect I wanted, as a place where the Chinese boys can sit between exercises, sip their tea, and sagely discuss world events. The solidity and permanence of the construction applied to the (Gym) though these qualities are really much more needed among the Japanese.

The interior of the Gym is most successful, also. The floor is of creosoled wood blocks (Somevillis countermanding of this material came to Lane too late) which appear to me very satisfactory, —though no use for a dance! The walls are dark brownish red brick—very handsome color and texture; the trusses and ceiling light gray (cement color, only lighter). The gallery front is continuous monolithic concrete, rubbed off to expose the aggregates, in which is a little pale color—pleasing in tone and texture.

The swimming pool seemed to me the most attractive I have ever seen, —even though the side-walls above pool are temporarily lined with cement, awaiting arrival from Italy of marble slabs. The upper walls are a very light gray sand-lime brick, in faint herringbone panels—very pleasing. The floor and the pool is of dark brown paving brick, scored to present shipping. The pool is lined with blue-veined white marble slabs about 15" × 30"—and filled with crystal clear [?]-wall water is most inviting. I enjoyed a swim, with little Charlie Lane—felt quite as though taking part in a sacred rite of some kind. No one who could see this lively, light pool, with the sun streaming in and flooding all with out-of-doors, would ever again be satisfied with the usual dark basement pool like that at Columbia. I sent Jim MacKenzie Care M.&.D. (whose wedding invitation I just read in your letter) 2 photos of the Gym, with congratulations on his part in its design.

Lane's Auditorium details are coming along well, though he has much still to do in the way of drawing. I urged Chao to approve the use of Guastavino's Akoustolith (the $17,500 estimate, with the additional stairs required, and the small credits detailed, really amounts to about $28,000 gold), for the inner ceiling of the dome, barrel vaults, and certain panels. Chao is favorable, and will put the style in the budget; but of course it may be turned down by the Ministry. The walls are to be of the sand lime brick, and the floor of Japanese cork-tile (@ 1/5 cost of [?] cork) delivered on job. Lane's idea is to spray the dome ceiling in colors-red and gold and blue, to get a motiled(?) effect; but I have warned him against getting it too dark, in trying to get it too rich. I favor more gold and lime red.

Apparently Lane has received nothing from us in the Auditorium lighting. I know we did not have H. C. Meyer make regular plans + specifications on these second buildings, as he did on the Library + gym; but thought we had sent Lane something. I remember talking fully with Barrett Jones about the Auditorium, and deciding to have a great ball of light hanging from the centre; opalescent glass from direct lighting, but relying mainly on indirect light shining up out of truncated top of sphere and reflecting dome from ceiling.

Please send Lane, at once, duplicates of all intentions, drawings, [?], letters, etc. that have ever been sent to him on this point; and if nothing has ever been sent, send him a copy of a letter you can get, now, from B. Jones on this subject. Lane's idea is that it would be finest to have the dome absolutely unbroken; and light is by shooting light up at the dome from the 4 corners of the interior, and reflectors to [?] source of light being seen. For galleries, I am afraid that would give a queer, non-spherical look to the illustrated dome. Write Lane full of both ideas, so he can decide intelligently. If the sphere is to be used, I suppose it would have to be made in USA and shipped out—a string(?) really(?) against it. I like the idea of the sphere, myself. Don't bother to send Lane anything [?] on lighting except this matter of the dome.

Better [?][?] be given to packing. 75% of the [?] came broken; then Louis Cafe for the Auditorium facade was [?] so as to be about [?], etc. Make a special point with everybody, shipping anything out.

There will be no more building at Tsing Hua for some time—perhaps some years; there must finally be a change in the Ministry, the President is [?] wishing to use the Borea Feud for their own purpose. Lane expects to get the Science Building done about January, and the Auditorium about May 1919.

(P.S.)

Cable just [?] saying Forsyth sails for the Scatile June 8 on "Atenla Marn" (?). Best news I have had since I landed! I certainly need him. Will write again today or tomorrow.

—H.K.

3.6 托伯特·哈姆林(Talbot Hamlin)关于罗斯福总统铜牌信（1922年12月2日）

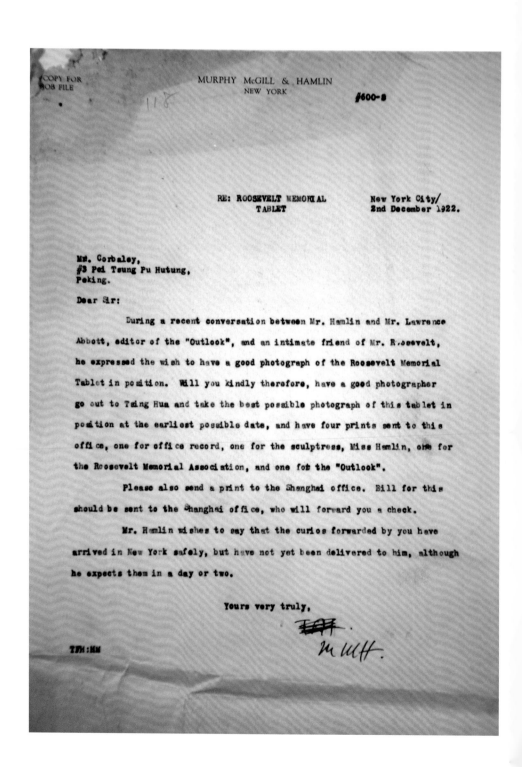

RE: ROSSEVELT MEMORIAL TABLET

New York City
2nd December 1922.

Mr. Corbaley,
#3 Pei Tsung Pu Hutung,
Peking.

Dear Sir:

During a recent conversation between Mr. Hamlin and Mr. Lawrence Abbott, editor of the "Outlook", and an intimate friend of Mr. Roosevelt, he expressed the wish to have a good photograph of the Roosevelt Memorial Tablet in position. Will you kindly therefore, have a good photographer go out to Tsing Hua and take the best possible photograph of this tablet in position at the earliest possible date, and have four prints sent to this office: one for office record, one for the sculptress, Miss Hamiln, one for the Roosevelt Memorial Association, and one for the "Outlook".

Please also send a print to the Shanghai office. Bill for this should be sent to the Shanghai office, who will forward you a check.

Mr. Hamlin wishes to say that the curies forwarded by you have arrived in New York safely, but have not yet been delivered to him, although he expects them in a day or two.

Yours very truly,
T. Hamlin

TFH: HM

90.5.1.
清华学堂. 筑七

第 4 章　周诒春和曹云祥的相关资料

4.1 The Tsinghua Question. The Far Eastern Review, vol. XV, no. 5, 1918（p.169）. 有关周诒春校长去职之文章全文转录及翻译

Much anxiety has been expressed lately in American educational circles and among Chinese returned students about the present status of Tsing Hua College, the institution founded and maintained by the Chinese Government for the preparation of Chinese students to whom the Government grants scholarships in America. The accusations brought against Dr. Tsur Ye-tsung, once Dean and, since 1913, president of the College, have resulted in his resignation, and the entire Peking community is not only convinced that an injustice was done Dr. Tsur, but also that his removal was suggested to those who accused him by a political clique with ulterior motives which it is not diplomatic to define. Apart from the character of these motives, any attempt to involve the administration of Tsing Hua College in Chinese petty politics must rob the institution of much of its prestige and usefulness, and must therefore be resented by all Americans and Chinese students returned from America who are either directly or indirectly interested in education in this country.

In 1908 the United States Government announced to China its intention to reduce the Boxer Indemnity claim from $24,440,728 to $13,655,472. The Chinese Government then undertook to use the annual remittances to educate Chinese students in America, and later announced its intention of establishing a preparatory school at Tsing Hua to train the successful candidates for scholarships in America. It was understood that a member of the Legation staff should have a voice in the direction of the institution. The agreements entered into were not precisely hard and fast in a legal sense, but were binding upon China in a moral sense.

Recently, when the United States Government, together with the Governments of the Allied Powers, agreed to the suspension of Boxer Indemnity payments for a period of five years the Ministry of Foreign Affairs gave a clear understanding that China would continue to send students to America and to maintain Tsing Hua College on the same financial basis that has prevailed heretofore.

In the maintenance of Tsing Hua College, in its freedom from political meddling and the arbitrary reduction of its subsidies, Americans have an intense interest and a sound moral right to demonstrate that interest. The removal of President Tsur, the Waichiaopu's control of expenditures in minute detail and the recent attempt to appoint to the Presidency a man who is not an American returned student show a new tendency towards a kind of control of Tsing Hua which every one is convinced is opposed not only to the interests of the College but also to American influence and interests in China. The checking of this tendency must primarily be left in the hands of our diplomatic representatives in China, but it is the business of every American organization in China to see that all Americans clearly understand the situation and that they are prepared to support our diplomatic representatives by concerted action or propaganda when called upon. A similar duty devolves upon the Chinese returned students and their various organizations.

— Peking American News Bulletin

———————————— ❋ ❋ ❋ ————————————

近来，美国教育界和中国归国留学生对清华学校的现状表达了极大的担忧。该校由中国政府为获美国政府奖学金的留美预备学生建立并维系。对曾任系主任，并从1913年起任校长的周诒春博士的指控最终致使其辞去职务，整个北京社会不仅相信这是对周博士的不公，而且认为他的去职是某些持卑劣动机的政治团体操纵的结果。除了这些动机的特点外，任何将清华管理层卷入中国可悲的政治斗争的企图，都必将使清华学校的声誉与影响力大打折扣，因此必将受到所有直接或间接对这个国家的教育感兴趣的美国人和中国归国留学生的憎恨。

1908年，美国政府向中国宣布，打算将庚子赔款从24440728美元减至13655472美元。随后，中国政府承诺将每年的退款用以派遣中国学生赴美接受教育，并在不久后宣布计划在清华园设立一所预备学校，以培养获得赴美奖学金的成功候选人。据了解，公使馆的工作人员应该在该学校的决策上有发言权。双方达成的协议并不具有严格的法律意义，不过在道德上对中国具有约束力。

最近，当美国政府和协约国政府共同声明同意庚子赔款暂停支付5年时，（中国）外务部明确表示中国将继续向美国派遣学生，并将提供与以往相同的财政基础来保证清华大学的运行。

在清华学校的维系问题上，在保证其不受政治干预、不随意削减补助金的问题上，美国人有着强烈的关注，并以一种坚实的道德立场来加以证明。周校长的去职、外务部对开支事无巨细的控制，还有最近试图委任一名非美国留学生出身的校长的举措，都显示出一种控制清华的新趋势。大家相信，这不仅违背学校的利益，也违背美国在中国的影响和利益。对这种倾向的遏制手段主要掌握在驻华外交代表们的手中，但在中国的每一个美国组织都应该看到，所有美国人都清楚地了解形势，并已做好在接到要求时通过一致的行动或宣传来支持我们的外交代表的准备。类似的职责也落在中国归国留学生及他们的各类组织肩上。

——北京《美国新闻公报》

（刘翘楚转录、翻译，杜林东校对）

4.2　Tsao, Y S. The China Press (1925—1938), Sep 22, 1928 (p.4). 曹云祥对去职问题之回应

THE TSING HUA CASE

Editor, The China Press：

Sir, —Recently, so much damaging information has been circulated by several news agencies concerning Tsing Hua College, that the undersigned feels in duty bound to correct; and I sincerely hope you will be good enough to grant me the necessary space in your valuable columns at this late date.

It has been reported that American influence has been negligible for a considerable time at this "American Indemnity College", that recently the students took a violent dislike to the president, so they discharged him, that they compelled five professors to resign, that since 1919 only two examinations have been held, that the students expect degrees whether they attended classes and passed examinations or not thus reducing their education to a farce.

While educational work in China is difficult to conduct and should be improved upon, there is no necessity to circulate unauthenticated information and to paint the picture darker than what it really is. In all fairness to Tsing Hua College, despite its faults and shortcomings, the college administration, the faculty and the students have not enacted such a "farce" as has been reported.

The president that served from April, 1922, to February, 1928, resigned cheerfully and of his own accord when he failed to receive the necessary support from his superior authorities after serving under no less than fourteen changes of foreign ministers during his administration. He was given farewell parties by both the faculty and the student body, while he requested them all to support the incoming president for the welfare of the college as a whole. The next president proved to be a fill-gap only, but for the two months he served, it was to the satisfaction of all. Eventually, he was told to yield his place to a Mukden nominee, who left hastily with his party soon afterwards. Sine then two deans have been acting temporarily pending the appointment of a new president by the Nationalist Government. Hence, "recently" there has been no violent disliking and discharging of presidents by the students.

Concerning attendance to classes and examinations, when a preparatory student fails to attend classes three times without reason, he is given a demerit, and nine demerits during his eight years at college constitutes a dismissal from the institution. When a university student absents himself from classes twenty times without permission during his four years, he is also dismissed. These rules are rigidly enforced. With regard to the keeping of record of work done and examinations held, there are systems of daily marks, the monthly tests and half-yearly examinations. It was only when there was war waging in the neighborhood or a general closing down of schools due to grave political unrest that the Faculty Council decides to postpone the examinations and close the college. Last summer, the graduating class passed their examinations and were sent abroad. The university students, as a whole are offered excellent courses by the professors, and in appreciation, they study very diligently and eagerly, as can be attested by the anxiety of the athlete directors and the over-crowded condition of the library all day long. The very fact that all Tsing Hua students who graduate have uniformly done very creditable work in American colleges, belles the statement of reducing education to a farce by "holding only two examinations in ten years."

Tsing Hua College was established in 1910 with the original intention of preparing students for study in America, and it was only four years age that the policy was changed by raising the standard to that of a university. Hitherto, no degrees have been granted at all to the preparatory students who have been sent to American colleges to complete their studies. It is safe to say that of the thousand odd students who have gone and returned, every one of them has received degrees and quite a number with great distinction. However, beginning with the summer of 1920, the university students will graduate with degrees conferred upon them, and they will have to compete with the graduates of other colleges for the fellowship examinations in order to be sent as post-graduates to the United States. When America remitted the indemnity for educational work, there was no bargaining for the extension of American influence in Tsing Hua College, far less a "yoke of American influence." I cannot, for one moment, imagine that any country that remits an indemnity for educational or cultural enterprise, harbors the ulterior motive of extending its national influence in China. Good-will, there will naturally be provided it can be maintained socially and politically.

Since last May, Tsing Hua has had no regular president, and the faculty, the alumni and the students have been hoping earnestly for the appointment of a good president from among the thirty odd aspiring candidates. It appears that the majority of the alumni would like to see a Tsing Hua alumnus appointed, as he is likely to be familiar with the conditions of the College, while there are others who believe that anyone who is conversant with the American educational conditions would be acceptable in order to widen the range of choice. Both these views represent an honest difference of opinion, and perhaps upon the reported appointment of Mr. Lo Chia-luen, the majority of the students rallied to his support while the alumni were not so ready to accept him. As the five professors recently denounced by the students are practically all alumni of the College, it is to be hoped that the differences will soon sink, especially in consideration of the several alumni members already appointed to the new Board of Directors.

Mr. Lo, I understand, is a good scholar and a favorite student of Dr. Tsai Yuan-pei, the Minister of Education, and Tsing Hua College has a good plant, a sound educational policy, a dependable source of income and an endowment fund. The faculty is well selected, the students carefully recruited and the alumni loyal to its Alma Mater. With adequate support from all sides, the new president will be able to develop the college into an educational institution of importance for the training of the future manhood of the nation.

As one who has served the College for about six years out of the seventeen years of its history, it is my fervent wish that the differences will soon be amicably settled, so that normal educational work could be resumed at an early date. Educational work must be regarded as a public trust, and anyone who is properly qualified and feels confident of being able to render creditable service, should be supported and be given a fair chance for the indication of his qualities.

I am, etc.,

Y. S. Tsao.

Ex-President of Tsing Hua College.

Kaifeng, Honan, Sept, 17, 1928.

（周翔峰转录）

4.3 Tsao, Y S. The China Press (1925—1938)， Apr 22, 1929 (p.4). 曹云祥在1929年对国立清华大学发展之评述，表达对罗家伦校长的支持

TSING HUA UNIVERSITY

Editor, THE CHINA PRESS：

Sir: —Once more we want to speak through your columns on an important issue now at stake. It is the detriment of the new policy of Tsing Hua University by the Board of Trustees at Nanking. As the public is fully aware that the fundamental policy of Tsing Hua has been changed since 1925. Before that time, the institution was known as the college "for preparation of advanced studies in America." Every summer, about fifty students sail away on stately President liners. But behind such a scene, thousands of dollars are squandered in many unnecessary expenditures. On the other hand, millions of youths in China never get the chance to enter universities, not to say going abroad. What a social injustice, what a ridiculous idea that money makes the education.

Moved by the spirit of the Age that there should be equal opportunities for all men, the Tsing Hua administration, in the year 1925, decided to convert the College into a university. Instead of bestowing the time-honored privileges to a few men, the American returned Indemnity shall be used for developing a permanent institution of learning where, students, in ever-increasing numbers, pursue higher education at comparatively little expense. And instead of keeping out the wretched poor, the Tsing Hua University opens its way for all students who want to spend strenuous years there, both mentally and physically. At present, Tsing Hua has developed into a full-fledged university which gives strong evidence to the fact that the new policy is a complete success.

Such is the story of the coming into being of the Tsing Hua University. But since its birth, the University has suffered from its inherent weakness. From the days of the College, this institution has been under the sole control of the Ministry of Foreign Affairs. Curiously enough, there has never been an educationalist who became the President of Tsing Hua. In some way or the other, the Minister sent one of his counselors to take charge of the school. The College itself was nothing more than a department of the Ministry. But since the founding of National Government at Nanking, the Ministry of Education was also admitted to a share in the administration. Thus the University was placed under a dual control. The practice is a disintegration of authority, the result of which is irresponsibility. There is no precedent of the same kind in the educational history of any country. And worst of all, the two Ministries vested their authority in a Board of Trustees, which has the supreme power of veto over the affairs of the University.

The preceding paragraph will not fail to give an impression about the complicated nature of the notorious Board of Trustees. Under such conditions, the Board is a "buffer zone", so to speak between the two Ministries. This dual character provides too much ground for motives other than educational to couch in, and when this occurs, the development plan of the University and the essential interests of the students are, of course, considered as secondary questions. Personal dislike and factional feeling play an important part in the meetings of the Trustees. Therefore it is not surprising that, at the recent Meeting of the Board, the development plan of the University, as submitted by President Lo Chialun, was in total rejected. Thus the new policy of Tsing Hua, for four years winning the confidence and support of all concerned, received a severe blow at the hands of the Trustees.

The recent resolutions of the Board of Trustees are contrary, in every sense, to the development of Tsing Hua. In the first place, the Trustees voted a reduction of $60,000 in the budget for the current year. They seem to ignore the fact that since last summer, University expenditures are strictly regulated by income. Salaries of staff members have been reduced to a considerable degree. The reduction of $60,000 from the budget would mean the complete deadlock of school finance. A second resolution worse than the one just mentioned is the sending of 30 more students to America by the Board of Trustees during the coming summer. This is an abrupt departure from the policy adopted since 1925 that the development of the University should be the sole and immediate concern of the governing authority. Considering that the Board is recognized as the faithful guardian of this policy, we deem that they are unworthy of such a great trust. A third resolution which virtually brings an end to the Municipal Engineering Department of the University, shows even further that the Board of Trustees has actually done more harm than good to this institution. The Engineering Department was established in 1925 with an equipment costing more than $200,000 and a total enrollment of 60 students. But these things made little impression, if any, in the mind of Trustees. Lastly, the Board did nothing proper for the investigation and careful investment of our sinking fund. The result of maladministration makes the total sum $8,000,000 reduced to $3,000,000. Thus, we, the students of Tsing Hua University, have come to the realization that the Board of Trustees will no longer subject to any toleration; and drastic measures should be taken for the abolition of this irresponsible organ.

Such is the chief object of our present movement. We are fighting for the strict observance of the development plan of this institution. Any obstacle that stands in the way of progress must be done away with. And to avoid misunderstanding on the part of the readers, we take this opportunity in making clear what the movement is and what it is not.

In the first place, the present movement is a struggle for principle, not for the man. We hold fast to the idea that Tsing Hua University should be under one responsible authority, one supreme rule, nothing of a dual character. We strive to abolish the Board of Trustees simply for the reason that it is a superfluous, and irresponsible organ, a disintegration of authority. It follows logically that Tsing Hua should be directly under the sole control of the Ministry of Education. As the movement is not personal one, we do not support the person of Mr. Lo Chia-lun, nor do we have any ill feeling towards any individual member of the Board of Trustees.

In the second place, the present movement is a manifestation on the part of the students in connection with the policy of the University. We deem that the urgent need of the hour is the development of the University, not the sending of 30 more students abroad at the sacrifice of 500 at home. The resolutions of the Board of Trustees are detrimental to the welfare of the University, which, for justice sake, we cannot remain silent. Therefore let the public be informed with this plain truth, which has too often been misinterpreted by the slanderers.

Lastly, we want to call the attention of the readers to the fact that Tsing Hua University is serving to preserve our memory of the Boxer Uprising in the year 1900 when the combined naval forces of the Powers occupied the City of Peking. To the American mind, this institution is a direct outcome of their generous action in returning on half of their share of the Indemnity. As the matter stands now, the remaining portion of the sinking fund, amounting only to $3,000,000, can hardly ensure the future support and development of the University. And it largely depends upon our endeavor in working for the abolition of the Board of Trustees that this institution will possibly become a mighty factor in the upbuilding of this country.

Thanking you for your valuable space,

We remain,

Yours etc.

The Tsing Hua University

April 13, 1929.

（周翔峰转录）

4.4 The Tenth Anniversary of Tsing' Hua College, Peking. Millard's Review of the Far East (1919—1921), May 21, 1921 (p.636). 全文转录及翻译清华建校十周年庆典活动与清华董事会之简要说明

THE TENTH ANNIVERSARY OF TSING. HUA COLLEGE

The tenth anniversary of Tsing Hua College, celebrated from April 3o, to May 2, was made memorable by a simple message which Dr. W.W.Yen, Minister of Foreign Affairs, announced in the course of his speech at the dedication of the $200,000 auditorium. The message was to this effect: "Steps have been taken for the creation of an endowment fund. In the accomplishment of this, two things will be achieved; the institution will be perpetuated and an immortal monument built to American liberality and friendship."

This decision of the Chinese government is of equal importance to the announcement of the intention of the American portion of the Boxer Indemnity and the Chinese decision to utilize the fund for the   of the education of young men and women in American colleges and universities. Finding it necessary to have a special school in China to prepare student for entering high institution of learning in the United States, the Chinese government established Tsing Hua College. Tsing Hua is an offshoot of the Chinese-American arrangement with no prospect for its perpetuity. It will cease to exist in 1940 when the Boxer Indemnity will have been completely returned. The creation of an endowment fund for the future of the college is an important measure. It is more or less surprising that the authorities of this college should not fund a special feature in the celebration of the tenth anniversary. Compared with that fund , the anniversary became a matter of small importance. It was an early as 1917, the question of an endowment fund for Tsing Hua received the attention of the Chinese authorities who were entrusted with the administration of the college.

A board of trustees has now been organized, including the Minister and Vice-Minister of Foreign Affairs, and the American Minister to China, and this board will take over the endowment fund in the hope that the fund would increase to such a size by the end of 1940 that the income therefrom would be sufficient to carry on the work in Tsing Hua together with the educational mission in the United States which is an essential part of the college activities .Regulations are being drawn to deal with the investment of the fund and will be of such a nature that it will be in safe hands.

A pamphlet containing the history of the institution was distributed on the campus and traced the events of the institution back to the time when they first started to employ American teachers on the staff.

Ceremonies for the tenth anniversary were numerous and varied, extending over a period of three days from Saturday to Monday. They included the opening of the new auditorium. Games and drills were given by the boy-scouts and the athletic classes on the field in front of the new Roosevelt Memorial Gymnasium. Plays both in Chinese and English were presented by the students. Speeches were given by President King, Minister Yen and Minister Crane in the auditorium which was dedicated with much ceremony. The American teachers of the college were conspicuous by their absence on its tenth anniversary. Here and there a few of them were seen on Sunday. Whether there were many Chinese teachers educated in America present, it was difficult to say. On such an occasion in the United States they would have worn academic caps and gowns. Neither did the Chinese teachers nor the Americans wear them last Saturday and Sunday. If they did so, the ceremonies might have been made more impressive. In spite of this, the outing to the historical place where Emperor Kang Hsi used to tread, coupled with the privilege to inspect the newly erected building at Tsing Hua, was thoroughly enjoyable.

---------------------------- * * * ----------------------------

清华学校建校十周年纪念庆典

清华学校建校十周年纪念庆典于4月3日至5月2日举行，外务部部长颜惠庆博士在耗资20万美元的大礼堂落成典礼上发表了简单而令人难忘的致辞。其主要内容是这样的："基金捐赠机构已经启动。在实现这一目标的过程中，将完成两件事：一是该机构将长期存在，二是为美国的自由和友谊建立一座不朽的纪念碑。"

中国政府的这一决定，与美国宣布减免部分庚子赔款以及中国决定将这笔资金用于年轻男女学生进入美国高校学习同等重要。中国政府发现有必要在中国设立一所特殊学校，为学生进入美国高等学府做准备，于是成立了清华学校。 清华是中美协议的一部分，不可能永久存在。1940年，当庚子赔款退还完成时，它便将停止运营。为了学校的未来，设立捐赠基金是一项重要措施。学校主管部门不为庆祝十周年举办特别活动，这多少让人感到意外。与那笔基金相比，周年庆典变得无关紧要。早在1917年，清华捐赠基金的问题就引起了当时该校主管部门的重视。

现在已经组建了理事会，包括外务部的部长和副部长，还有美国驻华大使。理事会将接管捐赠基金，希望能在1940年年底将其增至足够的规模，使其收入足以继续清华的工作以及前往美国的教育使命，这是该校活动的一个重要部分。目前正在起草处理基金投资事宜的规定，它将确保基金被安全保管。

校园里分发了一本小册子，载有该校历史，并将其历史追溯到该校第一次聘请美国教师的时候。

十周年纪念仪式繁多，从星期六到星期一持续了三天。其中包括新礼堂的开幕，童子军与体育班在罗斯福纪念体育馆前进行的操练表演，还有学生们表演的中英文戏剧。金邦正校长、颜惠庆部长和克瑞思部长在礼堂发表了隆重的演讲。学校美国教师在十周年庆典上的缺席引人注目。星期日的时候却到处都能看到一些美国教师。很难说是否有许多在美国受过教育的中国教师在场。在美国这样的场合，他们会戴学术帽，穿学术礼服。中国教师和美国教师上星期六和星期日都没穿。如果他们这样做了，仪式可能会更加令人印象深刻。尽管如此，到康熙帝经常踏足的历史遗迹游览，再加上有幸参观清华新建成的大楼，这一经历非常愉快。

（孟可转录、翻译，杜林东校对）

第5章 罗家伦时代的相关资料

①
译者注：此处Northern rebellion应该
指的是中原大战（又称蒋冯阎战争），
指1930年（民国19年）5至10月蒋介石
与与北方军阀阎锡山、冯玉祥、李宗仁
等在河南、山东、安徽等省发生的一场
新军阀混战。

5.1 Why President Lo Chia-luen Resigned from Tsing Hua. The China Weekly Review (1923—1950)，1930-06-21: 106. 全文转录及翻译关于罗家伦去职之新闻报道

Why President Lo Chia-Luen Resigned
From Tsing Hua

"Among the immediate and pernicious effects of the Northern rebellion upon the Central Government's administration in the North, particularly in the field of education," says a recent Kuo Min dispatch from Nanking, "was the recent resignation of President Lo Chia-luen of the National Tsinghua University." The reason for president Lo's action, according to his petition of resignation submitted to the Ministry of Education, was the fact that "during a meeting of the students' representatives on May 20, the question of the president's tenure of office was brought up for discussion. But at a general meeting of the entire student body on the following day, however, the question was quashed by a majority vote."

Mr. Lo pointed out that while the action of the entire student body could be regarded as a vindication of himself, he could not, at the same time, ignore the fact that the question of the presidency was one to be decided by the Party and the Government and, therefore, lay outside the province of student activities. The action of the student representatives was, for this reason, ominous for the future of Government educational administration for the upholding of which he felt in duty bound to take a moral stand. Rather than impair the respect and the dignity of the Presidency of Tsing Hua, therefore, he decided to resign. In the course of a statement, Mr. Lo dealt with the matter of the increased amount of the monthly instalment funds, because of the increased exchange value of gold. He recommended that they be used exclusively for the new buildings. For this he had obtained authorization from the Ministry of Education a few months ago, so that the present four buildings under construction would be duly completed, and, by proper management, a sufficient amount of money would be saved to enable the new Chemistry building to be directed.

罗家伦校长为何从清华辞职

一则最近的南京国民党电讯表示："在北方叛乱①对中央政府在北方造成的众多直接和有害影响中，发生在教育领域的一起特别事件，是罗家伦校长最近从国立清华大学辞职。"根据罗家伦校长提交教育部的请愿书，他之所以辞职，是因为"在5月20日的学生代表会议上，关于校长任期的问题被提出讨论。但是，在第二天的全体学生大会上，这个问题却以多数票被否决"。

罗先生指出，虽然整个学生团体的行为都可以视为对他本人的辩护，但他不能忽略这样一个事实，即校长任职的问题是由党和政府决定的，故不属于学生活动的范畴。因此，学生代表的行动对于政府教育管理的未来是不利的，站在道德立场上，他认为自己有责任支持政府的决定。所以，为了不损害校长的尊严，他决定辞职。在声明中，罗先生提及了因黄金的兑换价值增加而导致的每月分期付款资金①增加的问题。他建议将其专用于新建筑的建设。为此，他于几个月前获得了教育部的授权，所以目前在建的四座建筑物②将按时竣工，并且通过适当的安排，将节省足够的资金用来建设新的化学馆③。

（张钰淳转录、翻译，杜林东校对）

① 编者注：monthly instalment funds指美国使馆按月交给基金会的款项，用于清华的日常开支，来源于1908年之后庚子赔款的退款。罗家伦任职期间曾对这笔基金进行整顿，请辞后交给中华教育文化基金董事会代管。详见本书上编第5章。

② 译者注：当时在建的包括图书馆（扩建部分）与气象台。

③ 译者注：清华大学化学馆兴建于1931年7月，至1933年夏竣工。

5.2 Tsing Hua President Resigns: Determination to Uphold the Dignity of His Office. The North-China Herald and Supreme Court & Consular Gazette (1870—1941), 1930-06-17: 448. 全文转录及翻译《北华捷报》关于罗家伦辞职之新闻

TSING HUA PRESIDENT RESIGNS
Determination to Uphold the Dignity of His Office：
A Review of Work Done in Past Two Years

Among the immediate effects of the Northern rebellion upon the Central Government's administration in the North, particularly in the field of education, was the recent resignation of President Lo Chialuen of the National Tsing Hua University, the maintenance of which, as is generally known, is wholly dependent upon an endowment fund derived from the American Boxer Indemnity refund.

The reason for President Lo's action. According to his petition of resignation submitted to the Ministry of Education, was due to the fact that "during a meeting of the students' representatives on May 20, the question of the President's tenure of office was brought up for discussion. But at a general meeting of the entire student body on the following day, however, the question was quashed by a majority vote."

Such a phenomenon in present circumstances, said Mr. Lo, was not extraordinary. But to have it happen in an institution for the welfare of which he had devoted two years of energetic effort was something which he could not bear to see.

Continuing, he pointed out that while the action of the entire student body could be regarded as a vindication of himself, he must not, at the same time, ignore the fact that the question of the presidency was one to be decided by the Party and the Government and, therefore, lay outside the province of student activities. The action of the student representatives was, for this reason, ominous for the future of Government educational administration for the upholding of which he felt in duty bound to take a moral stand. Rather than impair the respect and the dignity of the Presidency of Tsing Hua, therefore, he decided to resign the office.

A subsequent order from the Ministry of Education instructed Mr. Lo to continue in office. But he was determined in his decision and loft Peking after turning over the administration of Tsing Hua to the University Senate.

The following statement containing many interesting facts about the University's administration was also issued by Mr. Lo before his departure for Nanking:

Not only have I left Tsing Hua University, but I am leaving Peking this afternoon for my native province. I am determined: not to compromise on the moral stand which led me to resign from the presidency of Tsing Hua, and thereby to uphold the necessary dignity of that office. Any compromise would set an unfortunate precedent Tsing Hua for any of my successors who might attempt to serve the institution without impairing their self-respect and the moral prestige of the office.

I am glad that I can leave Tsing Hua with the feeling that I have served the institution honourably and faithfully for two years. I shall always be grateful to my colleagues for their co-operation without which the various reforms under my administration would not have been possible. The main progressive features of the work I have done are outlined in: pamphlet published a few months ago under the title, Tsing Hua: a University in the Making. On the eve of my departure, I merely want to express my sincere hope that some important policies may be continued and a few healthy traditions perpetuated.

1. The transfer of the Tsing Hua endowment funds from the virtual control of the Ministry of Foreign Affairs to the custody of the independent and self-perpetuating China Foundation, and that of the monthly instalment funds from the actual control of the same Ministry to the Foundation, are the results of a hard struggle to which I contributed no small part, and for which I feel especially indebted, among many other friends to Mr. J. V. A. MacMurray, the then American Minister to China, for his kind co-operation. I hope, particularly, that the endowment funds will be safeguarded according to the policy which I outlined in the Ten Guiding Principles approved by the Ministry of Education. Both the capital and the interest of the funds should not be touched before 1940, and in the meantime we should strictly limit the number of students to be sent abroad, so that after income from the Indemnity Refund ceases, the existence of Tsing Hua will not be jeopardised.

2. The increased amount of our monthly instalment funds, because of the increased exchange value of gold, should be used exclusively for the new buildings. For this I obtained authorization from the Ministry of Education a few months ago, so that the present four buildings under construction will be duly completed, and, by proper management, a sufficient amount of money will be saved to enable the new chemistry building to be erected

3. As the four new buildings—the biology building, the new students' dormitory, the new library and the meteorological tower—are now in active process of construction. I hope the original plans, which were so painstakingly worked out, will be adhered to closely

under vigilant technical and public supervision. The system which I started, under which cheques for the payment of construction work must be counter-signed by the Chairman of the Financial Committee on Building Construction, must be continued, as almost half of the ultimate cost of the biology building was generously endowed by the Rockefeller Foundation.

4. The practice inaugurated during my administration, whereby all accounts are open to the public, and, before their submission to the Ministry of Education, are first inspected by the University Senate and then carefully audited by chartered accountants, should always be observed for the benefit of Tsing Hua.

5. It was only by practicing strict economy and by cutting down administrative expenses that I was enabled to save $105,955 in the fiscal year of 1928 in comparison with the final budget of the previous year, although both professors and students' were largely increased in number and more and better equipment was added to the library and laboratories during the first year of my administration. Since I took over office, 24 administrative positions were abolished, representing an annual saving of $41,700 in salaries alone. But, on the other hand, during less than two years, $287,939 were spent for books and scientific apparatus, and in the budget for 1930 $240,000 are appropriated for the same purposes. I believe that efficiency is best served by enlisting fewer, but better qualified staff members and, and I firmly believe that, in a real institution of higher learning, every possible effort should be made to reduce administrative expenditures to the lowest minimum in order that academic equipment can be adequately provided.

6. During the past two academic years, 41 new professors and 21 new lecturers, including several distinguished visiting professors from Cambridge (England), Columbia, Chicago, Princeton (U.S.A.), and Kyoto (Japan) Universities, were added to the Tsing Hua faculty. They, together with 24 professors who were in Tsing Hua before my administration, constitute one of the most select faculties in the whole of China. I hope that Tsing Hua will always continue to enlist the service of eminent scholars both in China and abroad.

7. The graduate school ought to be the nucleus for the future intellectual development of Tsing Hua. Its foundation has been laid, and its programme should be persistently carried out. Quantity in the admission of students. I expect that Tsing Hua will take a leading part in promoting research word in China, just as Johns Hopkins did among American universities several decades ago.

8. The budget for 1930, which is essential to the immediate development of Tsing Hua, was prepared carefully and submitted to the Ministry of Education. It allots 20 per cent. of the total amount to books and laboratory apparatus, and another 20 per cent.

toward the cost of new buildings. Strict adherence to this budget will bring Tsing Hua to a new stage if progress.

In my humble opinion, the systematic development of an academic institution depends very much upon a far-sighted plan. In order to steer the ship of Tsing Hua through the stormy seas, I hope that the programme as embodied in the "Ten Guiding Principles" may still serve the institution as a set of working maps and charts.

"I am glad that I am leaving Tsing Hua only after having put it on a fairly secure basis. I trust my colleagues and friends who all have the welfare of Tsing Hua at heart will be able to pilot Tsing Hua to a safe haven, to bear its academic standard aloft, and to uphold the prestige necessary to any academic institution deserving of respect."

———————————— ＊＊＊ ————————————

清华校长辞职
维护职位尊严的决心：过去两年的工作回顾

6月26日，南京

在北方叛乱对中央政府在北方的众多直接影响中，发生在教育领域的一起特殊事件是最近国立清华大学校长罗家伦的辞职。众所周知，国立清华大学的维护完全依赖于美国庚子赔款的捐赠基金。

根据罗校长向教育部递交的辞呈，之所以采取这一行动是因为"在5月20日的学生代表会议上，校长的任期问题被提出来进行讨论。然而，第二天在全体学生大会上，这个问题被多数人投票否决了。"

罗校长说，这种现象在目前情况下发生并不奇怪。但是，在一个他为之奋斗两年的教育事业中发生这种事，是他不忍心看到的。

他继续指出，尽管整个学生会的举动可以被视为对自己的辩护，但他同时也不能忽略这样一个实事，那就是校长的任职问题是党和政府决定的，因此这不属于学生活动的范畴。基于以上，学生代表的行动对政府未来的教育管理是不利的，站在道德立场上，他认为自己有责任来支持政府的决定。所以为了不损害清华校长的尊严，他决定辞职。

教育部随后的命令指示罗校长继续任职。但他下定决心，并将清华行政管理权移交给校务委员会后离开了北京。

罗先生在动身前往南京前，亦发表载有许多有关大学管理趣闻的声明：

我不仅离开了清华大学，今天下午还要离开北京回到我的家乡。我表决：在导致我辞去清华校长职位的道德立场上不妥协，并以此维护该职位的必要尊严。任何妥协都会为我那些试图不损害自尊与职位道德的继任者们树立一个不好的先例。

我很高兴能伴着为清华正直而忠实地服务了两年的感受离开清华。我将永远感谢我的同事们的合作，没有他们的合作，我领导下的各种改革就不可能实现。几个月前，我出版了一本名为《清华：一所正在建设的大学》的小册子概述了我最近工作的进展。在我离开前夕，我只是想表达我真诚的希望：一些重要的政策可以继续下去，一些良好的传统可以得到延续。

1. 把清华的捐赠基金从外交部的虚拟控制权转移到独立自主且能够自我持续的中国基金会托管，并且将每月分期付款的资金的实际控制权由外交部转移到基金会是经历艰苦斗争的结果，其中我做了不少的贡献，我尤其感激包括当时的美国驻华大臣J.V.A. MacMurray先生在内的许多友人的合作。我特别希望，捐赠基金将根据我在教育部批准的十项指导原则中概述的政策得到保障。在1940年以前不应动用资金的本金和利息，同时我们应严格限制留学生人数，以免在补偿退款收入停止后，清华的生存就不会受到威胁。

2. 由于黄金交易价值的增加，我们每月分期增加的分期基金额度应该专门用于新建筑。为此，我在几个月前获得了教育部的批准，使目前正在建设的四栋大楼能够按时完工，并且通过适当的管理可节省足够的资金来建造新的化学馆。

3. 生物学馆、明斋、新图书馆（二期）和气象台四座新建筑正在积极建设中，我希望其施工能不惮劳思，通过完善的技术和公共监督保证严格按照原设计加以实现。我所开创新制度——支付建筑工程的支票必须由建筑工程财务委员会主席会签，必须延续下去，因为生物学馆的总建造费用中几乎有一半是由洛克菲勒基金会慷慨捐助的。

4. 在我任职期间，所有财政均向公众开放，在提交给教育部之前，先提交本校校务委员会审阅，然后由注册会计师加以仔细审计。这一做法应当始终被本校遵守。

5. 在我任职的第一年里，虽然教授和学生的人数大幅增加，且图书挂和实验室也增加了更多更好的设备，但在严格节约和减少行政开支下，我才能在1928年在最终财政预算较前一年节省105955美元。自从我上任以来，取消了24个行政职位，每年仅薪金就节省了41700美元。但是，另一方面，在不到两年的时间里，书籍和科学仪器的支出为287939美元，在1930年的预算中为同样目的拨出了240000美元。我相信，招募较少但素质更高的工作人员可以最好地提高效率，而且我坚信，在一个真正的高等教育机构中，应尽一切可能的努力将行政支出减少到最低限度，以便可以充分提供研究设备。

6. 在过去的两个学年里，清华大学增加了41位新教授和21位新讲师，其中包括多位来自剑桥大学、哥伦比亚大学、芝加哥大学、普林斯顿大学和京都大学的杰出的客座教授。他们和我上任前在清华任职的24位教授，构成了全中国最优秀的高校之一。我希望清华大学将一如既往地争取海内外杰出学者的支持。

7. 研究生院应成为清华未来知识分子发展的核心。它的基础已经奠定，其纲领应该坚持执行。录取学生的数量（应有所保证）。我希望清华大学能够像约翰霍普金斯大学几十年前在美国大学中所做的那样，在促进中国的科研事业中发挥带头作用。

8.（学校）已经认真制定了对清华的迅速发展至关重要的1930年度预算，并已提交给教育部。其中，总额的20%分配给书籍和实验室仪器，另外20%用于监造新校舍。如果这一预算得到严格执行，清华大学的发展将迈入一个新的阶段。

依我之拙见，一个学术机构的系统发展，在很大程度上取决于一个有远见的计划。为了指引"清华号"驶过波涛汹涌的海面，我希望"十项指导原则"所包含的计划，可以继续作为一套工作路线图为清华服务。

"我很高兴在有一个相当稳固的基础后才离开清华。我相信所有的同事和朋友都在内心关心清华，一定能把清华引到一个安全的避风港，维持其高水平的学术水准，并维护任何值得尊重的学术机构所必要的声望。"

（王语涵转录、翻译，杜林东校对）

5.3　Big Building Plan is Made at Tsing Hua. The China Press, 1931-09-03. 全文转录及翻译罗家伦去职后之清华建设情况

BIG BUILDING PLAN IS MADE AT TSING HUA
Peiping University Will Be Englaged By New Improved Erections

Construction of a Chemistry Hall at an aggregate cost of $270,000 and of a new electricity plant costing $250,000 are included in the huge three-year program announced recently by the Tsing Hua University board of the directors. Tsing Hua University is one of the leading educational institutions in Peiping.

Work will be started on these two projects as well as on the new engineering hall, water power experiment laboratory and new gymnasium early this month and it is expected that by next Spring the buildings will be ready for occupation.

The program in its entirety provides for the construction of 10 brand new buildings, an extension of the present college campus and a remodeling of the present old structures of the university.

Use Boxer Funds

The funds used will be appropriated from a part of the returned American Boxer Indemnity Funds which have been granted to Tsing Hua University.

At a meeting of the university directions last year the present construction program was adopted. A fund of only $760,000 was decided upon at the meeting last year but an extension of the activities was decided upon by Mr. Ung Wen-hou, Present of the university.

The complete program calls for the construction or remodeling of the following buildings over a three year period:

Chemistry Hall

1.　Construction of a Chemistry Hall at an estimated cost of $270,000.

2.　Extension of the Engineering Hall and the water power experiment laboratory, $150,000.

3.　Extension of the Gymnasium, $100,000.

4.　Remodeling of the Science Hall, $15,000.

5. Construction of a Geography Hall, $80,000.

6. Construction of a new hospital building, $50,000.

7. Construction of a new electricity plant, $250,000.

8. Betterment of the telephone service and plant, $15,000.

Bigger Campus

9. Enlargement of the school campus, $50,000.

10. Construction of the Executive Building for the University Department, $120,000.

11. Construction of a Literature Hall, $200,000.

12. Construction of a Law Hall, $200,000.

13. Construction of a new girl student dormitory, $100,000.

14. Construction of a new boy student dormitory, $200,000.

15. Enlargement of faculty dormitory, $35,000.

16. Construction of a new faculty dormitory, $150,000.

The total expenditure, according to the present estimates, will be $1,985,000.

--- * * * ---

清华制订重要的建筑计划

清华大学董事会最近宣布了一个为期三年的大计划，包括建造一个耗资27万美元的化学馆以及一个耗资25万美元的新发电厂。清华大学是北平最顶尖的教育机构之一。

这两个项目以及新的工程馆、水力实验室和新体育馆将于本月初开工，预计明年春天竣工。

整个项目包括建设10座全新的建筑，扩建现有的大学校园以及改造现有的旧建筑。

使用庚子赔款

所用经费将从退还给清华大学的部分美国庚子赔款中划拨。

在去年的一次大学指导会议上采用了目前的建设方案。该会议决定拨出仅76万美元的基金，但出席会议的翁文灏先生决定增加拨款。

完整的计划要求在三年内建造或改造以下建筑物：

化学馆

1. 建造一个化学馆，预计费用为27万美元。

2. 扩建工程厅馆及水电实验室，15万美元。

3. 扩建体育馆，10万美元。

4. 改造科学馆，1.5万美元。

5. 建造一个地理馆，8万美元。

6. 建造一座新的医院，5万美元。

7. 建造一个新的发电厂，25万美元。

8. 改进电话服务和设备，1.5万美元。

校园扩建

9. 扩建校园，5万美元。

10. 为大学部建造行政楼，12万美元。

11. 建造一个文学馆，20万美元。

12. 建造一个法律学馆，20万美元。

13. 建造一个新的女生宿舍，10万美元。

14. 建造一个新的男生宿舍，20万美元。

15. 扩大教员宿舍，3.5万美元。

16. 建造一个新的教师宿舍，15万美元。

根据本预算，支出总额将为1985000美元。

（安芃霏转录、翻译，杜林东校对）

第6章 **梅贻琦时代的相关资料**

Tsing Hua to Take over Old Summer Palace. The China Press, 1933-11-18: 2. 全文转录及翻译清华将接管圆明园之新闻

Tsing Hua To Take Over Old Summer Palace
Historic Grounds Will Be Used for Agricultural Experimental Station

PEIPING, Nov.17.

According to information from Chinese educational circles, the Ministry of Education in Nanking has granted the application of the National Tsing Hua University to take over the Old Summer Palace known to the Chinese as Yuan Ming Yuan.

It is understood that the university will convert the palace grounds into an agriculture experimental station for the use of its proposed school of agriculture to be established next year.

The Old Summer Palace was destroyed by the troops of the Anglo-French Expedition in 1860 in retaliation for the action of the Manchu court in ordering the execution of a number of foreign prisoners of war. The site of the palace is situated in the vicinity of Tsing Hua.

———————————— * * * ————————————

清华大学将接管圆明园
历史遗址将被用于农业试验站

北平，11月7日。

根据中国教育界的消息，教育部（南京）已批准国立清华大学接管圆明园的申请。

据了解，该大学将把圆明园遗址改建为农业实验站，用于拟在明年建立的农学院。

为了报复满清政府下令处决若干外国战俘的行为，圆明园在1860年被英法联军摧毁。该园的所在地位于清华附近。

（张钰淳转录、翻译）

下编 校园近代重要建筑测绘图集

清华大学礼堂
91.3.9. 冀

清华学堂
80.5%

2020年·25. 时年83岁.

第①组　建校初期的重要建筑

1. 二校门（重建）

测绘人	金茶璇 金兑镒 刘炫育	图纸名称	二校门南立面图 东立面图 顶视图
制图人	金兑镒		
班级	建22 建23	打印日期	2015年7月16日
指导教师	刘亦师	图纸编号	14-1

清华大学
建筑学院　　清华大学二校门测绘图

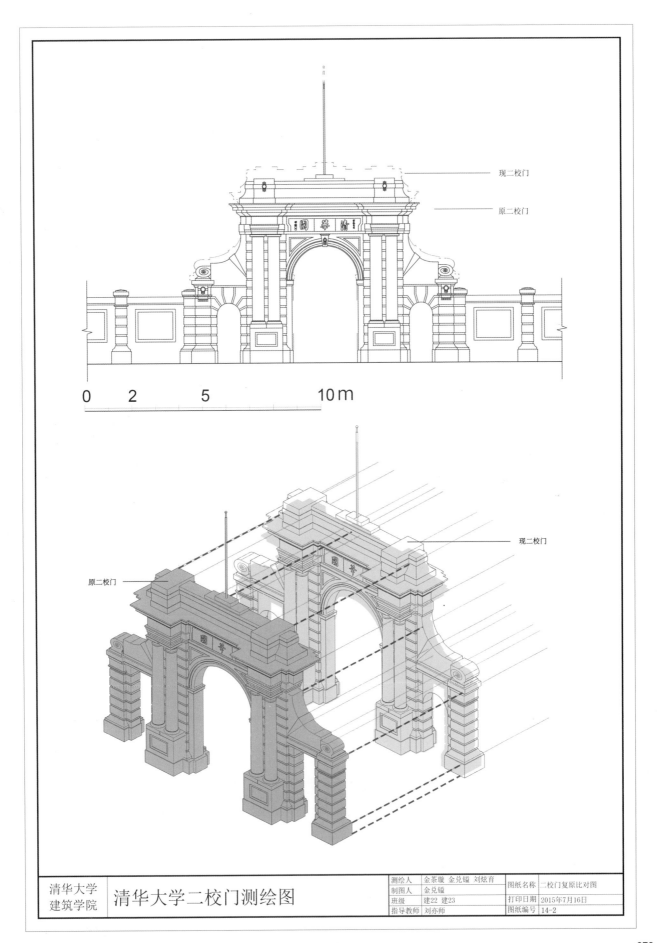

现二校门

原二校门

0　2　5　　　　　10 m

现二校门

原二校门

测绘人	金茶璇 金兑镒 刘炫育	图纸名称	二校门复原比对图
制图人	金兑镒		
班级	建22 建23	打印日期	2015年7月16日
指导教师	刘亦师	图纸编号	14-2

清华大学
建筑学院　　清华大学二校门测绘图

2. 同方部

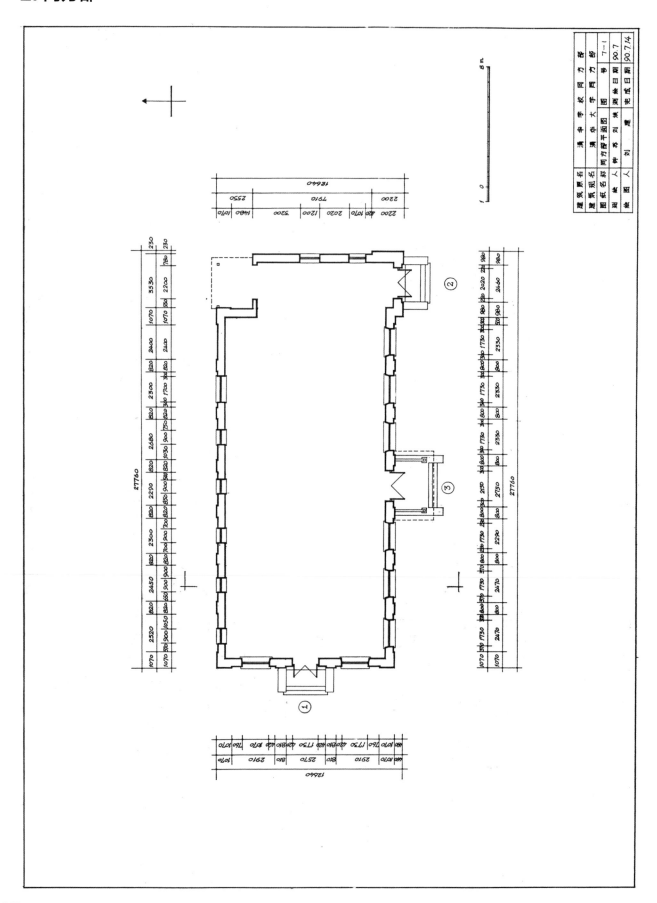

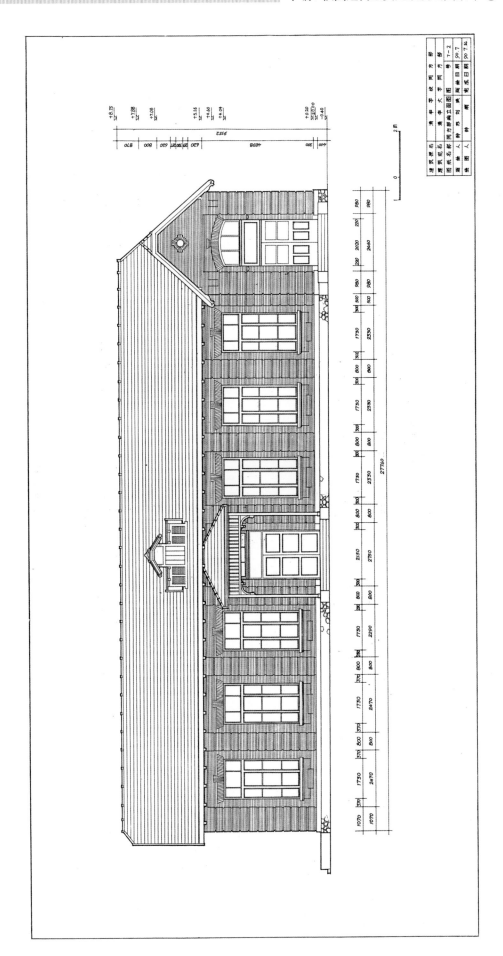

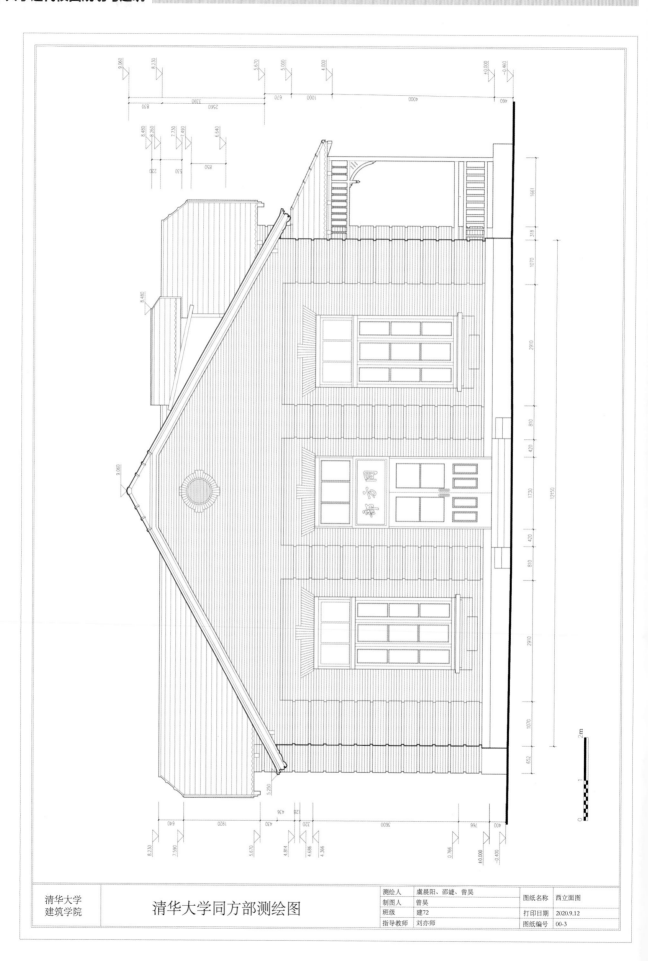

清华大学
建筑学院

清华大学同方部测绘图

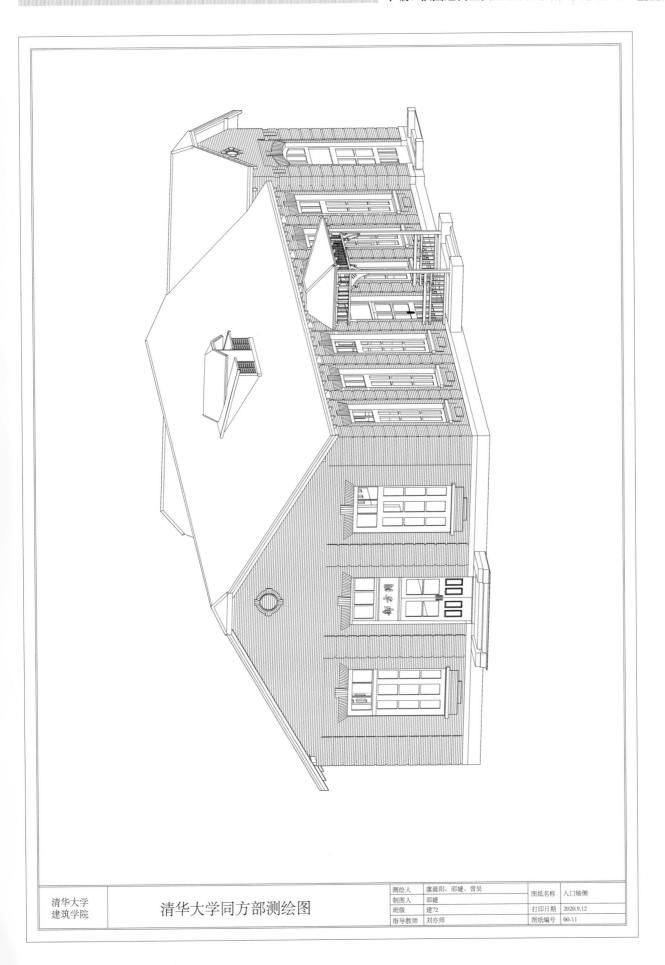

测绘人	虞晨阳、邵婕、曾昊		图纸名称	入口轴侧
制图人	邵婕			
班级	建72		打印日期	2020.9.12
指导教师	刘亦师		图纸编号	00-11

清华大学
建筑学院

清华大学同方部测绘图

3. 清华学堂

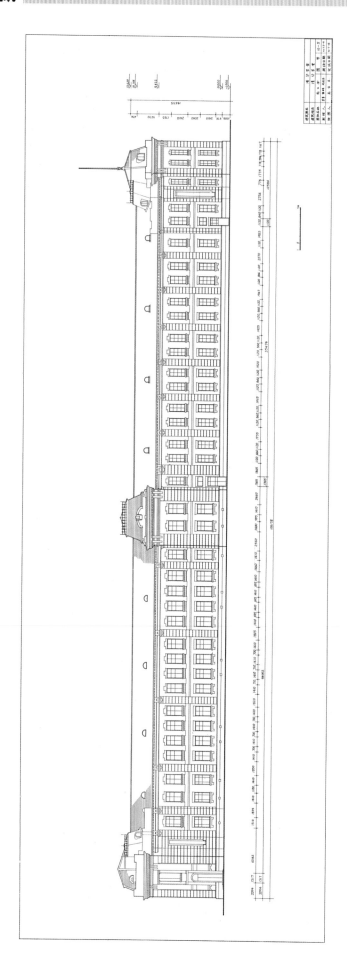

清华学堂

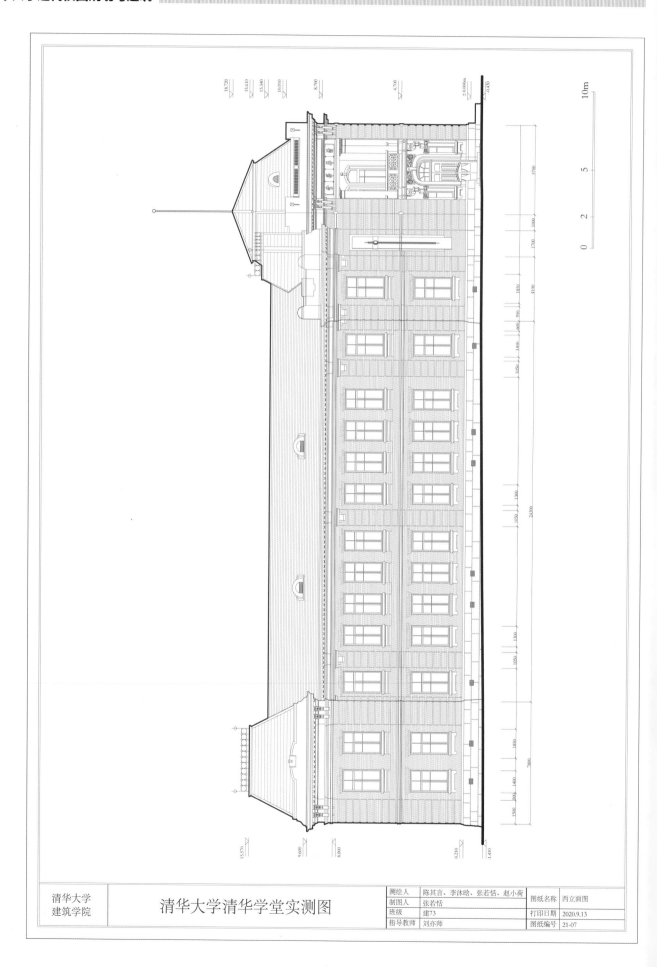

清华大学清华学堂实测图

测绘人	陈其言、李沐晗、张若恬、赵小荷	图纸名称	西立面图
制图人	张若恬		
班级	建73	打印日期	2020.9.13
指导教师	刘亦师	图纸编号	21-07

清华大学
建筑学院

清华大学
建筑学院

清华大学清华学堂测绘图

测绘人	陈其言、李沐晗、张若恬、赵小荷	图纸名称	西立面转角大样图
制图人	张若恬		
班级	建73	打印日期	2020.9.13
指导教师	刘亦师	图纸编号	21-14

387

4. 北院住宅

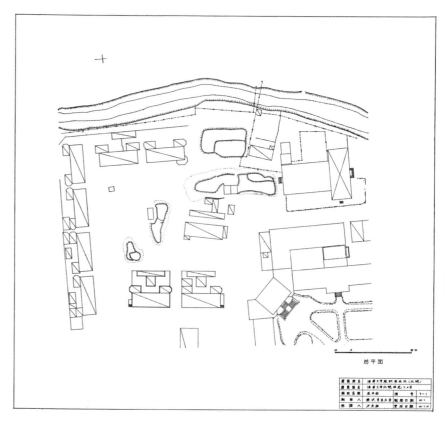

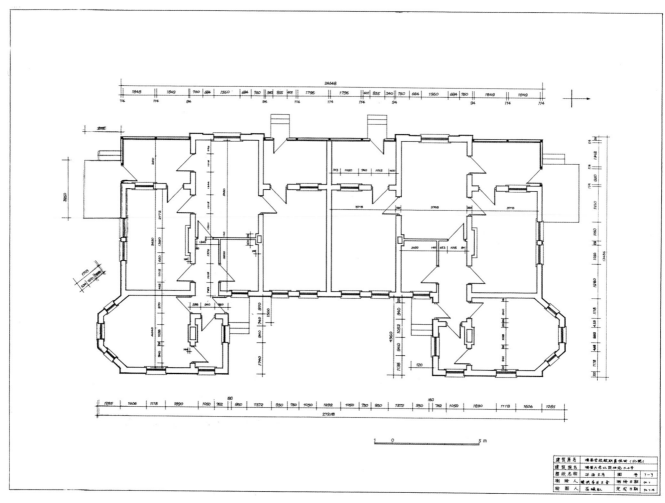

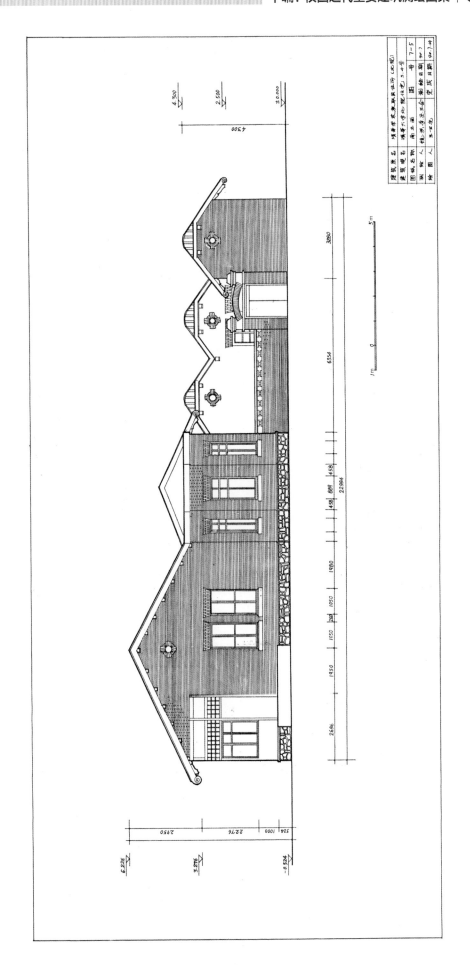

清华园清亭 98·10 冀生

第②组 "四大工程"

5. 大礼堂

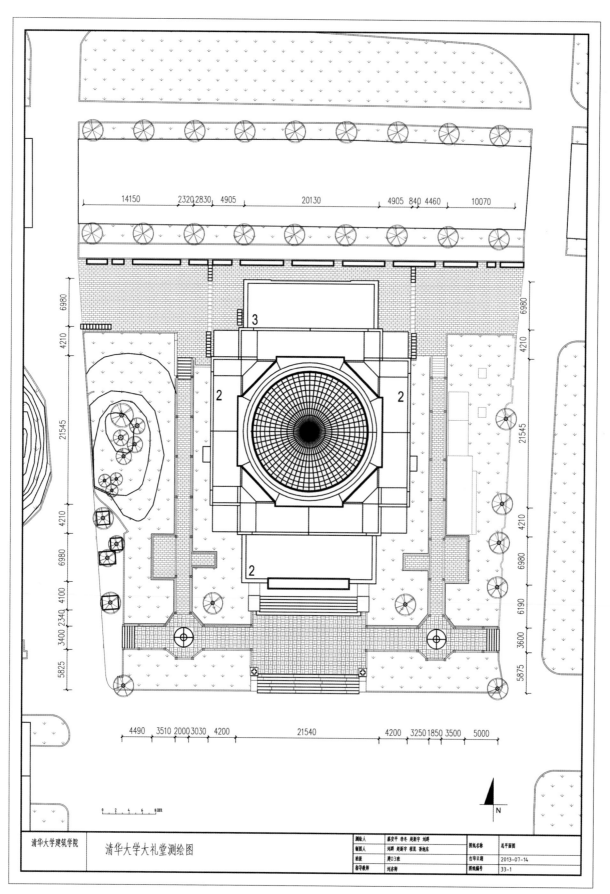

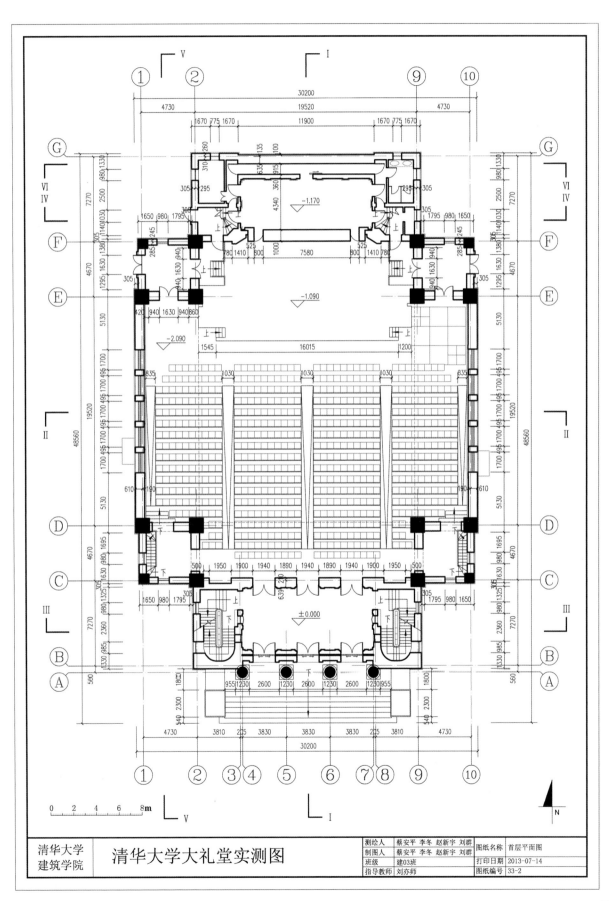

测绘人	蔡安平 李冬 赵新宇 刘群	图纸名称	首层平面图
制图人	蔡安平 李冬 赵新宇 刘群		
班级	建03班	打印日期	2013-07-14
指导教师	刘亦师	图纸编号	33-2

清华大学
建筑学院

清华大学大礼堂实测图

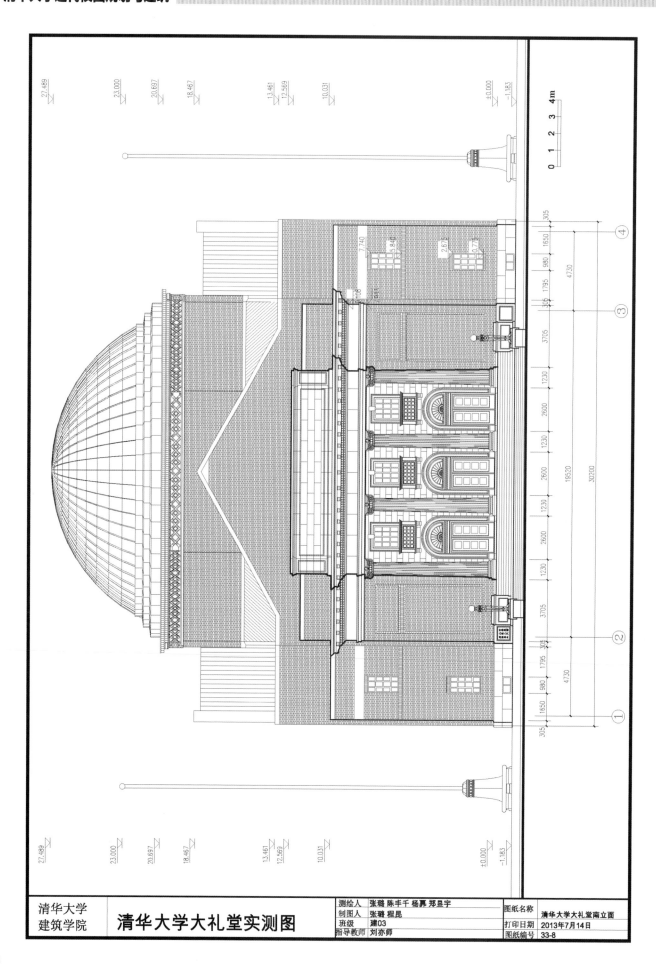

清华大学
建筑学院

清华大学大礼堂实测图

测绘人	张璐 陈丰千 杨焘 郑昱宇	图纸名称	清华大学大礼堂南立面
制图人	张璐 程昆		
班级	建03	打印日期	2013年7月14日
指导教师	刘亦师	图纸编号	33-8

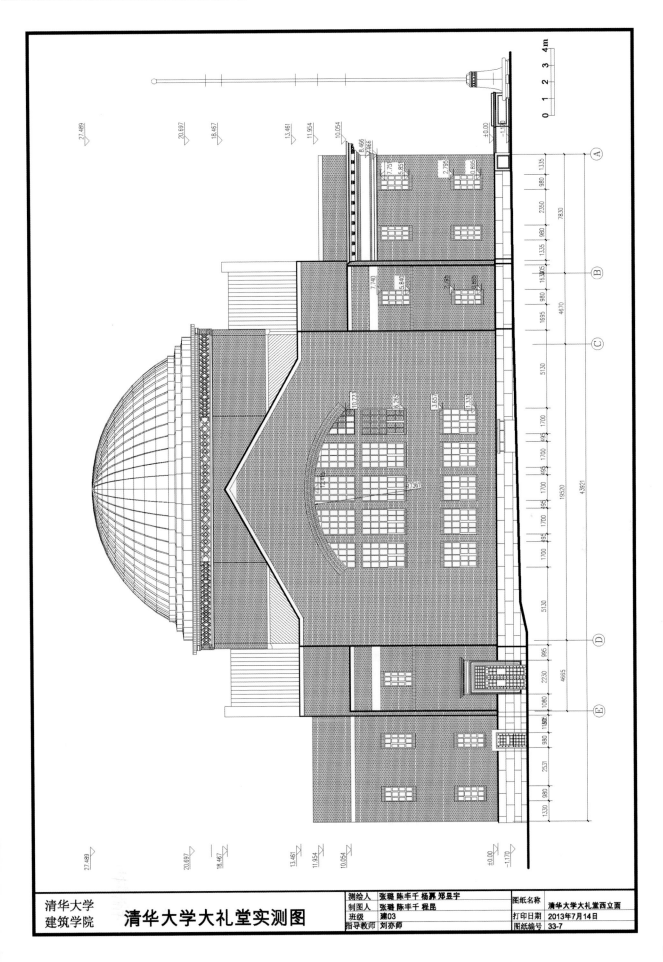

清华大学
建筑学院

清华大学大礼堂实测图

测绘人	张璐 陈丰千 杨昇 郑显宇	图纸名称	清华大学大礼堂西立面
制图人	张璐 陈丰千 程昆		
班级	建03	打印日期	2013年7月14日
指导教师	刘亦师	图纸编号	33-7

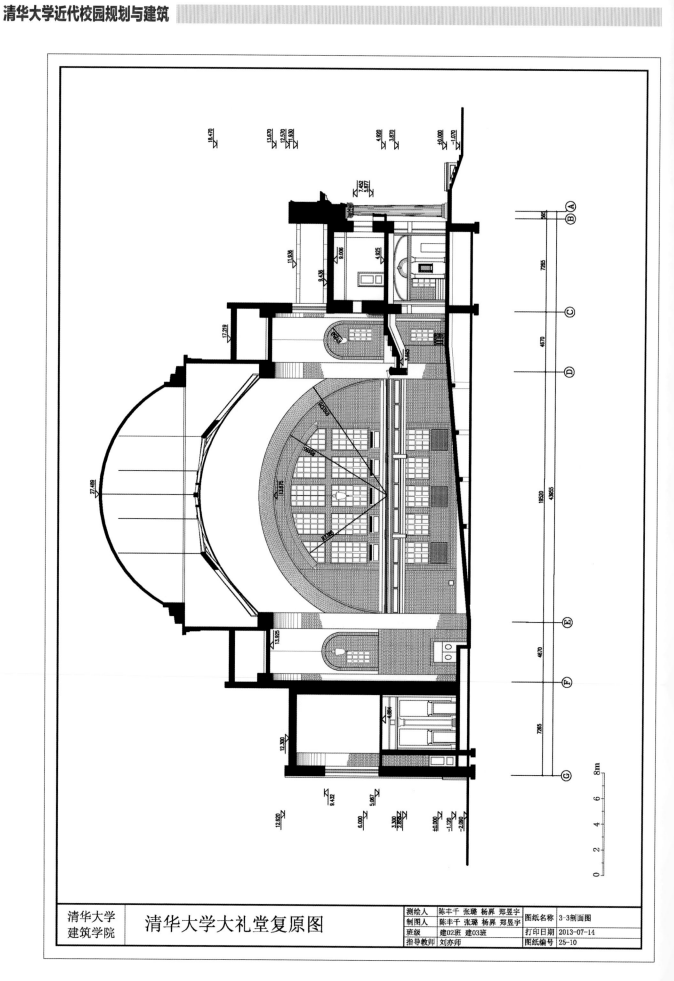

测绘人	陈丰千 张璐 杨昇 郑昱宇	图纸名称	3-3剖面图
制图人	陈丰千 张璐 杨昇 郑昱宇		
班级	建02班 建03班	打印日期	2013-07-14
指导教师	刘亦师	图纸编号	25-10

清华大学
建筑学院　清华大学大礼堂复原图

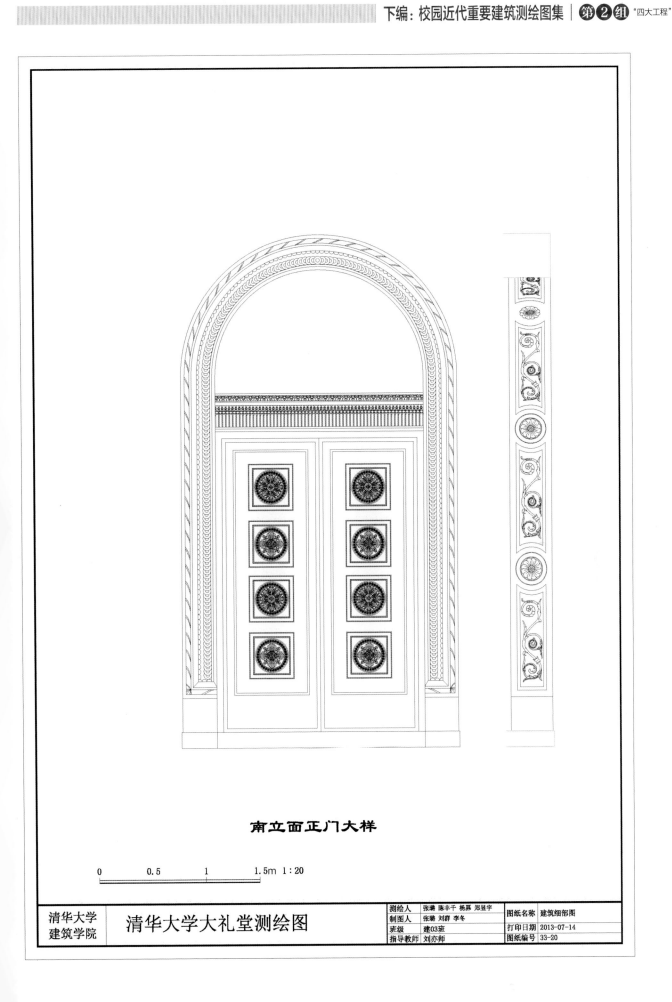

南立面正门大样

0　　　0.5　　　1　　　1.5m 1：20

测绘人	张璐 陈丰千 杨昇 郑昱宇	图纸名称	建筑细部图
制图人	张璐 刘群 李冬		
班级	建03班	打印日期	2013-07-14
指导教师	刘亦师	图纸编号	33-20

清华大学
建筑学院　　清华大学大礼堂测绘图

397

南立面阳台大样

路灯大样

旗杆大样

测绘人	张璐 陈丰千 杨昇 郑昱宇	图纸名称	建筑细部图
制图人	张璐 赵新宇		
班级	建03班	打印日期	2013-07-14
指导教师	刘亦师	图纸编号	34-17

清华大学大礼堂测绘图

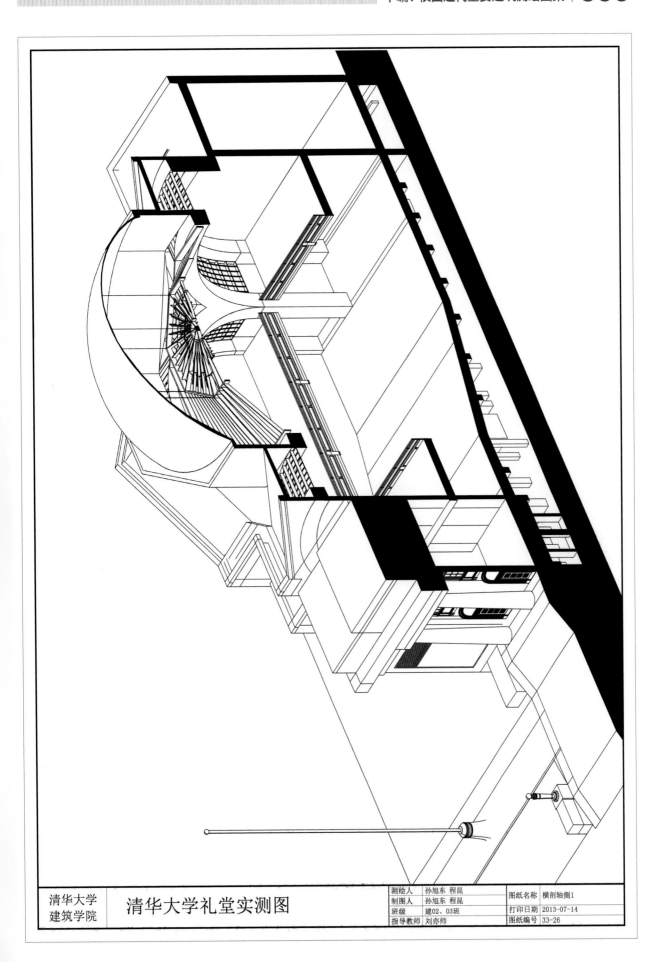

测绘人	孙旭东 程昆	图纸名称	横剖轴侧1
制图人	孙旭东 程昆		
班级	建02、03班	打印日期	2013-07-14
指导教师	刘亦师	图纸编号	33-26

清华大学
建筑学院

清华大学礼堂实测图

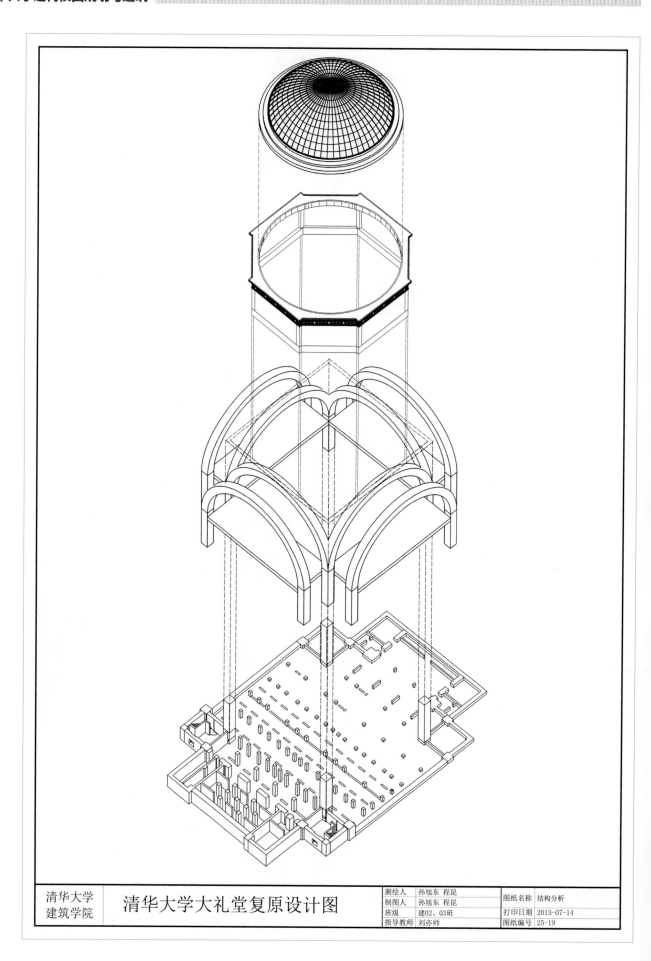

清华大学 建筑学院	清华大学大礼堂复原设计图	测绘人	孙旭东 程昆	图纸名称	结构分析
		制图人	孙旭东 程昆		
		班级	建02、03班	打印日期	2013-07-14
		指导教师	刘亦师	图纸编号	25-19

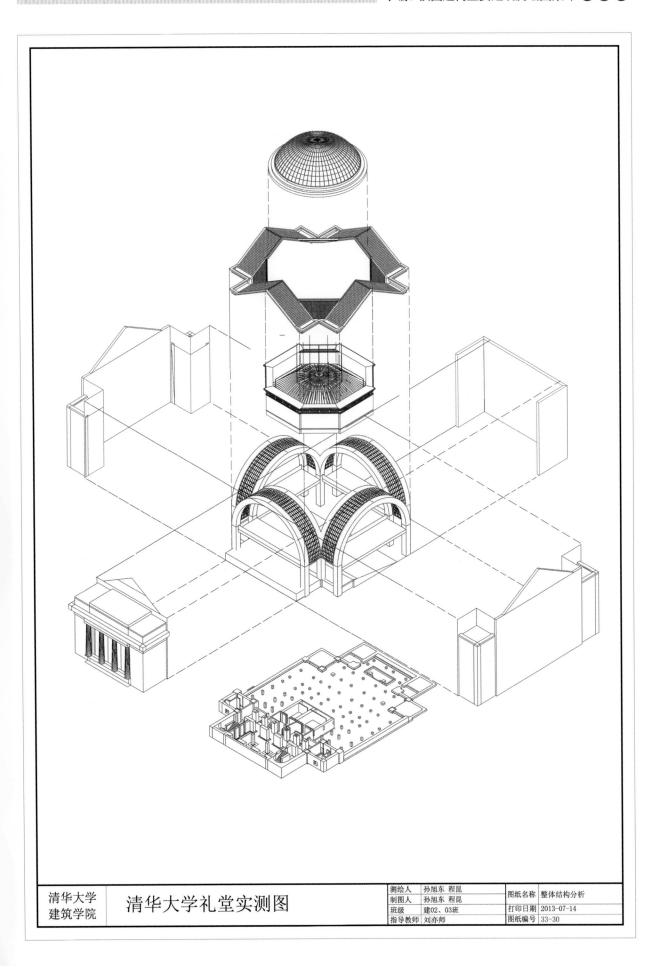

	测绘人	孙旭东 程昆	图纸名称	整体结构分析
清华大学 建筑学院	制图人	孙旭东 程昆		
	班级	建02、03班	打印日期	2013-07-14
	指导教师	刘亦师	图纸编号	33-30

清华大学礼堂实测图

6. 图书馆（一期）

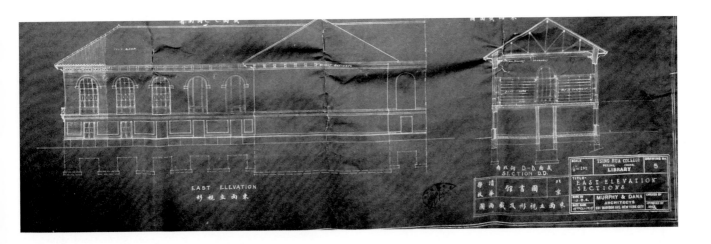

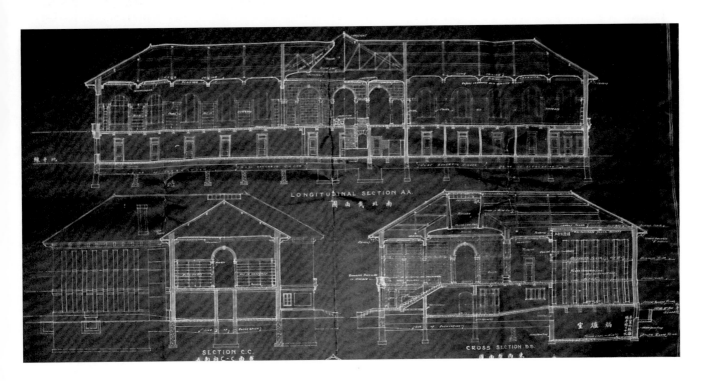

耶鲁大学墨菲档案之清华学校图书馆（一期）原始设计

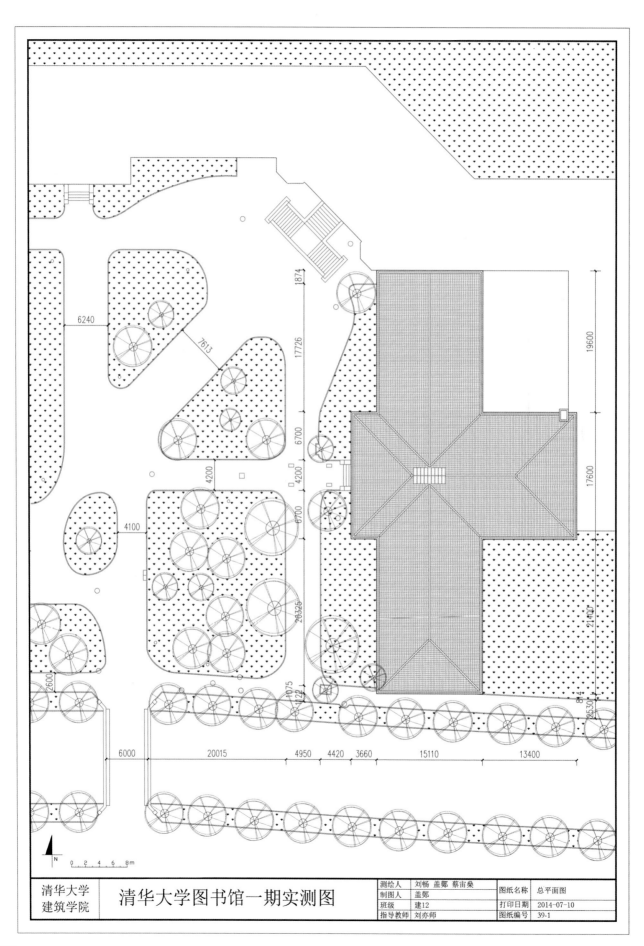

清华大学
建筑学院

清华大学图书馆一期实测图

测绘人	刘畅 盖郑 蔡宙燊	图纸名称	总平面图
制图人	盖郑		
班级	建12	打印日期	2014-07-10
指导教师	刘亦师	图纸编号	39-1

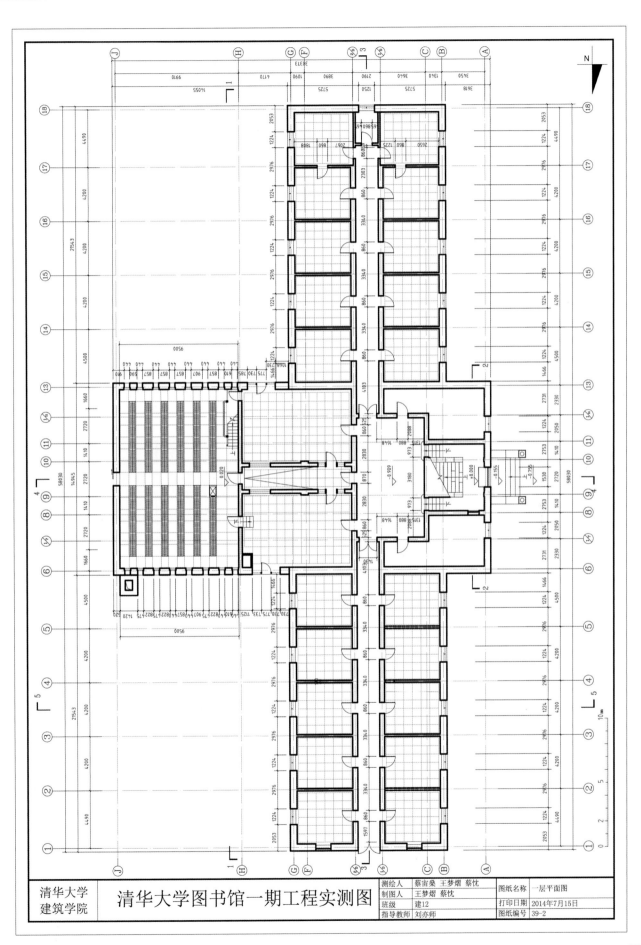

清华大学
建筑学院

清华大学图书馆一期工程实测图

测绘人	蔡宙桑 王梦熠 蔡忱	图纸名称	一层平面图
制图人	王梦熠 蔡忱		
班级	建12	打印日期	2014年7月15日
指导教师	刘亦师	图纸编号	39-2

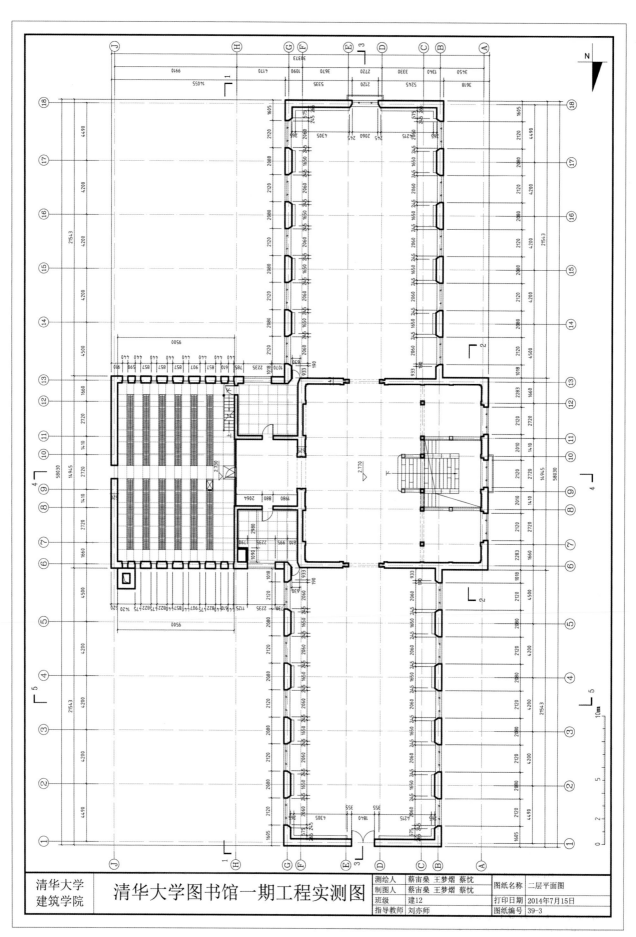

清华大学
建筑学院

清华大学图书馆一期工程实测图

测绘人	蔡宙桑 王梦熠 蔡忱	图纸名称	二层平面图
制图人	蔡宙桑 王梦熠 蔡忱	打印日期	2014年7月15日
班级	建12		
指导教师	刘亦师	图纸编号	39-3

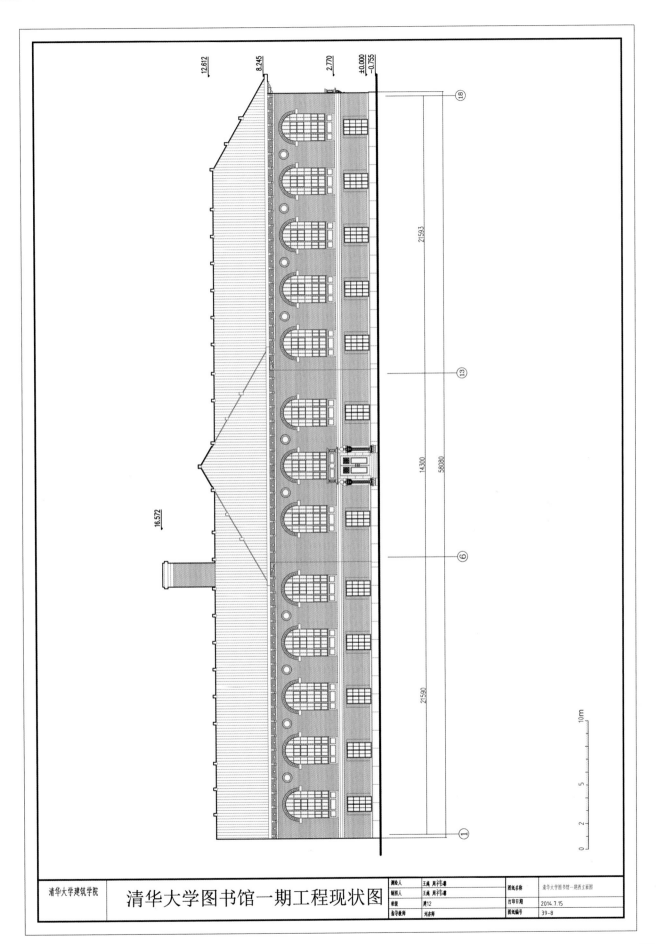

		图纸名称	清华大学图书馆一期西立面图
清华大学建筑学院	清华大学图书馆一期工程现状图	测绘人	王越 周祎馨		
		制图人	王越 周祎馨	打印日期	2014.7.15
		班级	建12	图纸编号	39-8
		指导教师	刘亦师		

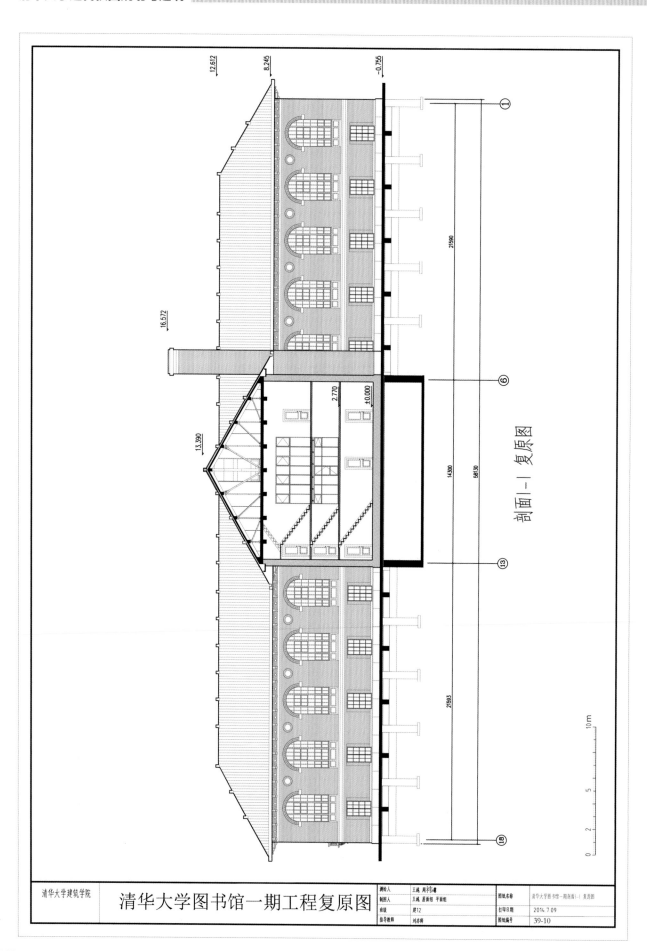

剖面1-1复原图

清华大学建筑学院

清华大学图书馆一期工程复原图

测绘人	王越 周祎曦		图纸名称	清华大学图书馆一期剖面1-1 复原图
制图人	王越 屋面组 平面组			
班级	建12		打印日期	2014.7.09
指导教师	刘志辉		图纸编号	39-10

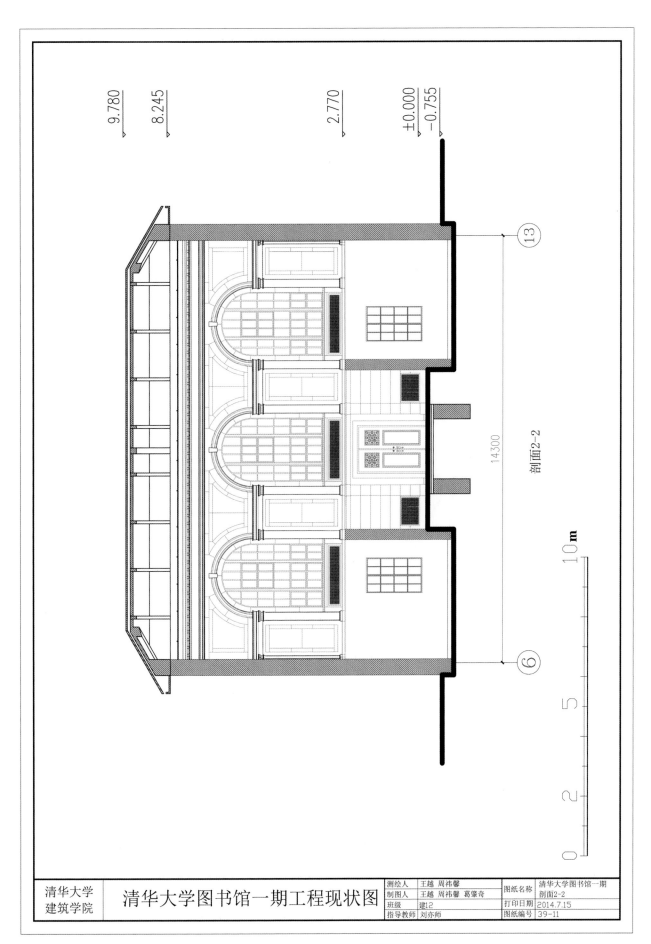

9.780
8.245
2.770
±0.000
−0.755

⑬

14300

剖面2-2

⑥

10 m

5

2

0

清华大学
建筑学院

清华大学图书馆一期工程现状图

测绘人	王越 周祎馨
制图人	王越 周祎馨 葛肇奇
班级	建12
指导教师	刘亦师

图纸名称	清华大学图书馆一期 剖面2-2
打印日期	2014.7.15
图纸编号	39-11

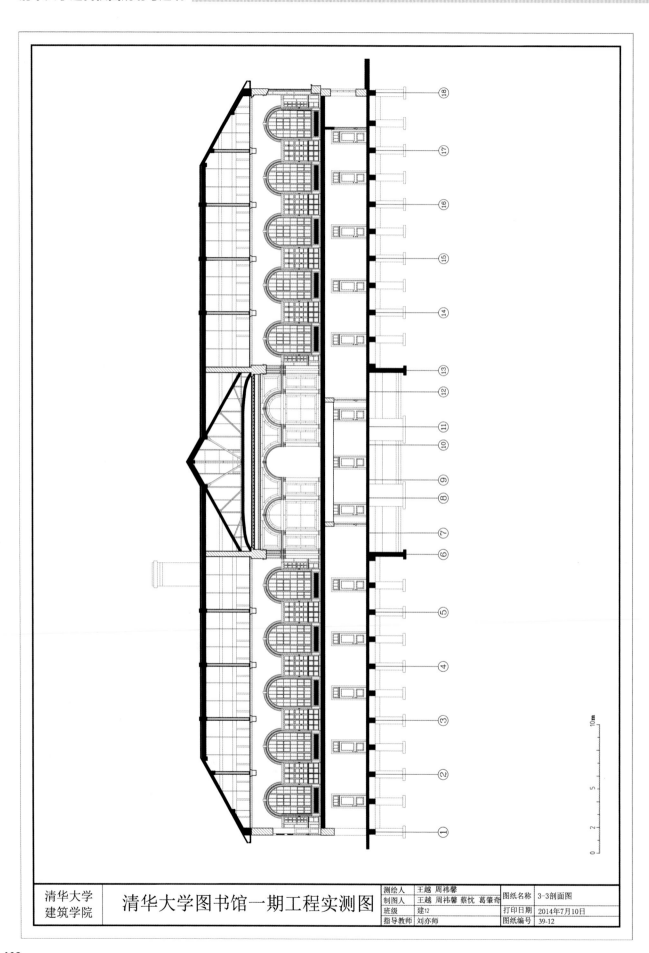

测绘人	王越 周祎馨		图纸名称	3-3剖面图
制图人	王越 周祎馨 蔡忱 葛肇奇			
班级	建12		打印日期	2014年7月10日
指导教师	刘亦师		图纸编号	39-12

清华大学
建筑学院

清华大学图书馆一期工程实测图

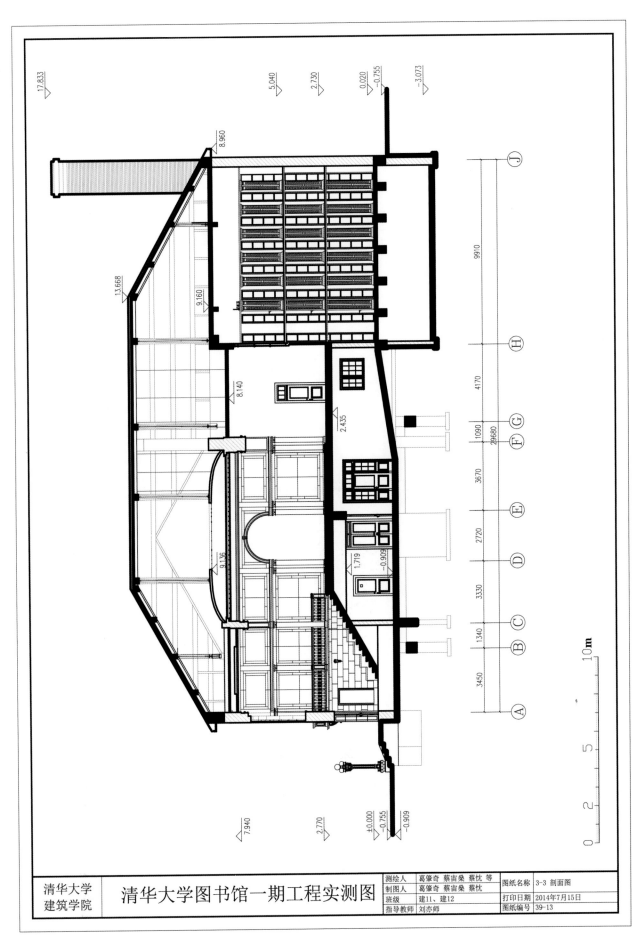

清华大学 建筑学院	清华大学图书馆一期工程实测图	测绘人	葛肇奇 蔡宙燊 蔡忱 等	图纸名称	3-3 剖面图
		制图人	葛肇奇 蔡宙燊 蔡忱		
		班级	建11、建12	打印日期	2014年7月15日
		指导教师	刘亦师	图纸编号	39-13

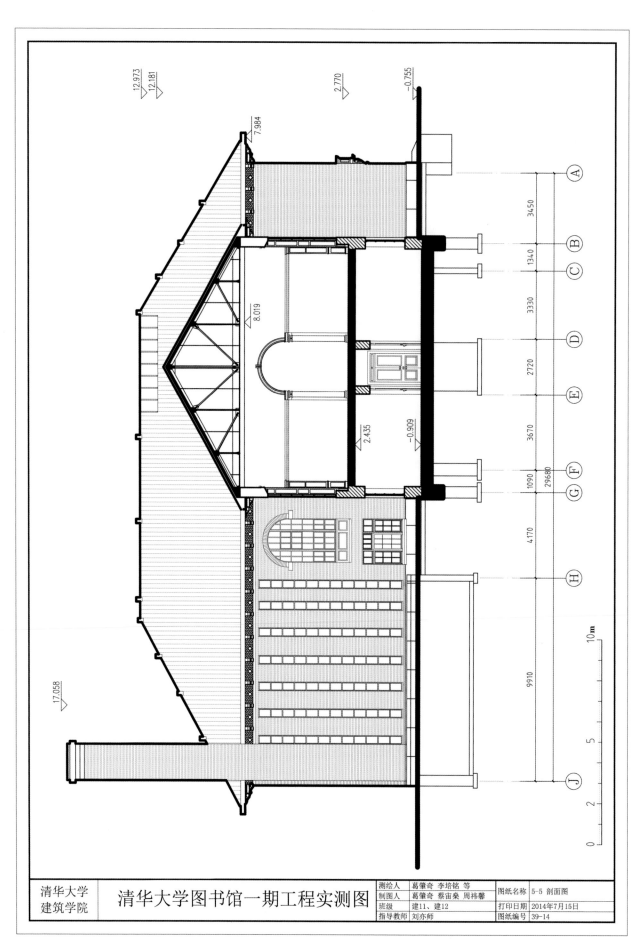

清华大学
建筑学院

清华大学图书馆一期工程实测图

测绘人	葛肇奇 李培铭 等	图纸名称	5-5 剖面图
制图人	葛肇奇 蔡宙桑 周沛馨		
班级	建11、建12	打印日期	2014年7月15日
指导教师	刘亦师	图纸编号	39-14

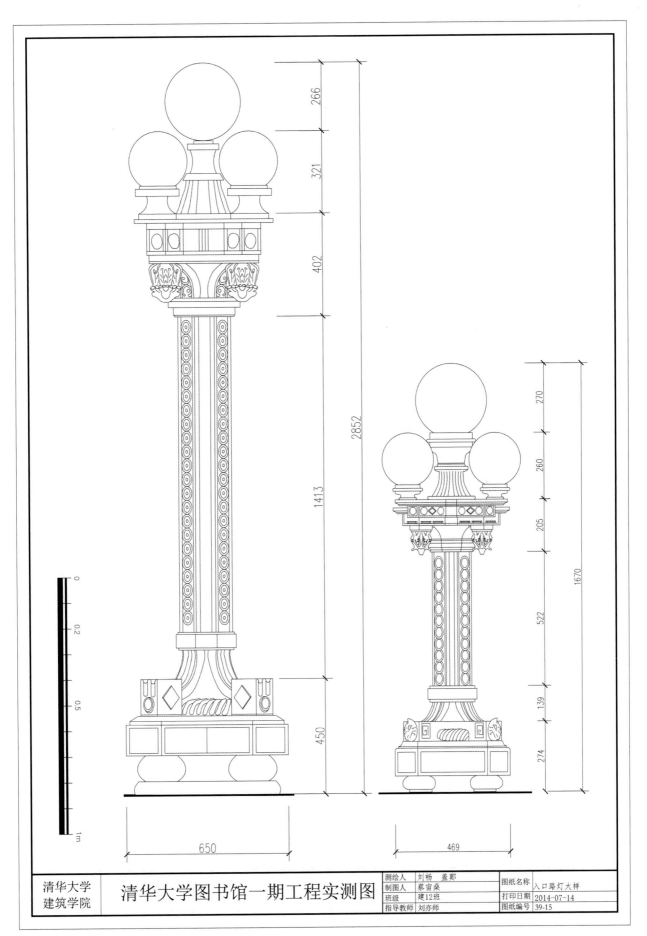

清华大学图书馆一期工程实测图

测绘人	刘畅 盖郑	图纸名称	入口路灯大样
制图人	蔡宙桑		
班级	建12班	打印日期	2014-07-14
指导教师	刘亦师	图纸编号	39-15

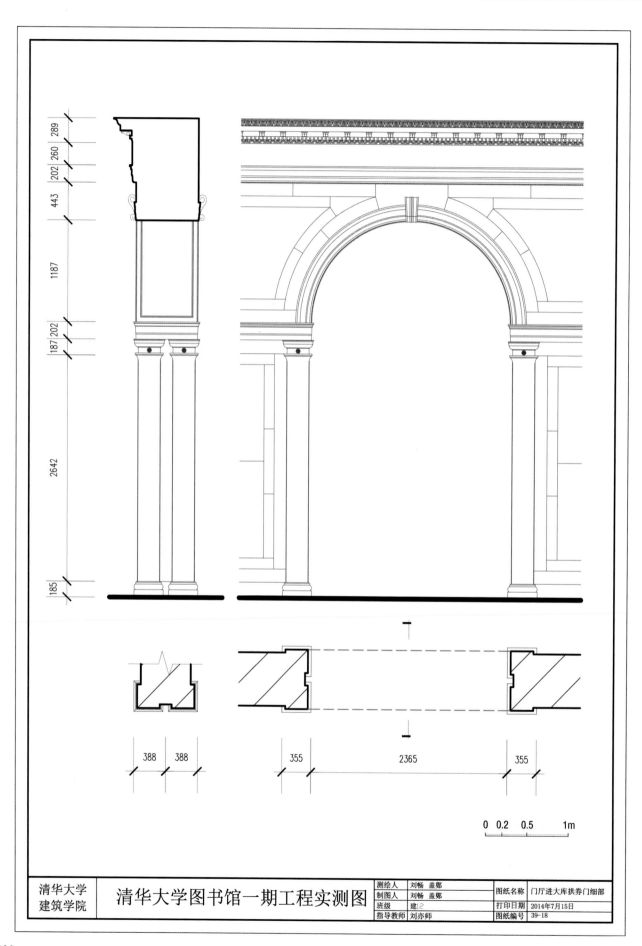

289
260
202
443
1187
187 | 202
2642
185

388 | 388

355 | 2365 | 355

0 0.2 0.5 1m

清华大学 建筑学院	清华大学图书馆一期工程实测图	测绘人	刘畅 盖颎	图纸名称	门厅进大库拱券门细部
		制图人	刘畅 盖颎		
		班级	建12	打印日期	2014年7月15日
		指导教师	刘亦师	图纸编号	39-18

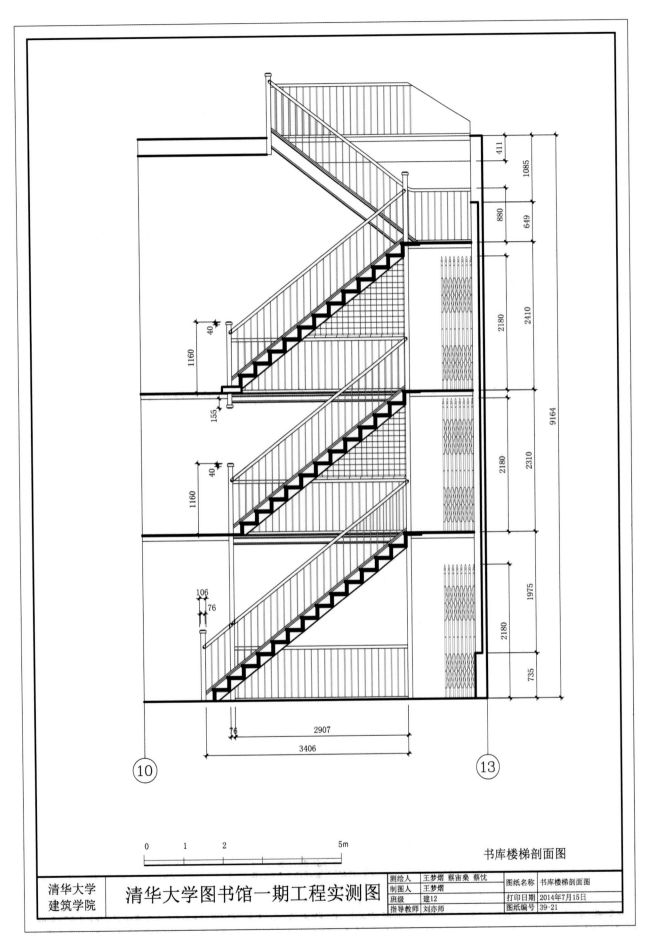

书库楼梯剖面图

0　　1　　2　　　　　　5m

清华大学 建筑学院	清华大学图书馆一期工程实测图	测绘人	王梦熠 蔡宙燊 蔡忱	图纸名称	书库楼梯剖面图
		制图人	王梦熠	打印日期	2014年7月15日
		班级	建12	图纸编号	39-21
		指导教师	刘亦师		

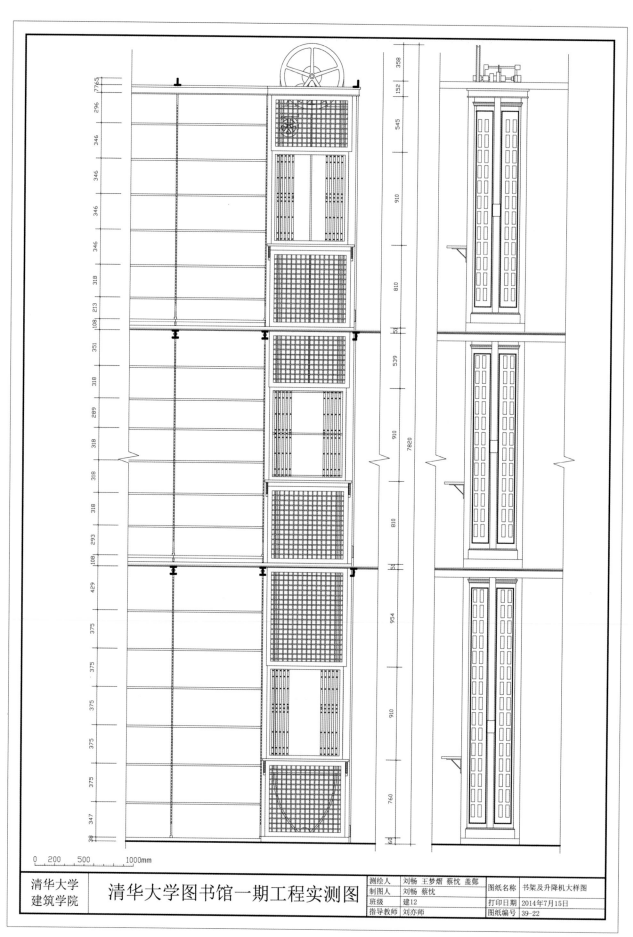

测绘人 刘畅 王梦熠 蔡忱 盖郧
制图人 刘畅 蔡忱
班级 建12
指导教师 刘亦师

图纸名称 书架及升降机大样图
打印日期 2014年7月15日
图纸编号 39-22

清华大学
建筑学院

清华大学图书馆一期工程实测图

0 200 500 1000mm

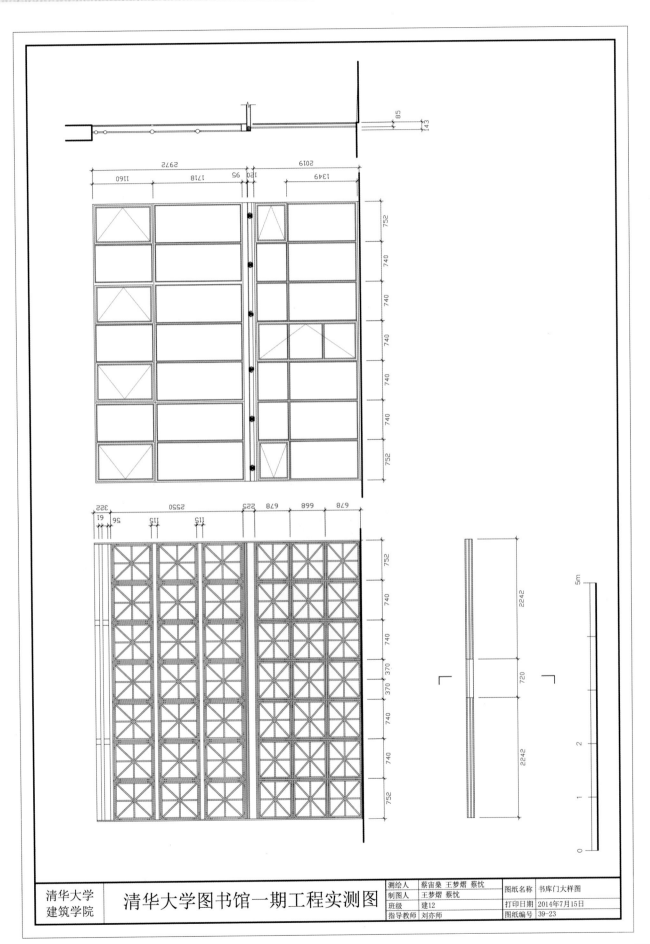

	图纸名称	书库门大样图
测绘人	蔡宙燊 王梦熠 蔡忱	
制图人	王梦熠 蔡忱	
班级 建12	打印日期	2014年7月15日
指导教师 刘亦师	图纸编号	39-23

清华大学图书馆一期工程实测图

清华大学
建筑学院

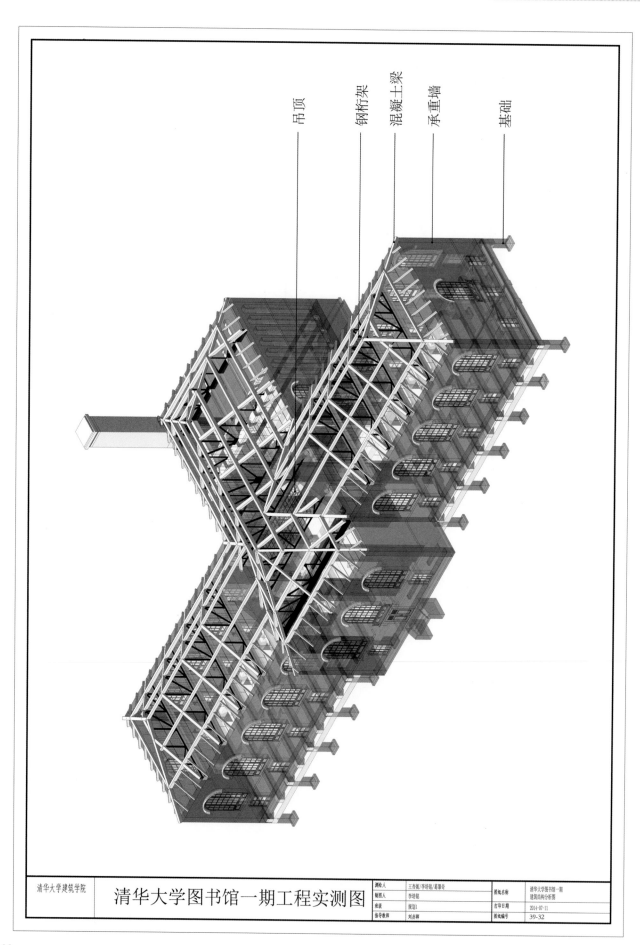

吊顶

钢桁架

混凝土梁

承重墙

基础

		测绘人	王杏妮/李培铭/葛肇奇	图纸名称	清华大学图书馆一期
清华大学建筑学院	清华大学图书馆一期工程实测图	制图人	李培铭		建筑结构分析图
		班级	建筑1	打印日期	2014-07-11
		指导教师	刘亦师	图纸编号	39-32

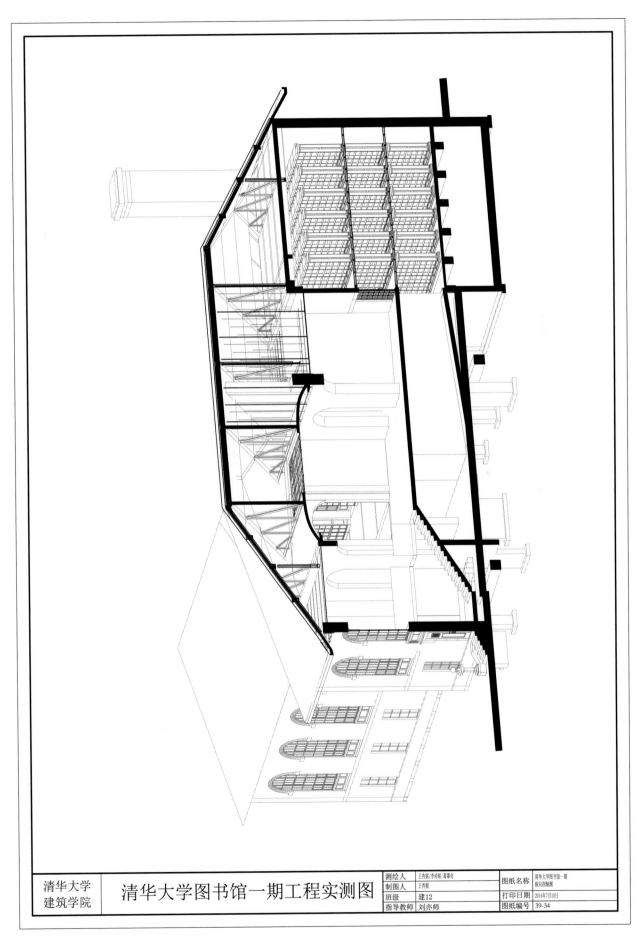

清华大学 建筑学院	清华大学图书馆一期工程实测图	测绘人	王杏妮/李培铭/葛肇奇	图纸名称	清华大学图书馆一期 纵向剖轴测
		制图人	王杏妮		
		班级	建12	打印日期	2014年7月10日
		指导教师	刘亦师	图纸编号	39-34

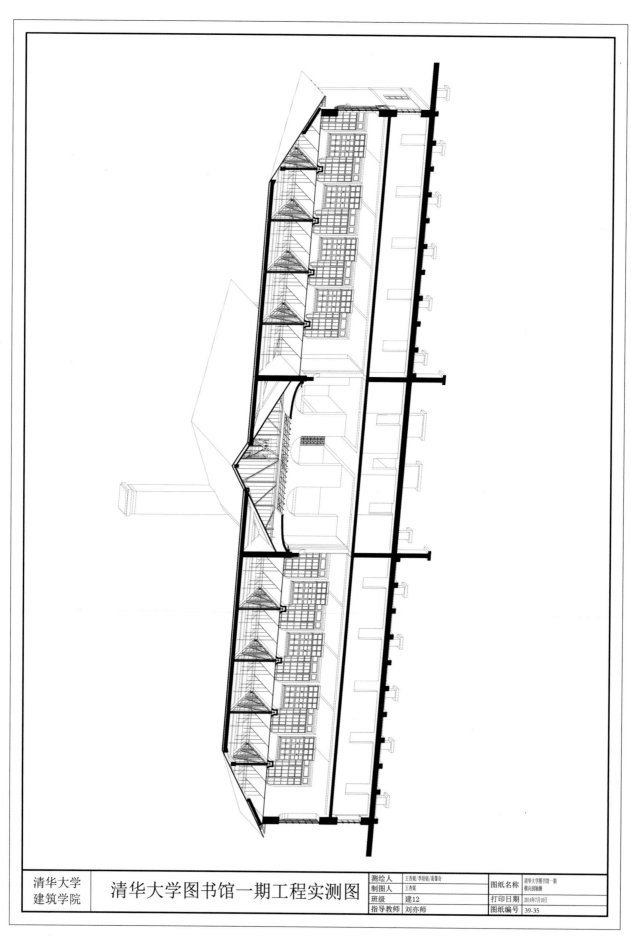

清华大学 建筑学院	清华大学图书馆一期工程实测图	测绘人	王杏妮/李培铭/葛馨奇	图纸名称	清华大学图书馆一期 横向剖轴测
		制图人	王杏妮		
		班级	建12	打印日期	2014年7月10日
		指导教师	刘亦师	图纸编号	39-35

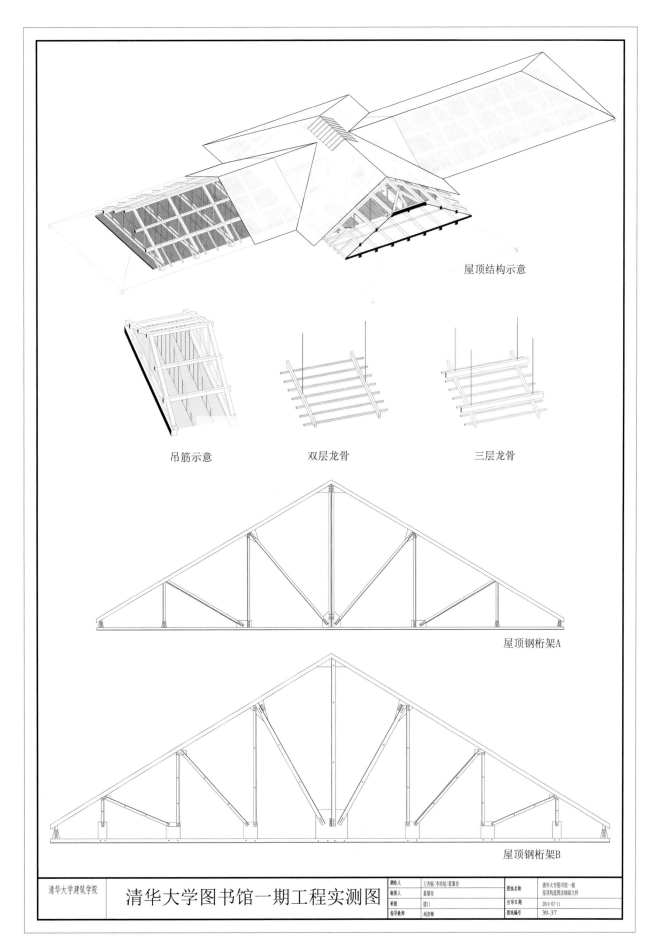

屋顶结构示意

吊筋示意　　　　双层龙骨　　　　三层龙骨

屋顶钢桁架A

屋顶钢桁架B

清华大学建筑学院	清华大学图书馆一期工程实测图	测绘人	王杏航/李培铭/葛肇奇	图纸名称	清华大学图书馆一期 屋顶构造图及细部大样
		制图人	葛肇奇		
		班级	建11	打印日期	2014-07-11
		指导教师	刘亦师	图纸编号	39-37

7. 科学馆

SCIENCE AND RECITATION BUILDING

PERSPECTIVES OF THE NEW BUILDINGS FOR TSING HUA UNIVERSITY, PEKING, CHINA

MURPHY & DANA, ARCHITECTS

耶鲁大学墨菲档案之清华学校科学馆效果图

科学馆建成的照片，1918 年
来源：墨菲档案

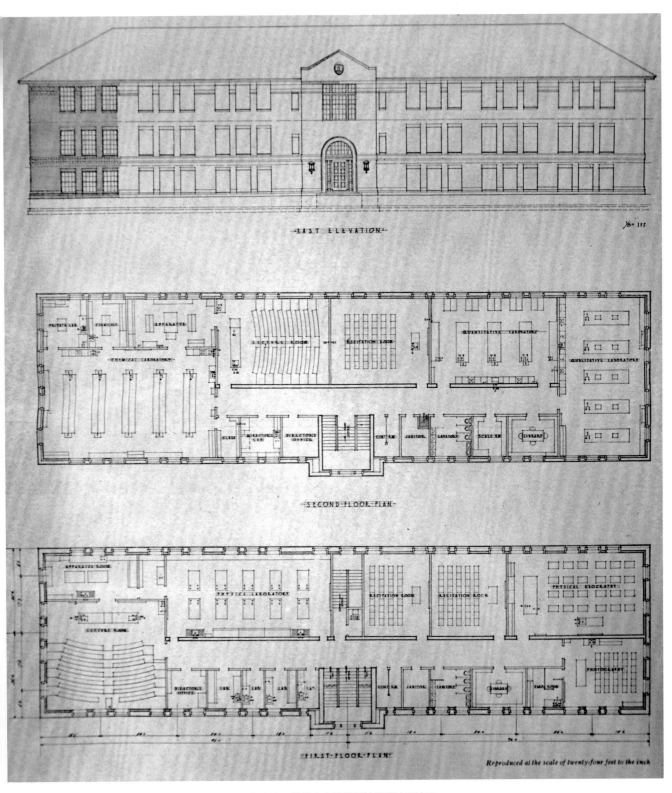

耶鲁大学墨菲档案之清华学校科学馆设计图

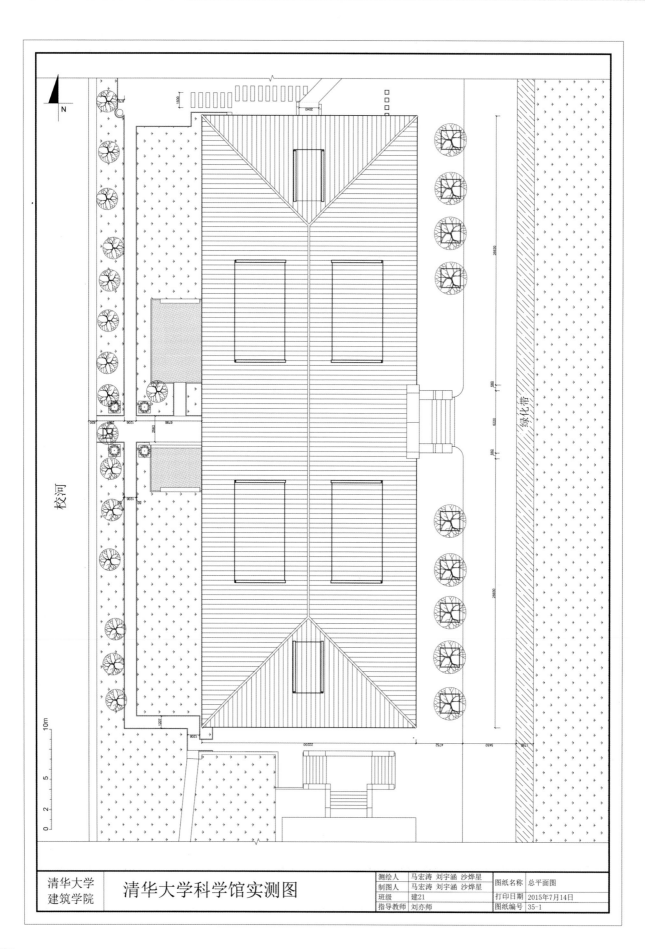

清华大学
建筑学院

清华大学科学馆实测图

测绘人	马宏涛 刘宇涵 沙烨星	图纸名称	总平面图
制图人	马宏涛 刘宇涵 沙烨星		
班级	建21	打印日期	2015年7月14日
指导教师	刘亦师	图纸编号	35-1

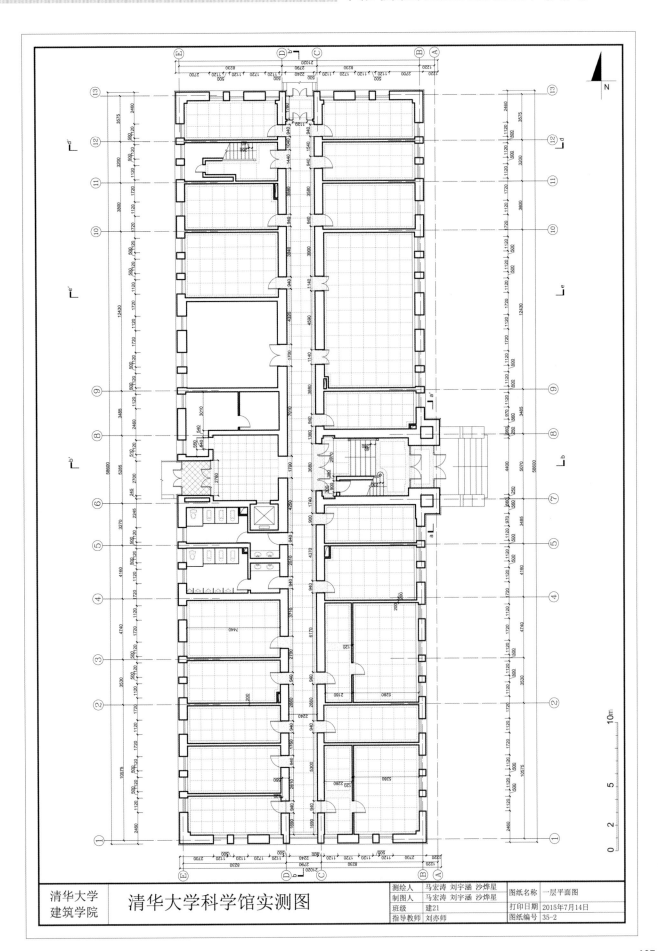

清华大学
建筑学院

清华大学科学馆实测图

测绘人	马宏涛 刘宇涵 沙烨星	图纸名称	一层平面图
制图人	马宏涛 刘宇涵 沙烨星		
班级	建21	打印日期	2015年7月14日
指导教师	刘亦师	图纸编号	35-2

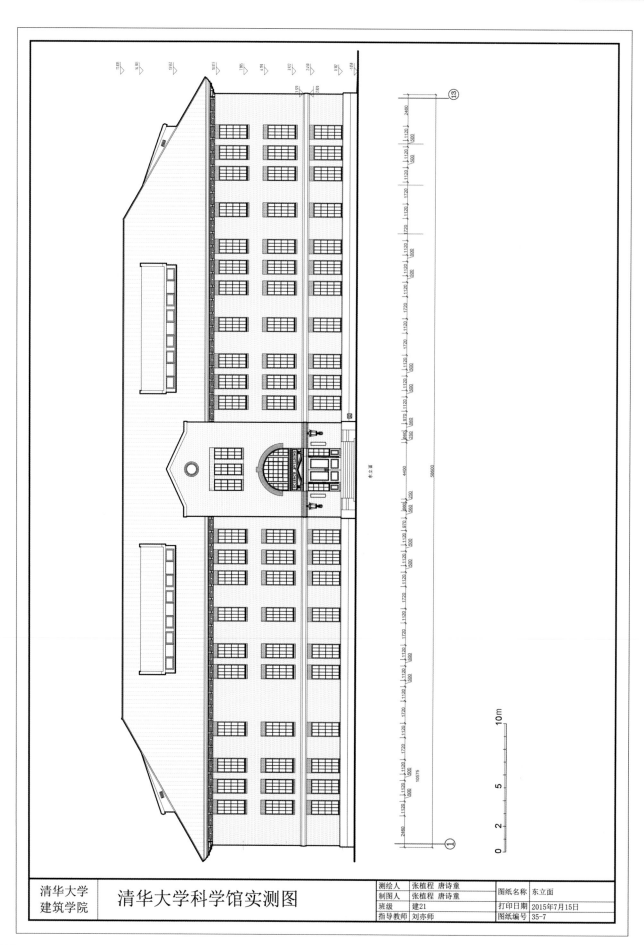

测绘人	张植程 唐诗童		图纸名称	东立面
制图人	张植程 唐诗童			
班级	建21		打印日期	2015年7月15日
指导教师	刘亦师		图纸编号	35-7

清华大学
建筑学院

清华大学科学馆实测图

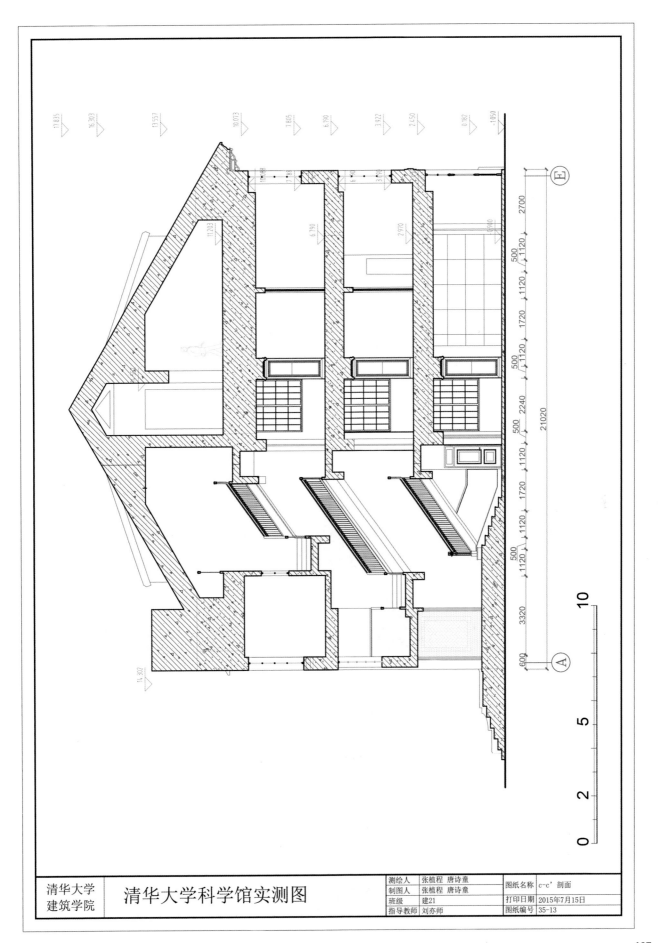

		图纸名称	c-c'剖面
测绘人	张植程 唐诗童		
制图人	张植程 唐诗童		
班级	建21	打印日期	2015年7月15日
指导教师	刘亦师	图纸编号	35-13

清华大学科学馆实测图

清华大学
建筑学院

测绘人	张植程 唐诗童	图纸名称	正立面山花大样
制图人	唐诗童	打印日期	2015年7月15日
班级	建21		
指导教师	刘亦师	图纸编号	35-16

SCIENCE BVILDING

0   0.4   1   2m

清华大学 建筑学院	清华大学科学馆实测图

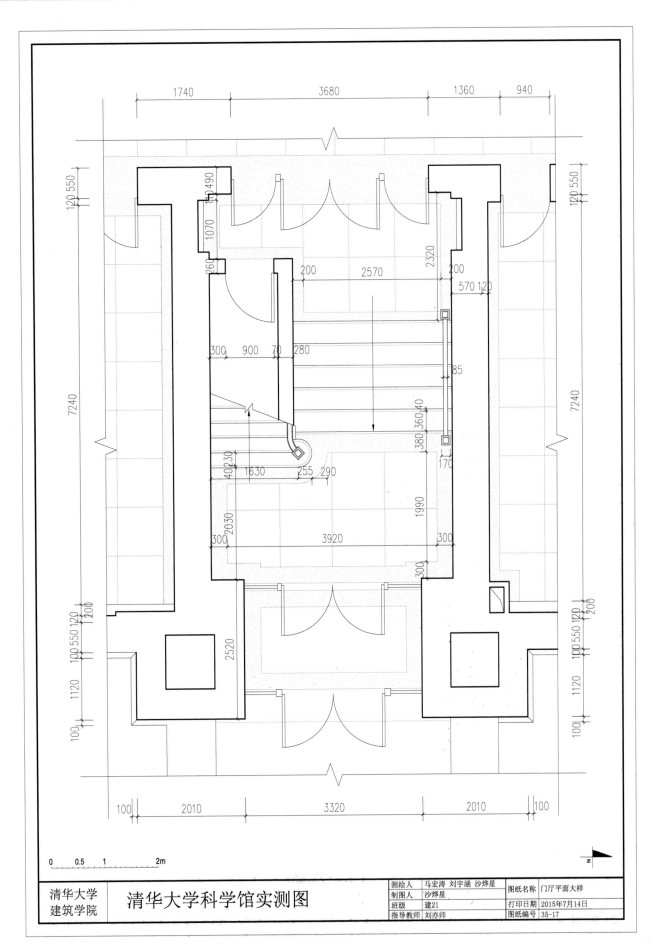

测绘人	马宏涛 刘宇涵 沙烨星	图纸名称	门厅平面大样
制图人	沙烨星		
班级	建21	打印日期	2015年7月14日
指导教师	刘亦师	图纸编号	35-17

清华大学
建筑学院　　清华大学科学馆实测图

测绘人	张植程 唐诗童	图纸名称	门厅轴测
制图人	张植程		
班级	建21	打印日期	2015年7月15日
指导教师	刘亦师	图纸编号	35-30

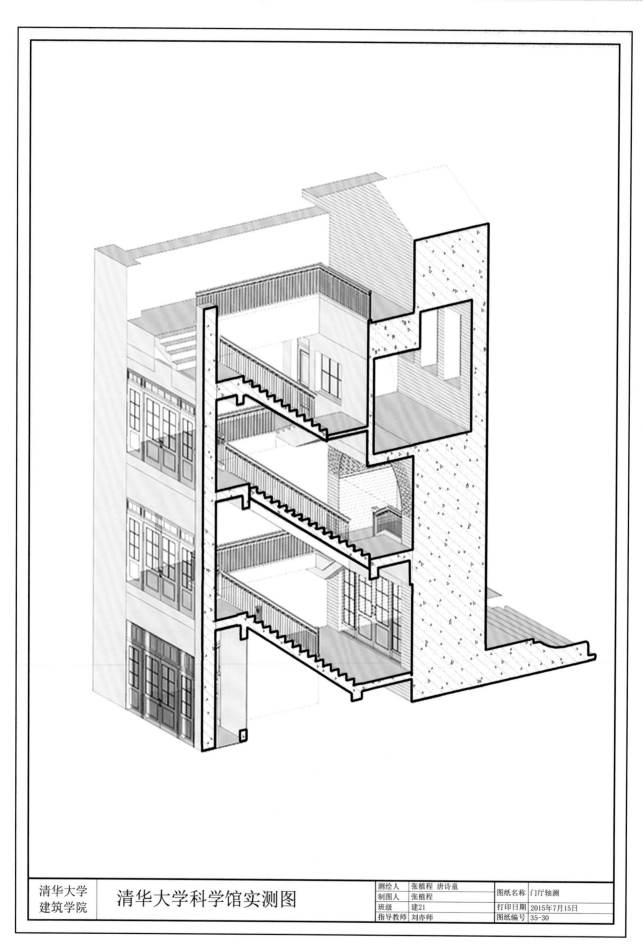

| 清华大学
建筑学院 | 清华大学科学馆实测图 |

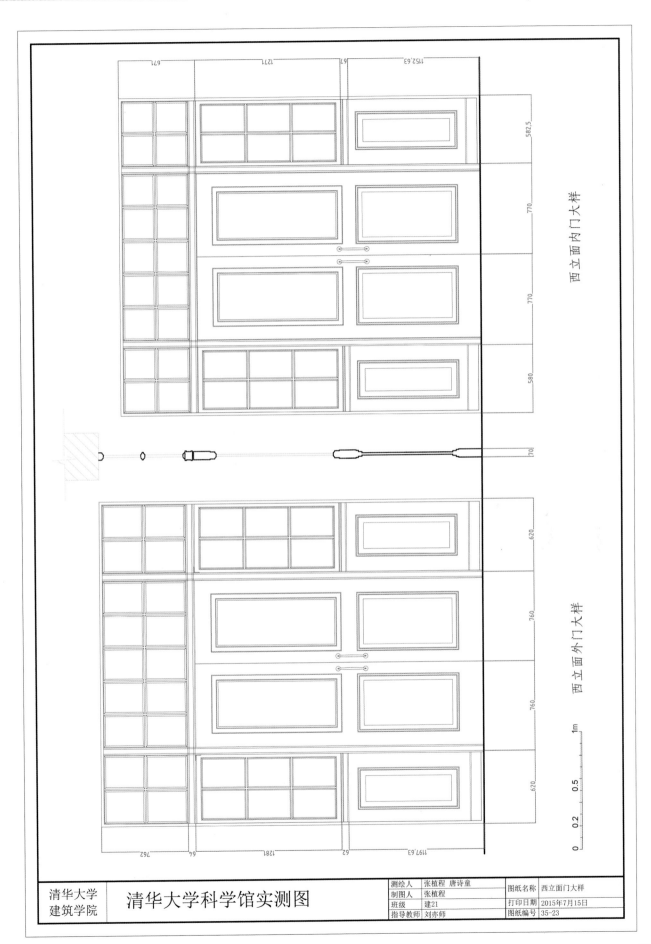

西立面内门大样

西立面外门大样

		图纸名称	西立面门大样
测绘人	张植程 唐诗童		
制图人	张植程		
班级	建21	打印日期	2015年7月15日
指导教师	刘亦师	图纸编号	35-23

清华大学
建筑学院

清华大学科学馆实测图

		测绘人	周皓 李智等	图纸名称	建筑轴测图

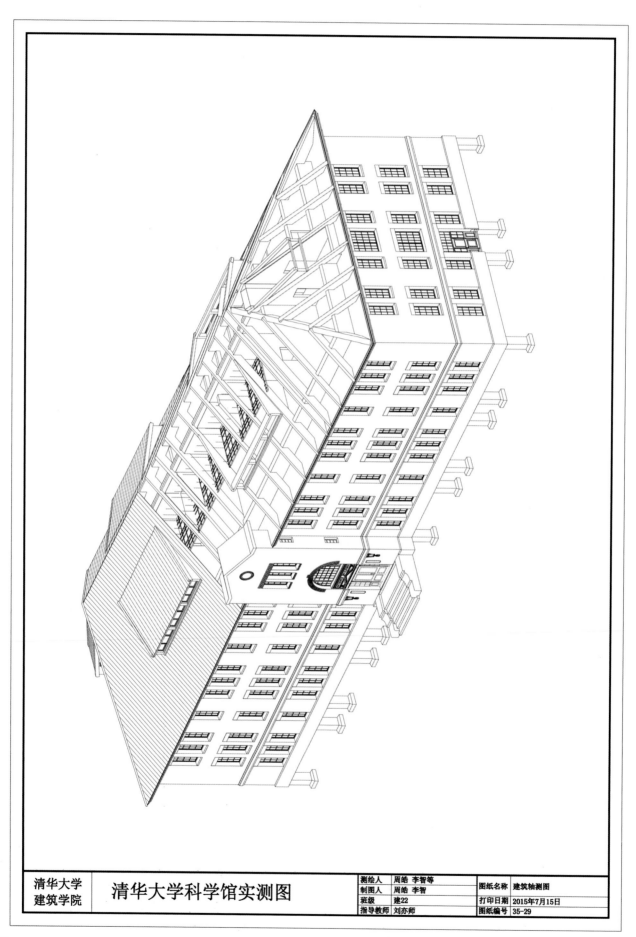

清华大学 建筑学院	清华大学科学馆实测图	测绘人	周皓 李智等	图纸名称	建筑轴测图
		制图人	周皓 李智		
		班级	建22	打印日期	2015年7月15日
		指导教师	刘亦师	图纸编号	35-29

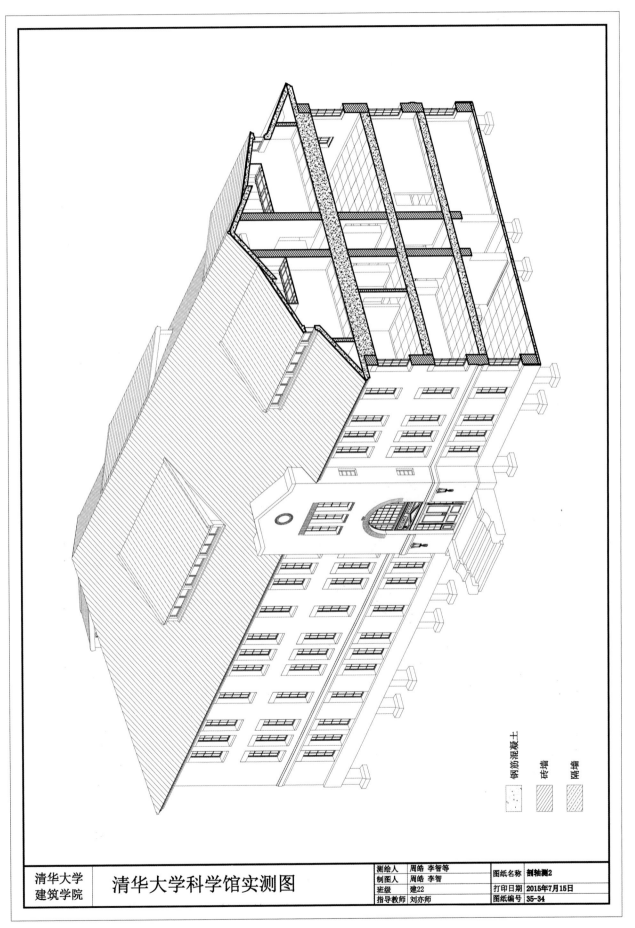

钢筋混凝土

砖墙

蜩墙

清华大学 建筑学院	清华大学科学馆实测图	测绘人	周皓 李智等	图纸名称	剖轴测2
		制图人	周皓 李智		
		班级	建22	打印日期	2015年7月15日
		指导教师	刘亦师	图纸编号	35-34

8. 西体育馆（含加建）

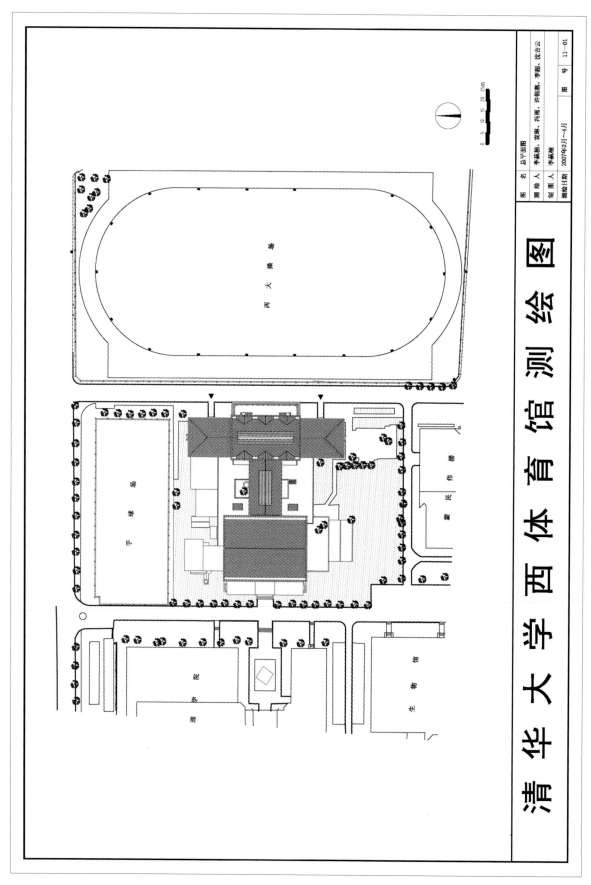

清华大学
建筑学院

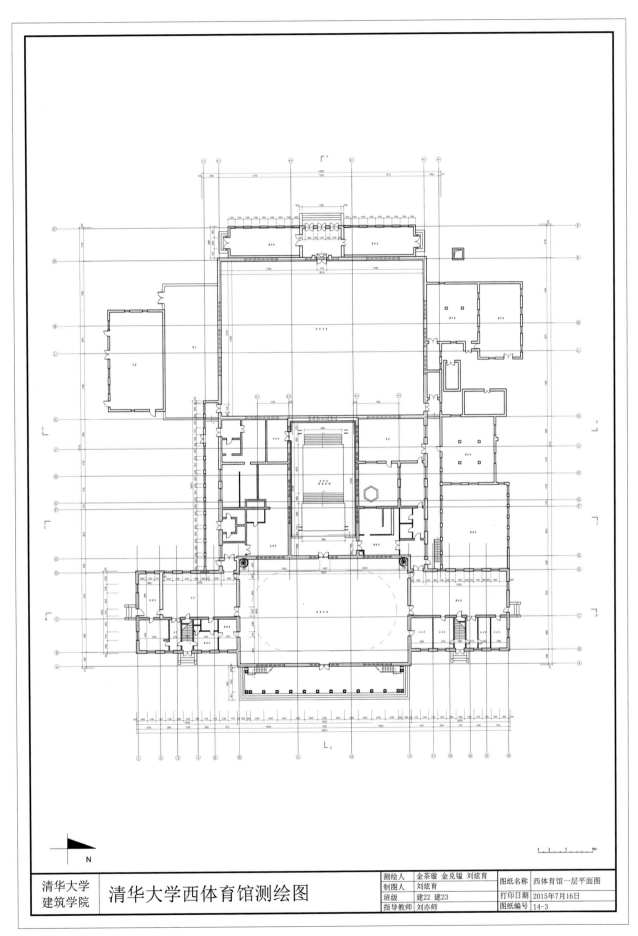

N

清华大学
建筑学院

清华大学西体育馆测绘图

测绘人	金茶璇 金兑镕 刘炫育	图纸名称	西体育馆一层平面图
制图人	刘炫育		
班级	建22 建23	打印日期	2015年7月16日
指导教师	刘亦师	图纸编号	14-3

		测绘人	金荼璇 金兑镪 刘炫育	图纸名称	西体育馆二层平面图

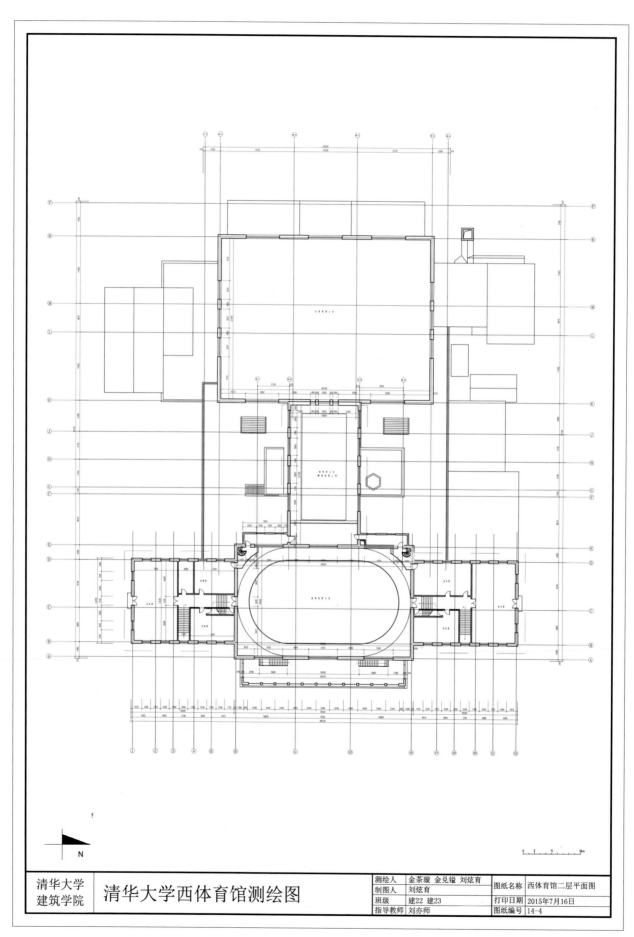

清华大学 建筑学院	清华大学西体育馆测绘图	测绘人	金荼璇 金兑镪 刘炫育	图纸名称	西体育馆二层平面图
		制图人	刘炫育		
		班级	建22 建23	打印日期	2015年7月16日
		指导教师	刘亦师	图纸编号	14-4

N

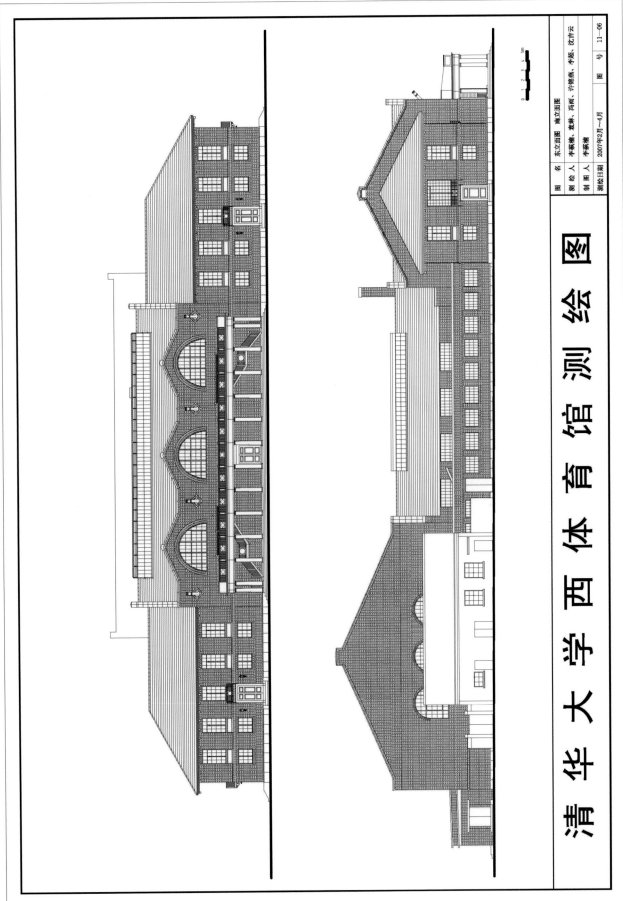

清华大学西体育馆测绘图

图 名	东立面图 南立面图
测绘人	李楠燕、袁琳、冯程、许娟燕、李悫、沈青云
制图人	李楠燕
测绘日期	2007年2月～4月

图 号 11—06

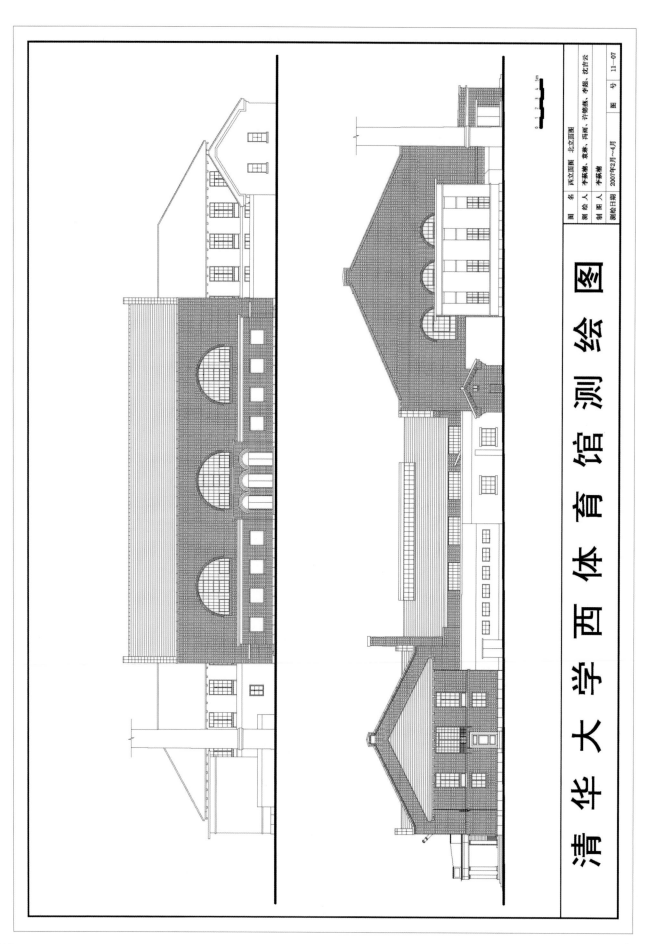

清 华 大 学 西 体 育 馆 测 绘 图

图　名	西立面图　北立面图		
测绘人	李慕楠、袁琳、许锦焕、李愿、沈吉云		
制图人	李慕楠		图　号　11—07
测绘日期	2007年2月～4月		

0 1 2 3 4 5m

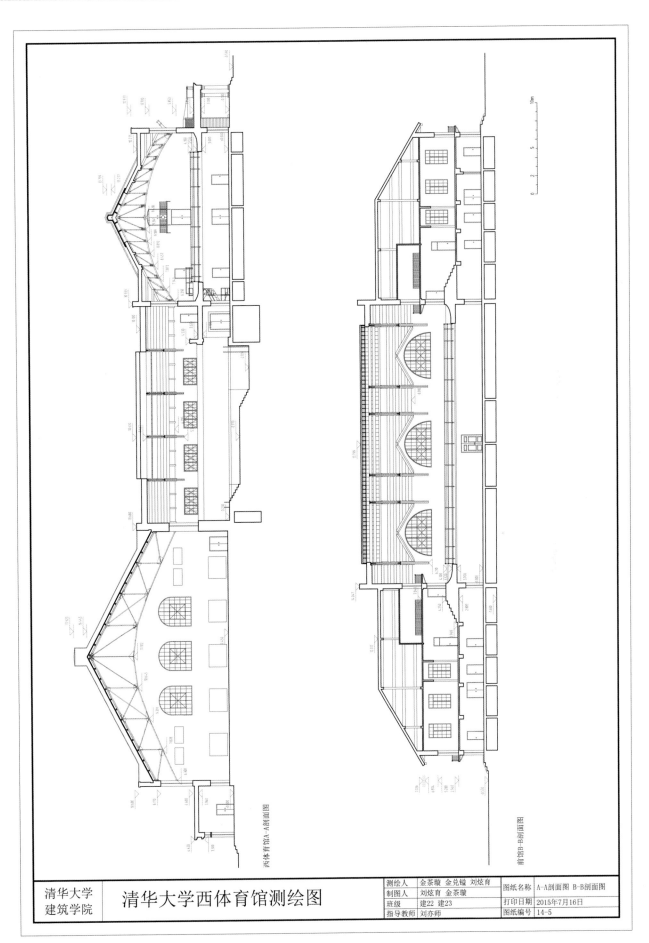

西体育馆A-A剖面图

前馆B-B剖面图

测绘人	金茶璇 金兑镒 刘炫育	图纸名称	A-A剖面图 B-B剖面图
制图人	刘炫育 金茶璇		
班级	建22 建23	打印日期	2015年7月16日
指导教师	刘亦师	图纸编号	14-5

清华大学
建筑学院

清华大学西体育馆测绘图

| 测绘人 | 金荟璇 金兑镒 刘炫育 | 图纸名称 | C-C剖面图 D-D剖面图 |
| 制图人 | 刘炫育 | | |

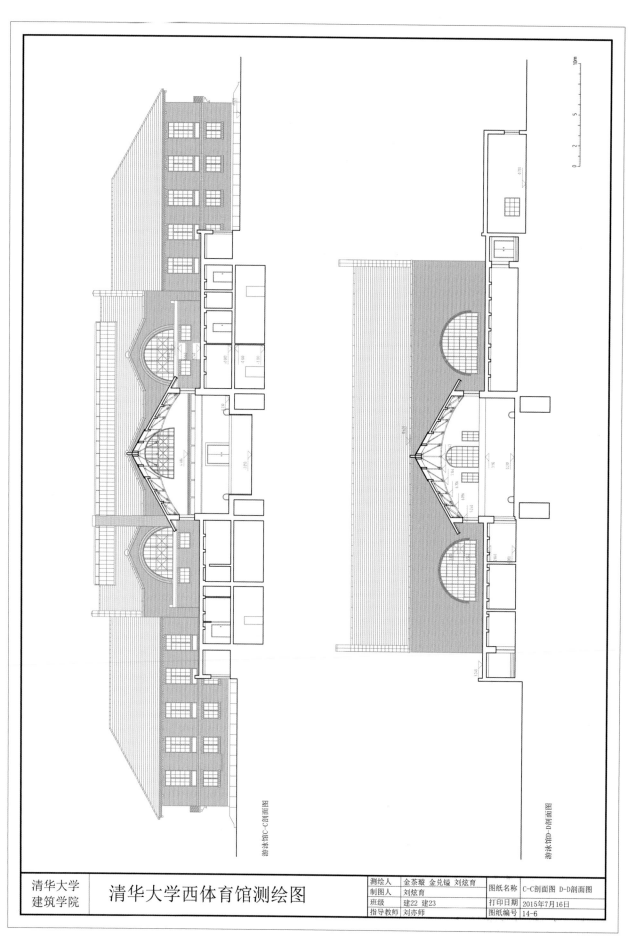

游泳馆C-C剖面图

游泳馆D-D剖面图

清华大学 建筑学院	清华大学西体育馆测绘图	测绘人	金荟璇 金兑镒 刘炫育	图纸名称	C-C剖面图 D-D剖面图
		制图人	刘炫育		
		班级	建22 建23	打印日期	2015年7月16日
		指导教师	刘亦师	图纸编号	14-6

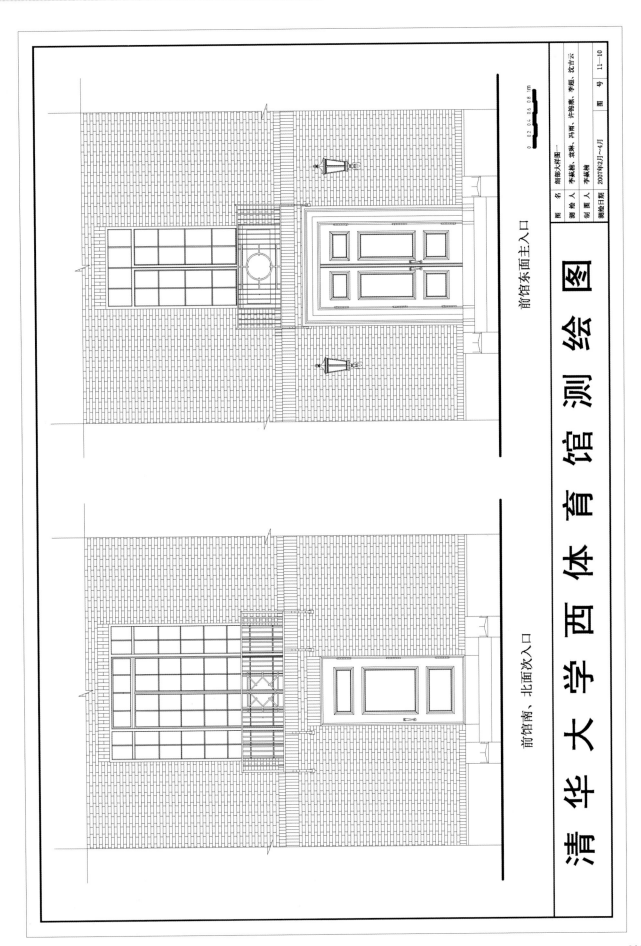

前馆东面主入口

前馆南、北面次入口

清华大学西体育馆测绘图

图 名	附部大样图一
测绘人	李森楠、姜淼、冯雨、许翰熹、李超、沈吉云
制图人	李森楠
测绘日期	2007年2月~4月

图号 11—10

0 0.2 0.4 0.6 0.8 1m

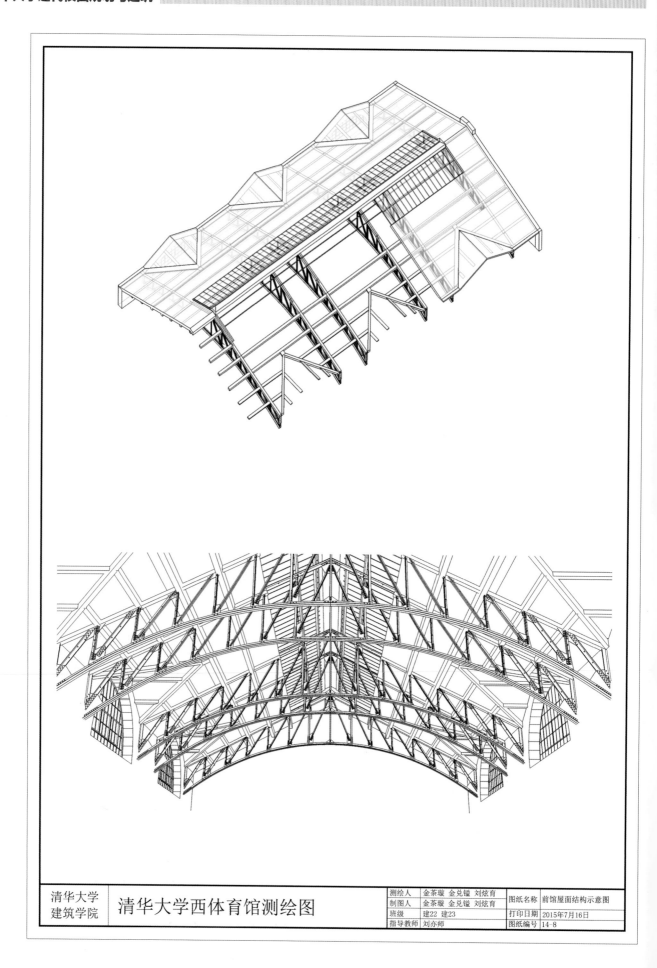

测绘人	金茶璇 金兑镒 刘炫育	图纸名称	前馆屋面结构示意图
制图人	金茶璇 金兑镒 刘炫育		
班级	建22 建23	打印日期	2015年7月16日
指导教师	刘亦师	图纸编号	14-8

清华大学
建筑学院

清华大学西体育馆测绘图

测绘人	金茶璇 金兑镠 刘炫育	图纸名称	游泳馆屋面结构示意图	
制图人	金茶璇 金兑镠 刘炫育			
班级	建22 建23	打印日期	2015年7月16日	
指导教师	刘亦师	图纸编号	14-9	

清华大学
建筑学院

清华大学西体育馆测绘图

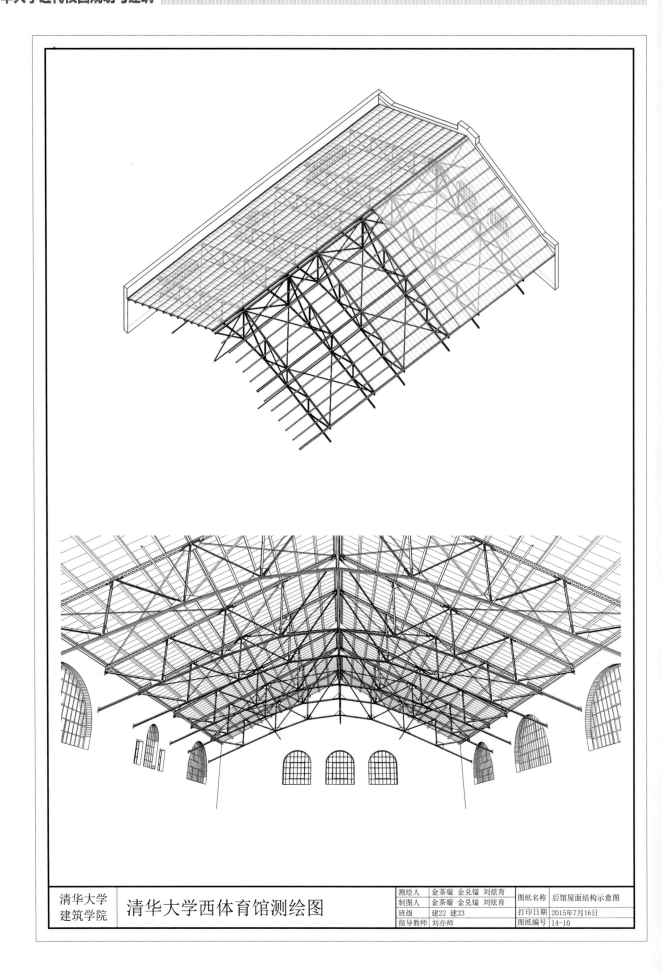

测绘人	金荃璇 金兑镒 刘炫育	图纸名称	后馆屋面结构示意图
制图人	金荃璇 金兑镒 刘炫育		
班级	建22 建23	打印日期	2015年7月16日
指导教师	刘亦师	图纸编号	14-10

清华大学
建筑学院

清华大学西体育馆测绘图

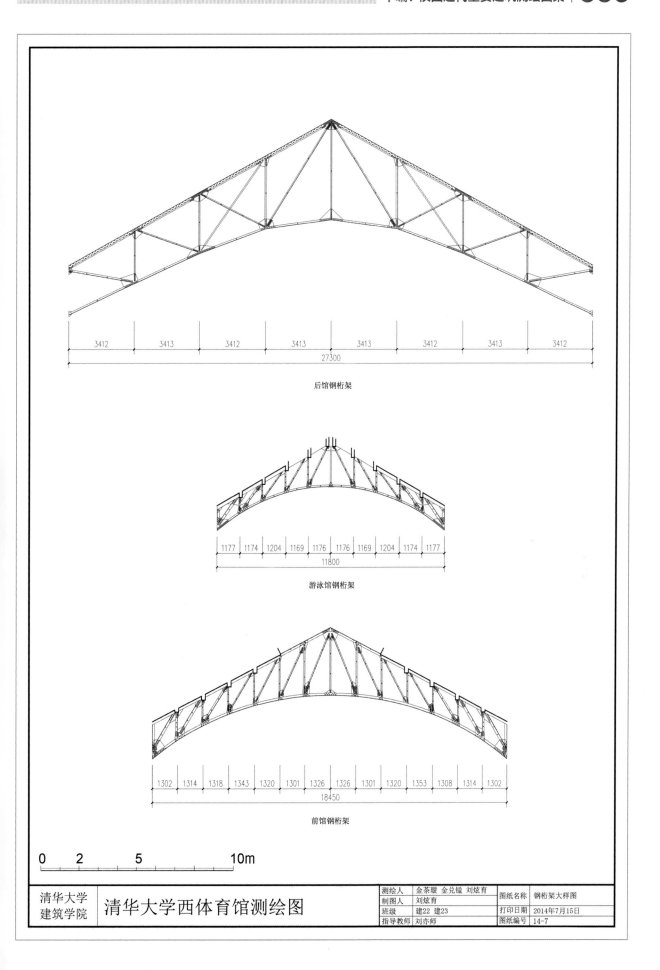

后馆钢桁架

游泳馆钢桁架

前馆钢桁架

0 2 5 10m

清华大学
建筑学院

清华大学西体育馆测绘图

测绘人	金茶璇 金兑镒 刘炫育	图纸名称	钢桁架大样图
制图人	刘炫育		
班级	建22 建23	打印日期	2014年7月15日
指导教师	刘亦师	图纸编号	14-7

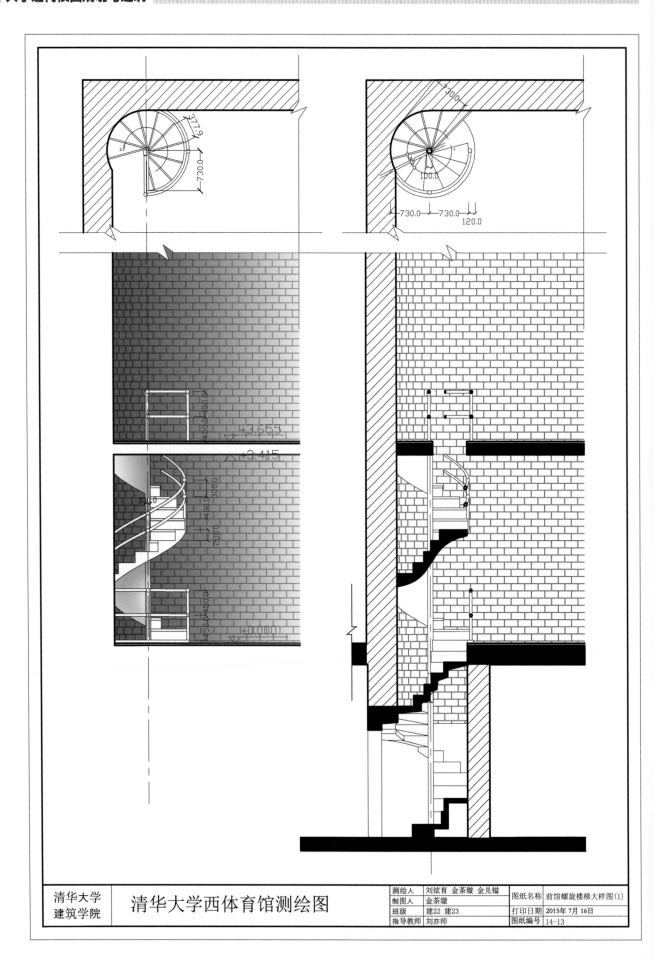

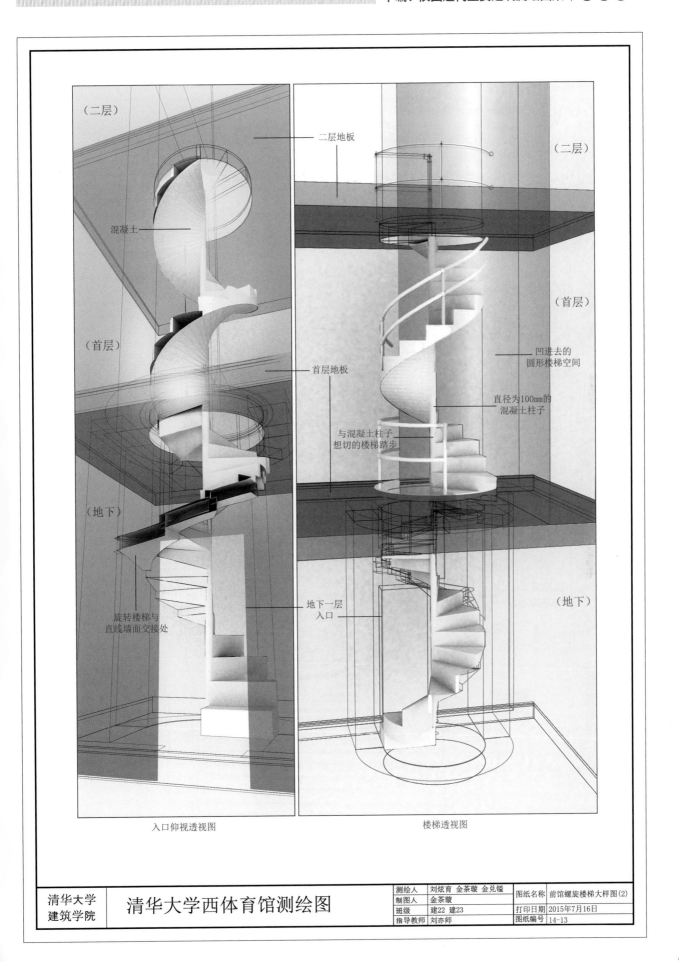

（二层）

二层地板

（二层）

混凝土

（首层）

（首层）

凹进去的
圆形楼梯空间

首层地板

直径为100mm的
混凝土柱子

与混凝土柱子
想切的楼梯踏步

（地下）

（地下）

旋转楼梯与
直线墙面交接处

地下一层
入口

入口仰视透视图

楼梯透视图

清华大学 建筑学院	清华大学西体育馆测绘图	测绘人	刘炫育 金茶璇 金兑镗	图纸名称	前馆螺旋楼梯大样图(2)
		制图人	金茶璇		
		班级	建22 建23	打印日期	2015年7月16日
		指导教师	刘亦师	图纸编号	14-13

清华大学图书馆
2004.6.4. 冀

第 3 组　　罗家伦、梅贻琦时代的重要建筑

9. 老土木馆

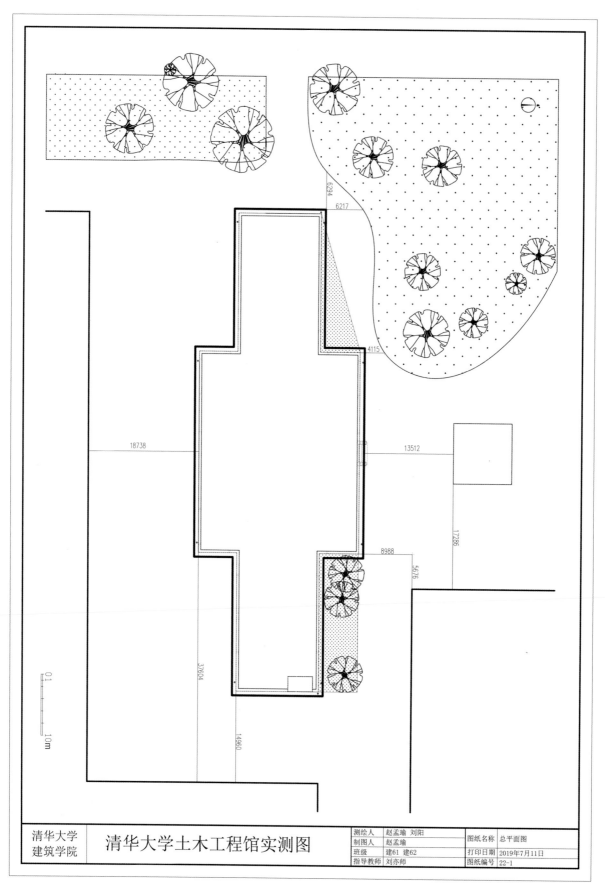

		测绘人	赵孟瑜 刘阳	图纸名称	总平面图
清华大学 建筑学院	清华大学土木工程馆实测图	制图人	赵孟瑜		
		班级	建61 建62	打印日期	2019年7月11日
		指导教师	刘亦师	图纸编号	22-1

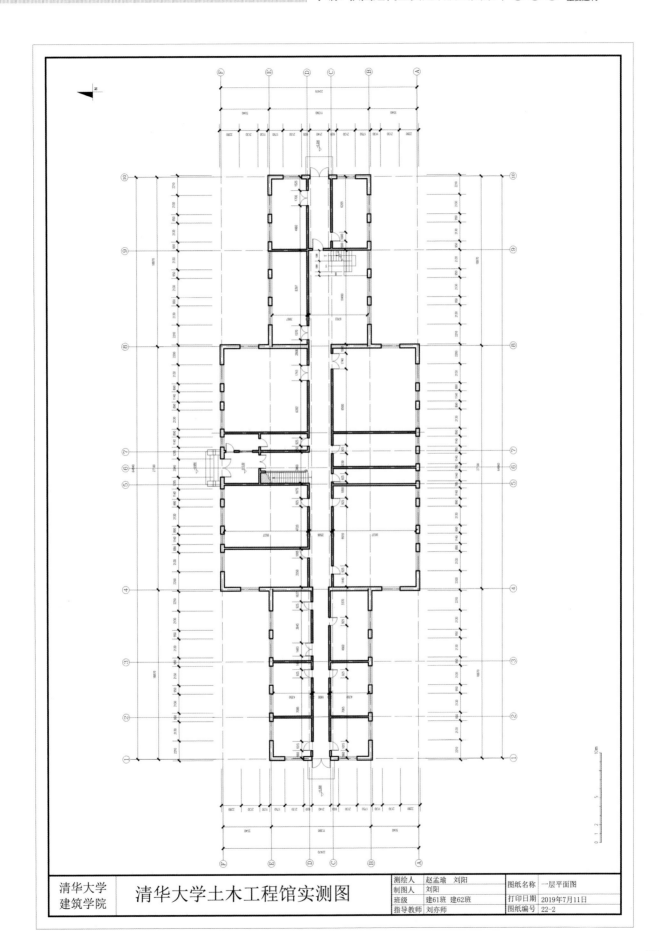

		测绘人	赵孟瑜 刘阳	图纸名称	一层平面图
清华大学 建筑学院	清华大学土木工程馆实测图	制图人	刘阳		
		班级	建61班 建62班	打印日期	2019年7月11日
		指导教师	刘亦师	图纸编号	22-2

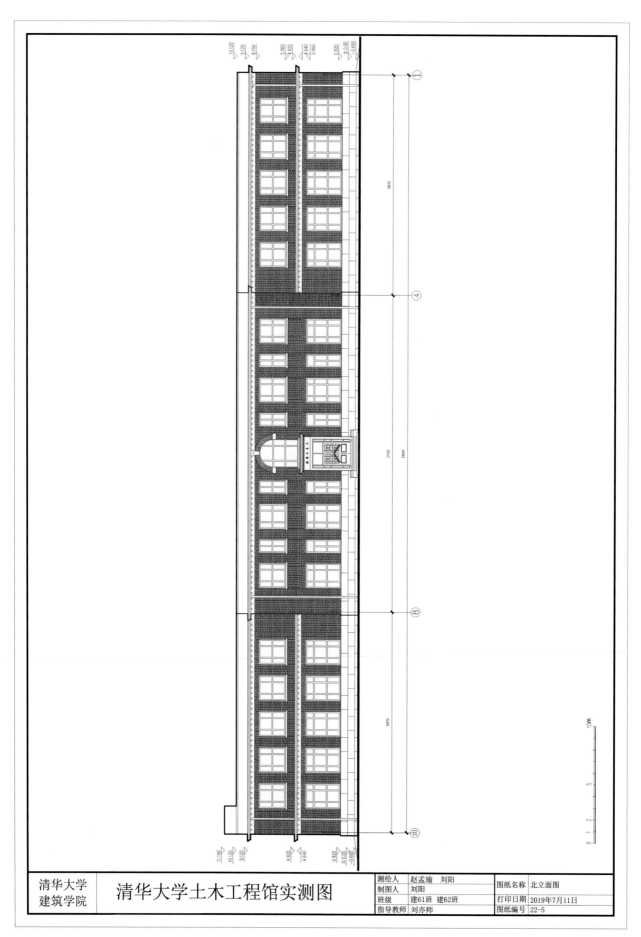

清华大学土木工程馆实测图

测绘人	赵孟瑜 刘阳	图纸名称	北立面图
制图人	刘阳		
班级	建61班 建62班	打印日期	2019年7月11日
指导教师	刘亦师	图纸编号	22-5

清华大学
建筑学院

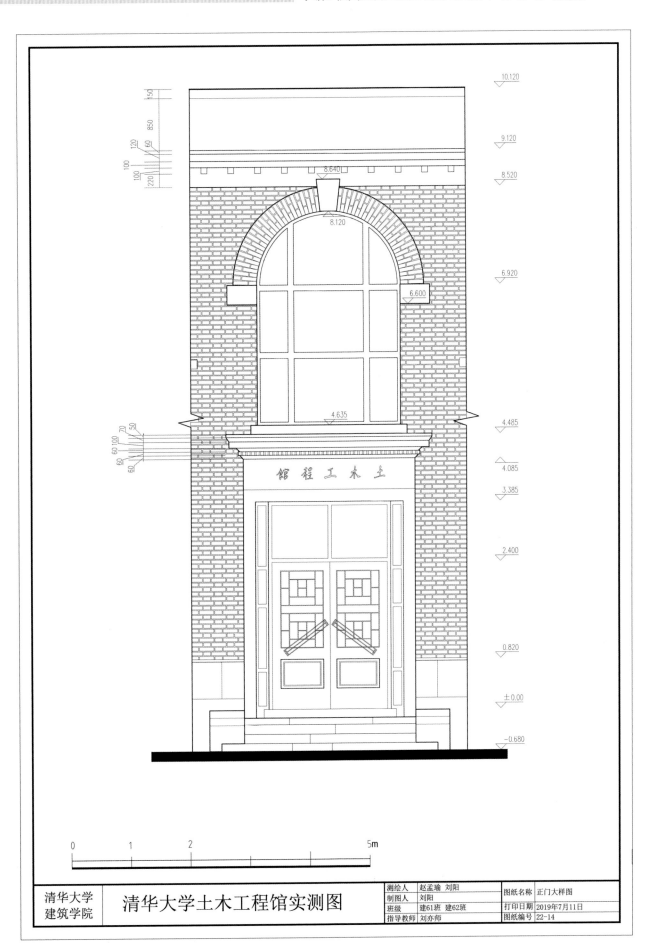

测绘人	赵孟瑜 刘阳	图纸名称	剖透视2
班级	建61班 建62班		
指导教师	刘亦师		

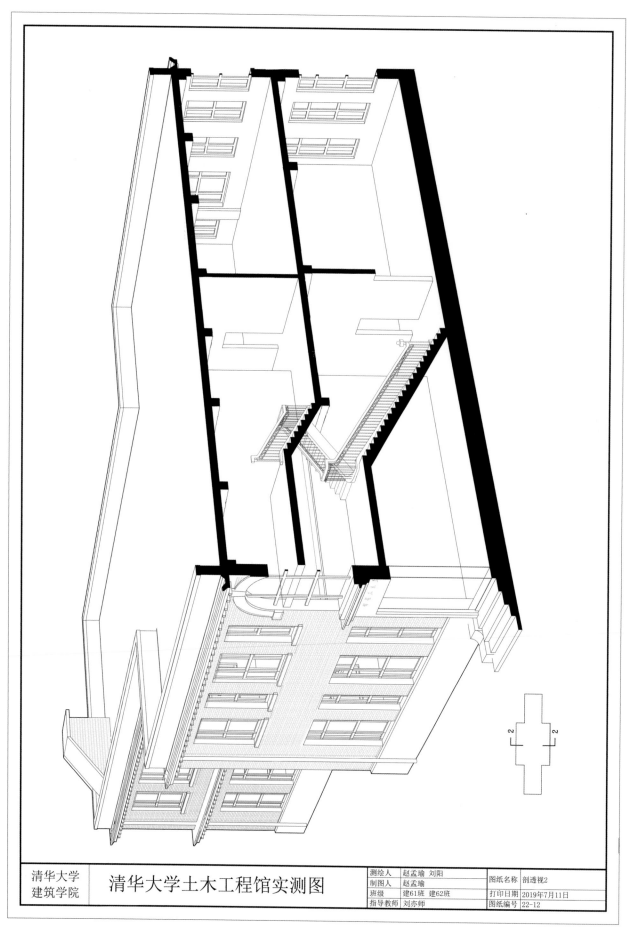

清华大学 建筑学院	清华大学土木工程馆实测图	测绘人	赵孟瑜 刘阳	图纸名称	剖透视2
		制图人	赵孟瑜		
		班级	建61班 建62班	打印日期	2019年7月11日
		指导教师	刘亦师	图纸编号	22-12

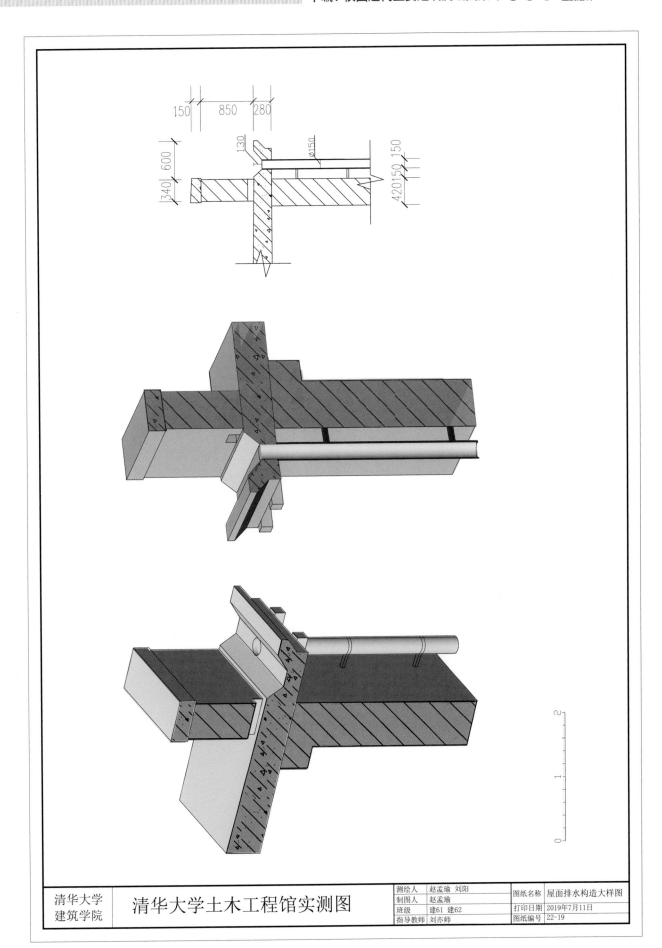

测绘人	赵孟瑜 刘阳		图纸名称	屋面排水构造大样图
制图人	赵孟瑜			
班级	建61 建62		打印日期	2019年7月11日
指导教师	刘亦师		图纸编号	22-19

清华大学建筑学院

清华大学土木工程馆实测图

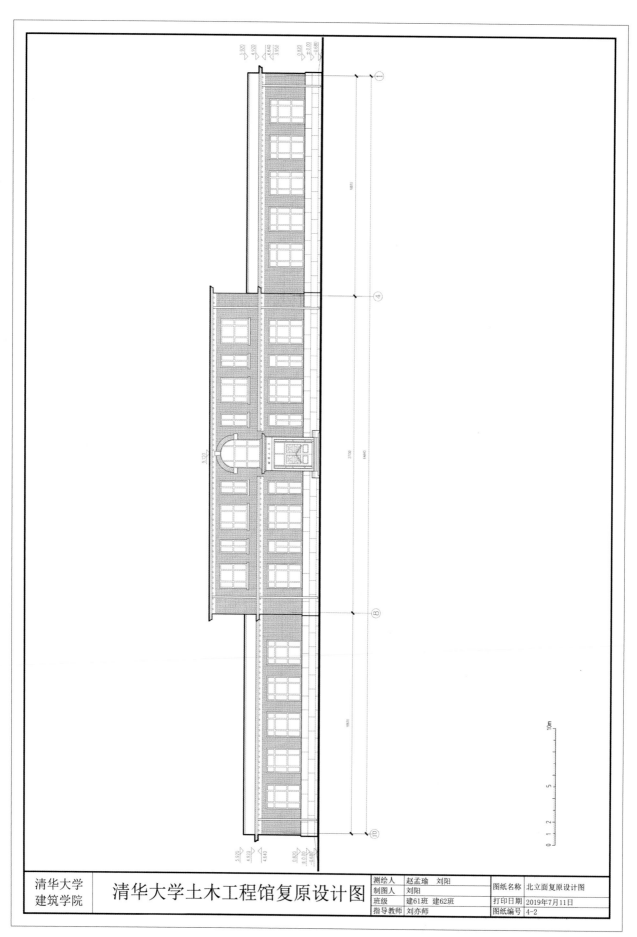

清华大学 建筑学院	清华大学土木工程馆复原设计图	测绘人	赵孟瑜　刘阳	图纸名称	北立面复原设计图
		制图人	刘阳		
		班级	建61班　建62班	打印日期	2019年7月11日
		指导教师	刘亦师	图纸编号	4-2

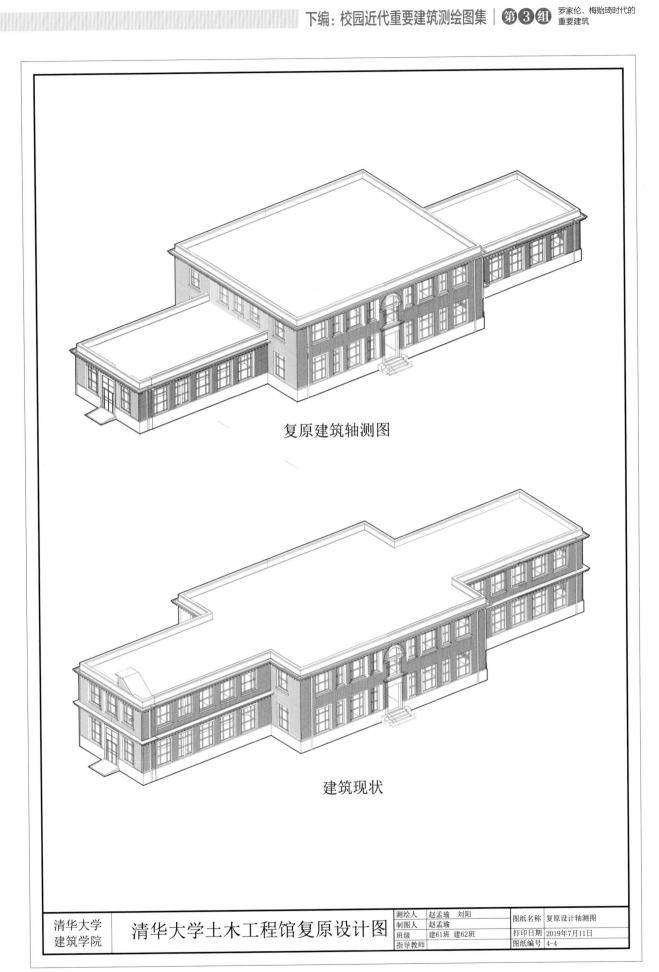

复原建筑轴测图

建筑现状

清华大学 建筑学院	清华大学土木工程馆复原设计图	测绘人	赵孟瑜　刘阳	图纸名称	复原设计轴测图
		制图人	赵孟瑜	打印日期	2019年7月11日
		班级	建61班 建62班	图纸编号	4-4
		指导教师			

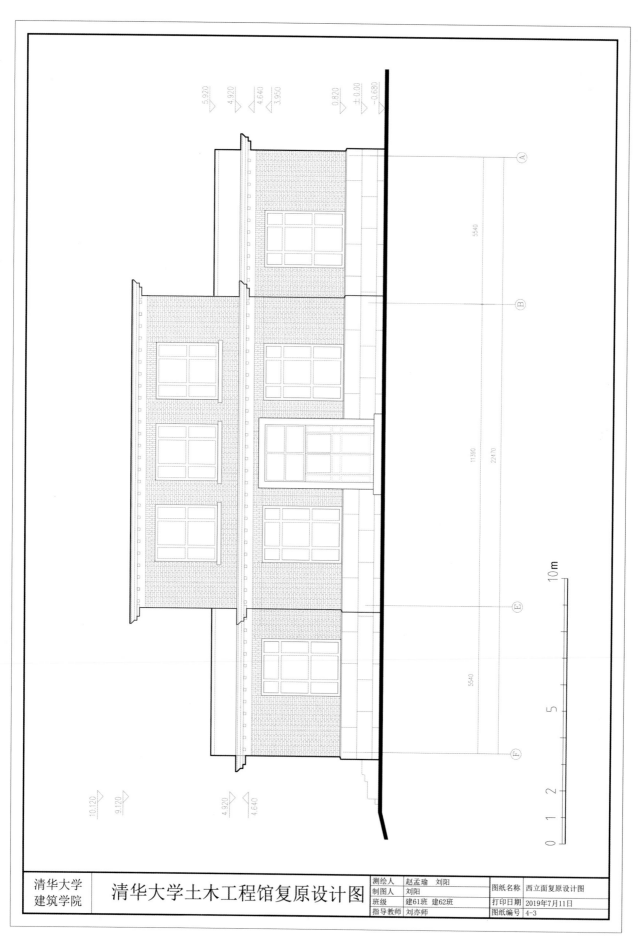

清华大学建筑学院	清华大学土木工程馆复原设计图	测绘人	赵孟瑜　刘阳	图纸名称	西立面复原设计图
		制图人	刘阳		
		班级	建61班　建62班	打印日期	2019年7月11日
		指导教师	刘亦师	图纸编号	4-3

458

清华大学 建筑学院	清华大学土木工程馆实测图	测绘人	赵孟瑜　刘阳	图纸名称	土木工程馆实景照片
		制图人	赵孟瑜　刘阳		
		班级	建61班　建62班	打印日期	2019年7月11日
		指导教师	刘亦师	图纸编号	22-22

10. 生物学馆

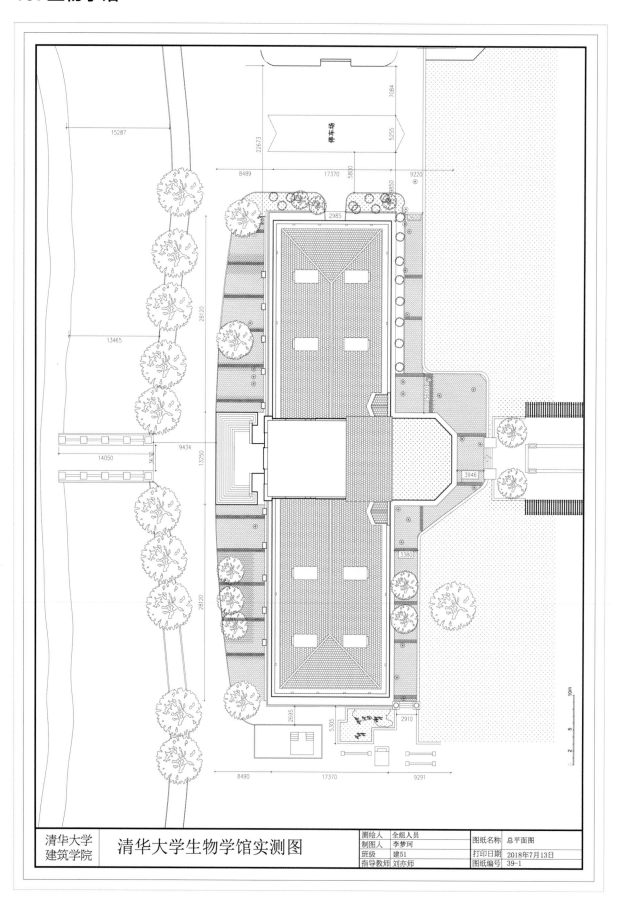

停车场

清华大学
建筑学院

清华大学生物学馆实测图

测绘人	全组人员	图纸名称	总平面图
制图人	李梦珂		
班级	建51	打印日期	2018年7月13日
指导教师	刘亦师	图纸编号	39-1

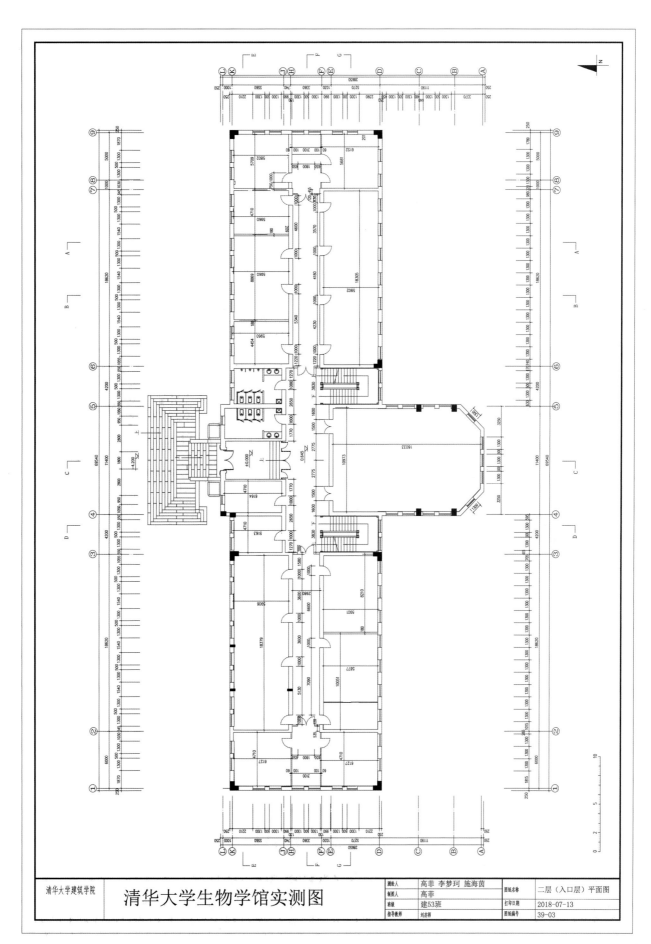

			测绘人	高菲 李梦珂 施海茵	图纸名称	二层（入口层）平面图
清华大学建筑学院	**清华大学生物学馆实测图**		制图人	高菲		
			班级	建53班	打印日期	2018-07-13
			指导教师	刘志君	图纸编号	39-03

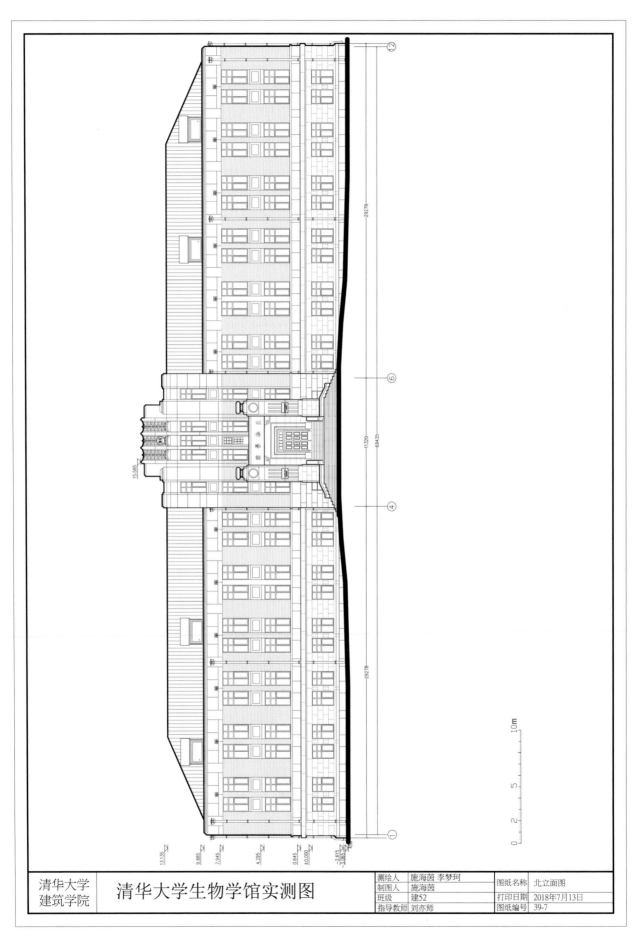

清华大学
建筑学院

清华大学生物学馆实测图

测绘人	施海茵 李梦珂	图纸名称	北立面图
制图人	施海茵	打印日期	2018年7月13日
班级	建52		
指导教师	刘亦师	图纸编号	39-7

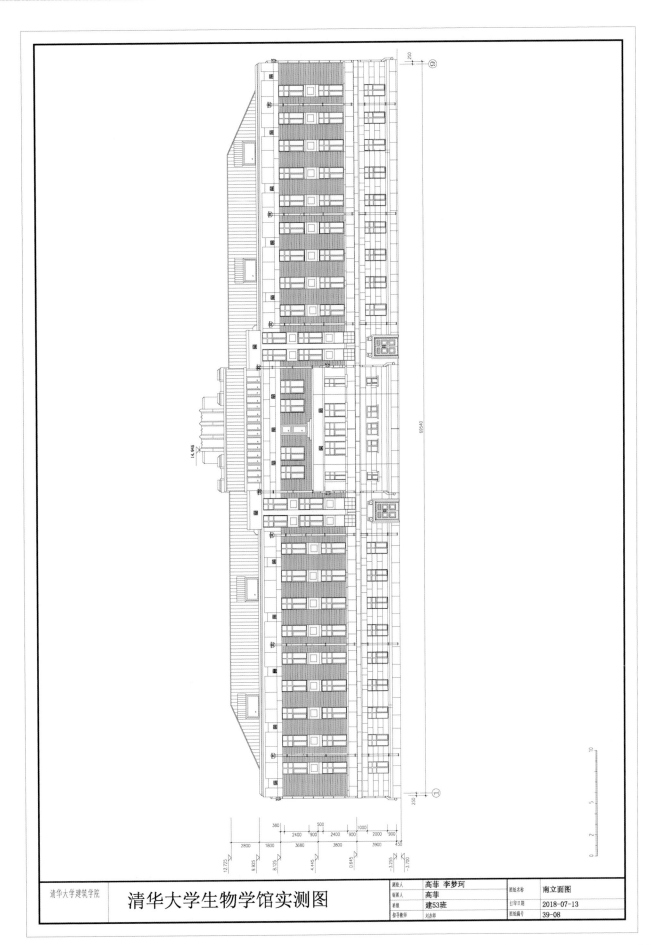

清华大学建筑学院

清华大学生物学馆实测图

	测绘人	高菲 李梦珂	图纸名称	南立面图
	制图人	高菲		
	班级	建53班	扫印日期	2018-07-13
	指导教师	刘志晖	图纸编号	39-08

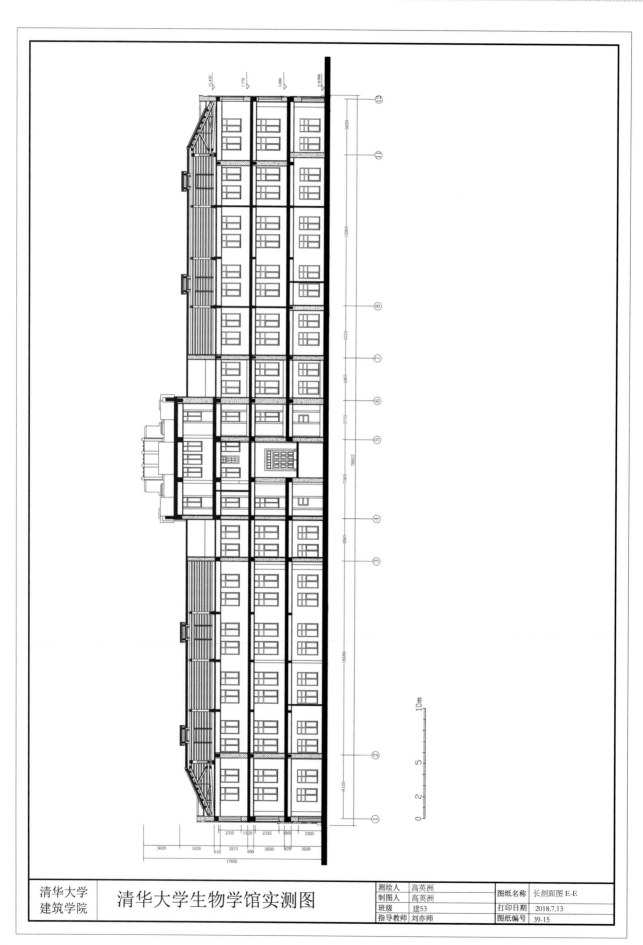

测绘人	高英洲	图纸名称	长剖面图 E-E
制图人	高英洲		
班级	建53	打印日期	2018.7.13
指导教师	刘亦师	图纸编号	39-15

清华大学
建筑学院

清华大学生物学馆实测图

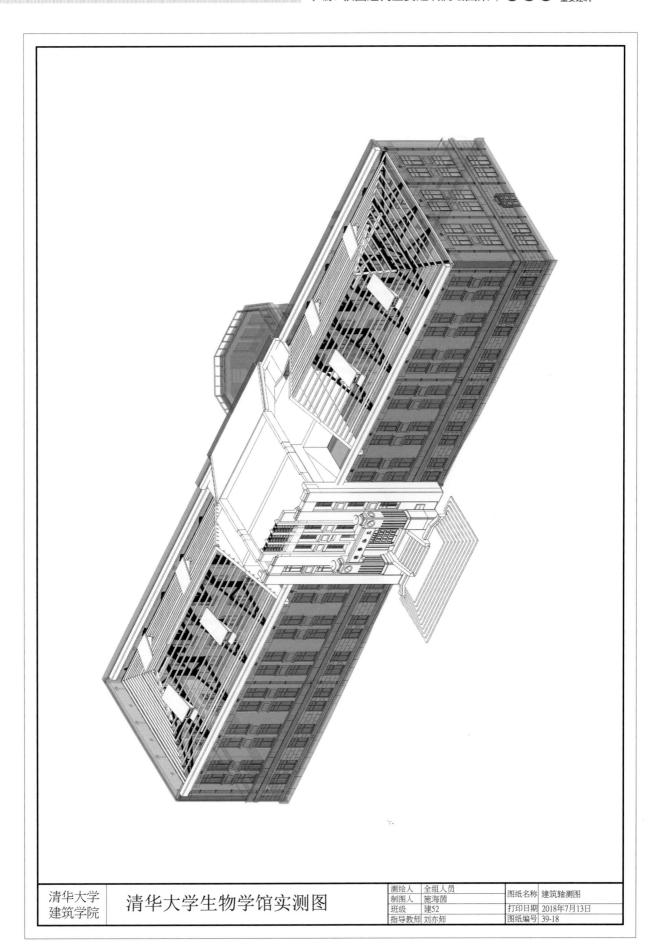

清华大学
建筑学院

清华大学生物学馆实测图

测绘人	全组人员	图纸名称	建筑轴测图
制图人	施海茵		
班级	建52	打印日期	2018年7月13日
指导教师	刘亦师	图纸编号	39-18

测绘人	郑妍彦 高英洲		图纸名称	楼梯大样1
制图人	高英洲			
班级	建52　建53		打印日期	2018.7.13
指导教师	刘亦师		图纸编号	39-29

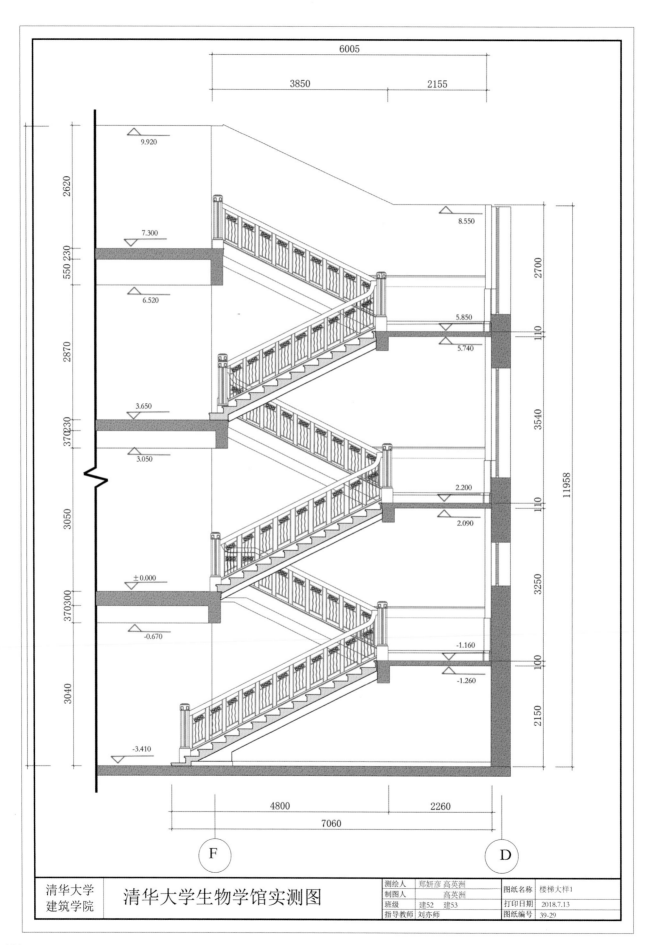

清华大学
建筑学院

清华大学生物学馆实测图

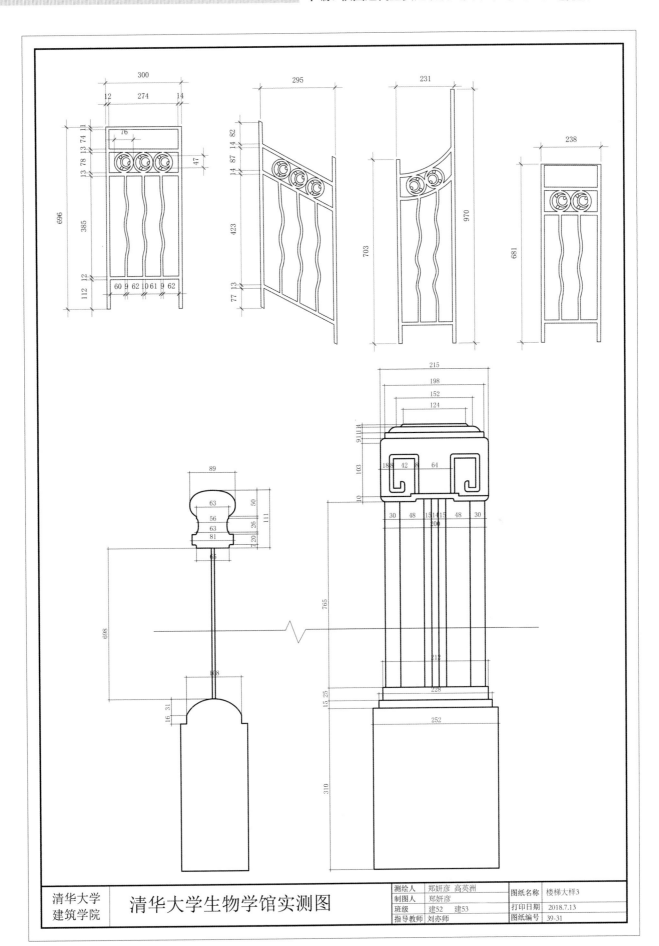

清华大学
建筑学院

清华大学生物学馆实测图

测绘人	郑妍彦 高英洲		图纸名称	楼梯大样3
制图人	郑妍彦		打印日期	2018.7.13
班级	建52　建53		图纸编号	39-31
指导教师	刘亦师			

467

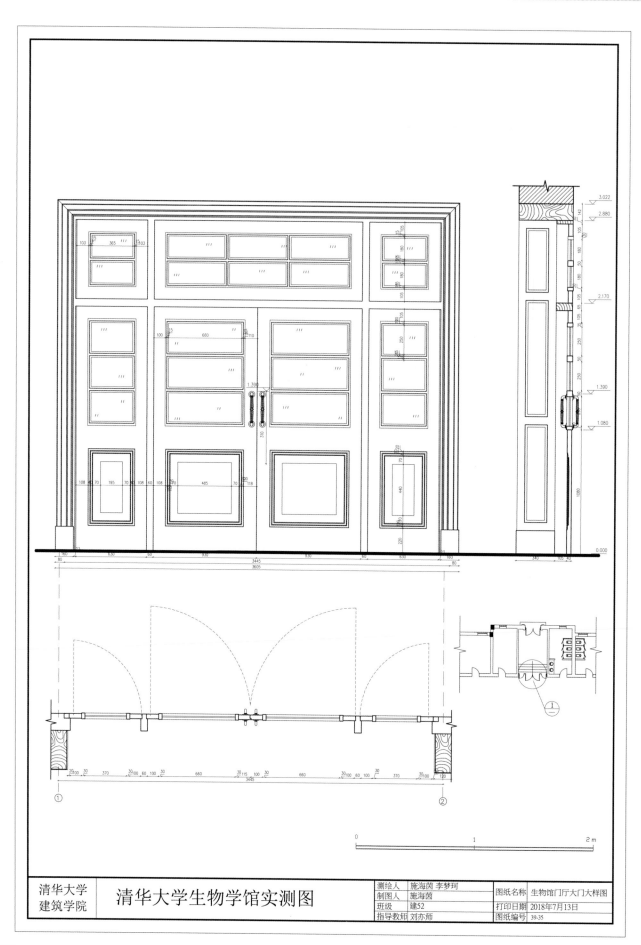

清华大学
建筑学院

清华大学生物学馆实测图

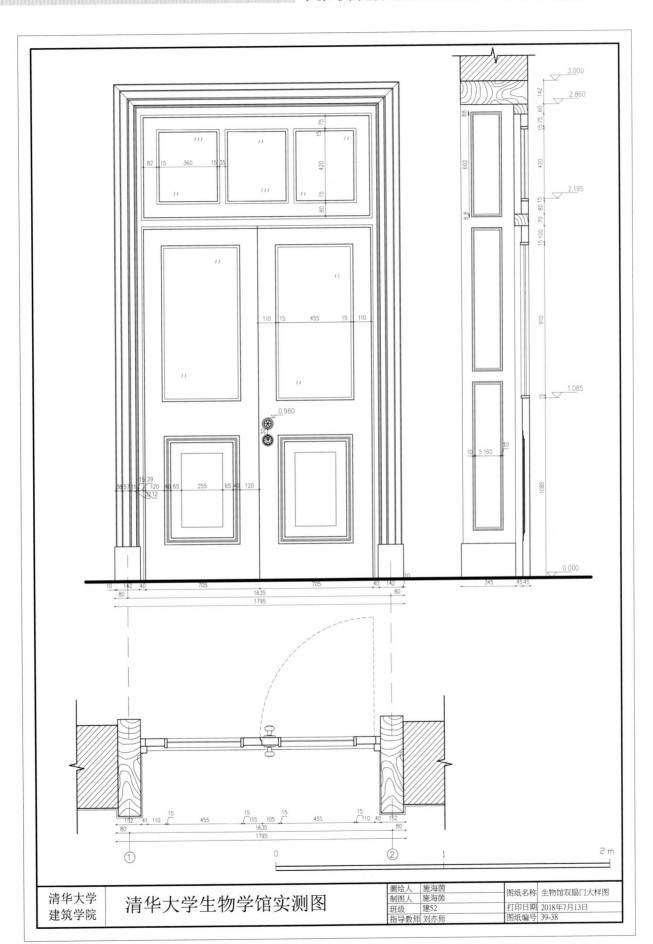

		图纸名称	生物馆双扇门大样图
清华大学 建筑学院	清华大学生物学馆实测图	测绘人	施海茵
		制图人	施海茵
		班级	建52
		打印日期	2018年7月13日
		指导教师	刘亦师
		图纸编号	39-38

469

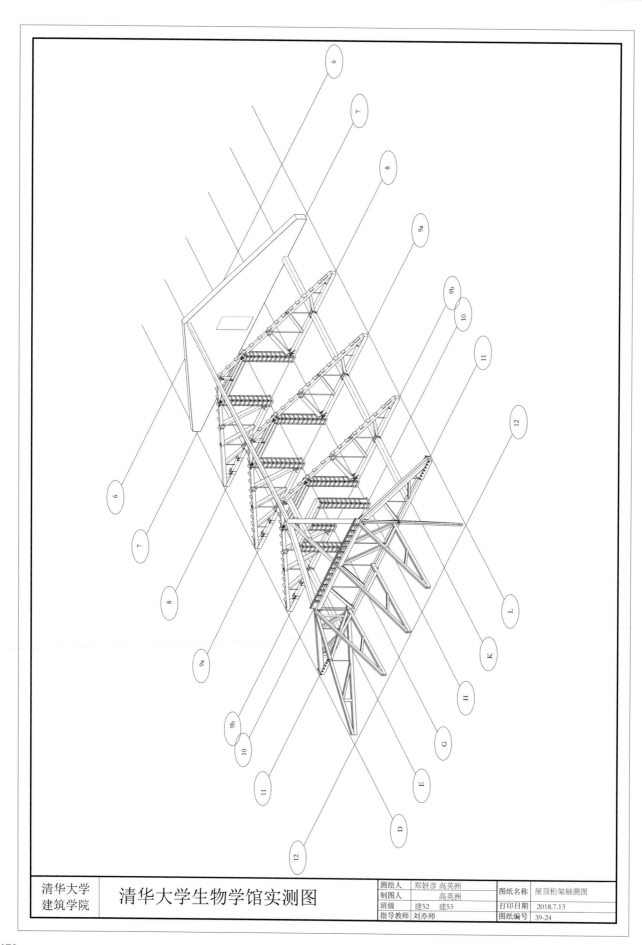

测绘人	郑妍彦 高英洲	图纸名称	屋顶桁架轴测图
制图人	高英洲		
班级	建52　建53	打印日期	2018.7.13
指导教师	刘亦师	图纸编号	39-24

清华大学
建筑学院

清华大学生物学馆实测图

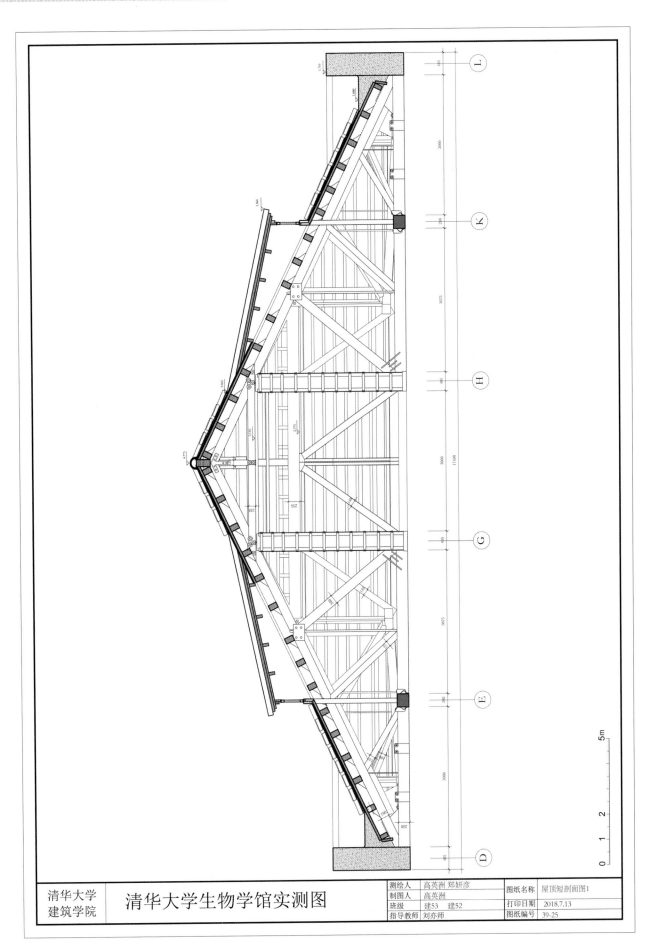

清华大学
建筑学院

清华大学生物学馆实测图

测绘人	高英洲 郑妍彦	图纸名称	屋顶短剖面图1
制图人	高英洲		
班级	建53　建52	打印日期	2018.7.13
指导教师	刘亦师	图纸编号	39-25

测绘人	高英洲 郑妍彦	图纸名称	屋顶短剖面图2
制图人	郑妍彦		
班级	建53 建52	打印日期	2018.7.13
指导教师	刘亦师	图纸编号	39-26

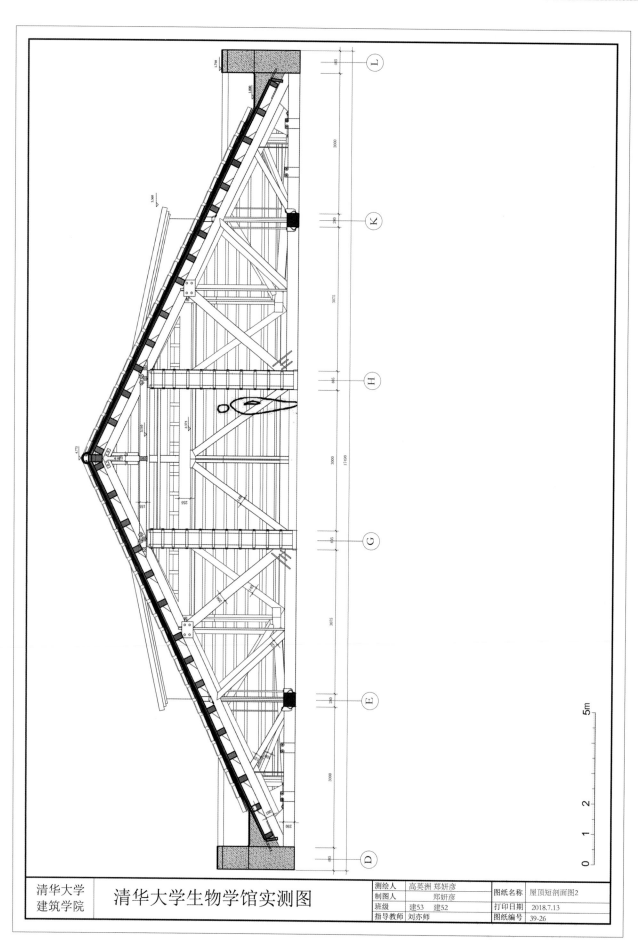

清华大学建筑学院	清华大学生物学馆实测图

清华大学
建筑学院

清华大学生物学馆实测图

测绘人		图纸名称	实测照片选
制图人			
班级		打印日期	2018.7.13
指导教师	刘亦师	图纸编号	39-39

11. 气象台

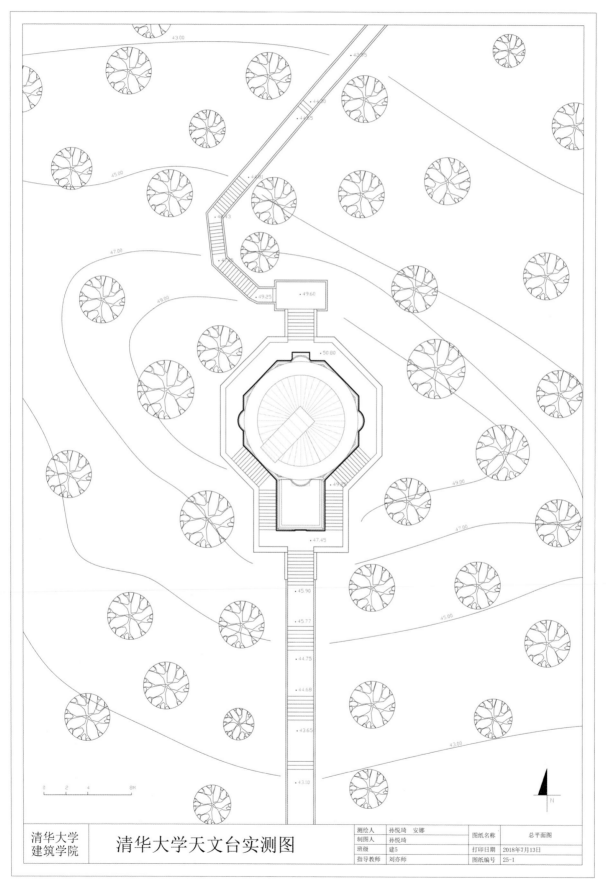

测绘人	孙悦琦 安娜		图纸名称	总平面图
制图人	孙悦琦			
班级	建5		打印日期	2018年7月13日
指导教师	刘亦帅		图纸编号	25-1

清华大学
建筑学院

清华大学天文台实测图

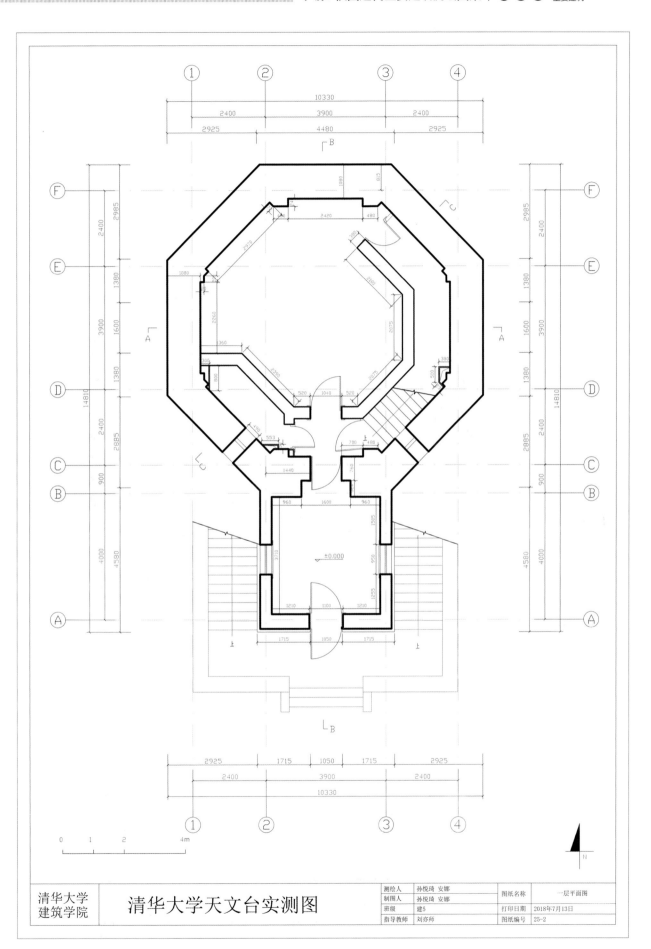

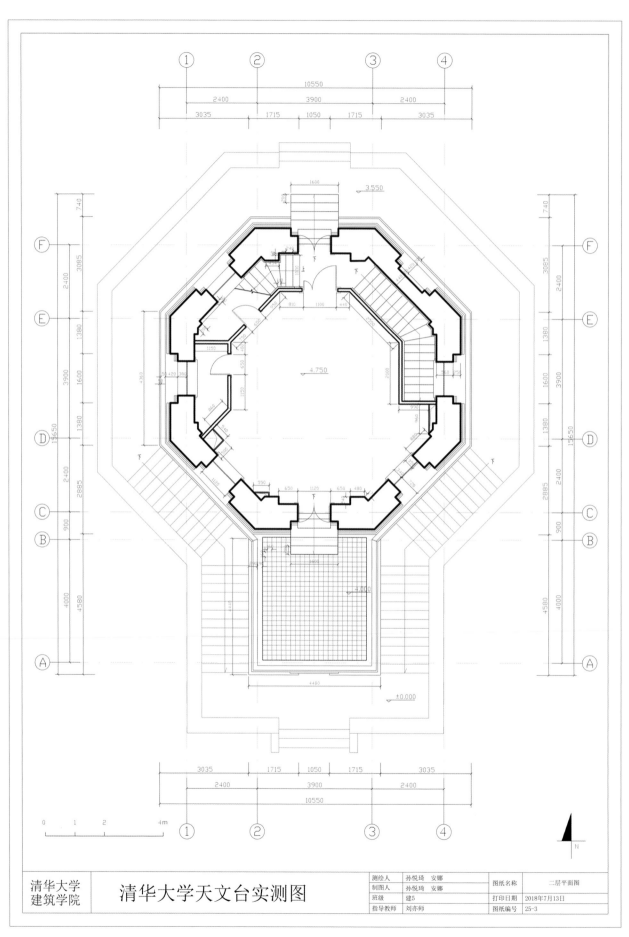

清华大学
建筑学院

清华大学天文台实测图

测绘人	孙悦琦 安娜	图纸名称	二层平面图
制图人	孙悦琦 安娜		
班级	建5	打印日期	2018年7月13日
指导教师	刘亦师	图纸编号	25-3

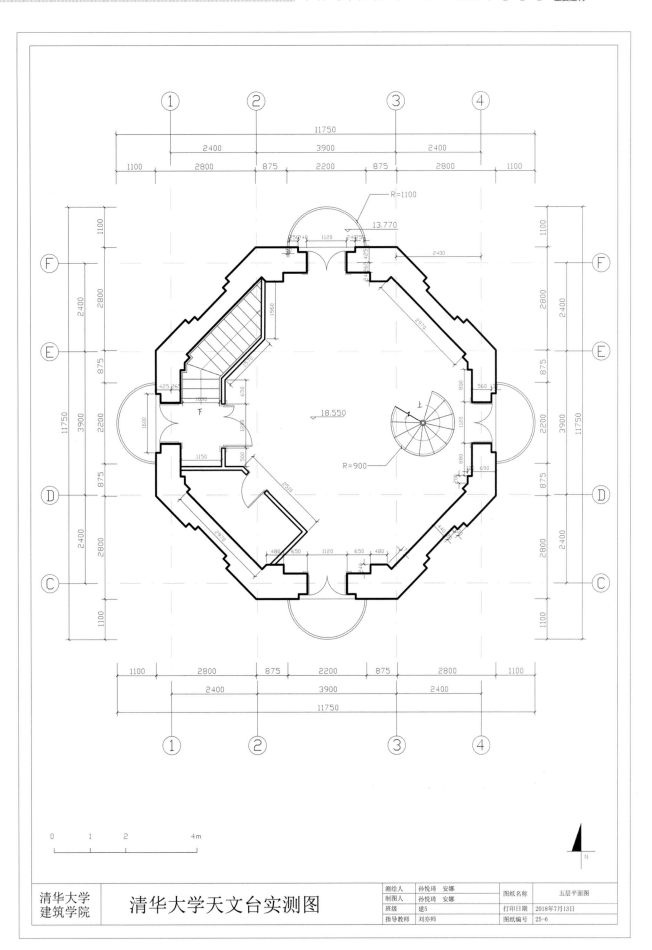

测绘人	孙悦琦　安娜	图纸名称	五层平面图
制图人	孙悦琦　安娜		
班级	建5	打印日期	2018年7月13日
指导教师	刘亦师	图纸编号	25-6

清华大学
建筑学院

清华大学天文台实测图

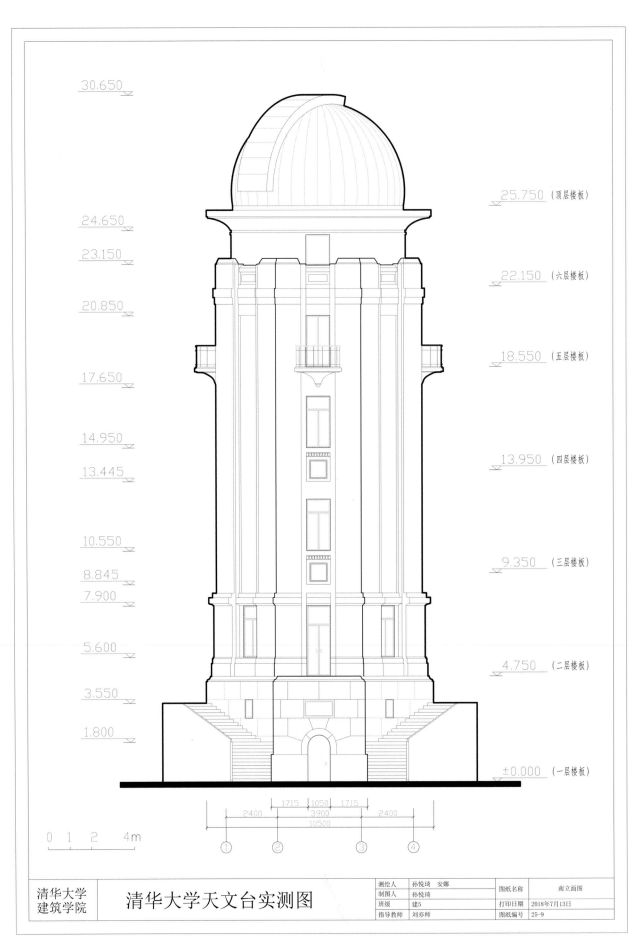

30.650

25.750 (顶层楼板)

24.650

23.150

22.150 (六层楼板)

20.850

18.550 (五层楼板)

17.650

14.950

13.950 (四层楼板)

13.445

10.550

9.350 (三层楼板)

8.845

7.900

5.600

4.750 (二层楼板)

3.550

1.800

±0.000 (一层楼板)

1715　1050　1715
2400　　3900　　2400
10500

① ② ③ ④

0 1 2 4m

	测绘人	孙悦琦 安娜	图纸名称	南立面图
清华大学 建筑学院	清华大学天文台实测图	制图人	孙悦琦		
		班级	建5	打印日期	2018年7月13日
		指导教师	刘亦师	图纸编号	25-9

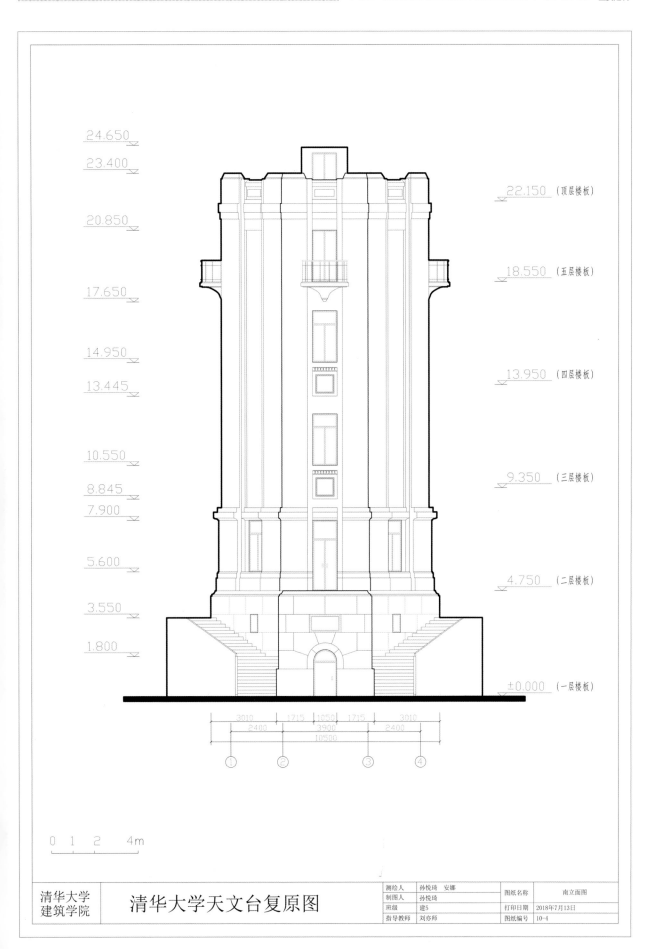

24.650

23.400

22.150 (顶层楼板)

20.850

18.550 (五层楼板)

17.650

14.950

13.950 (四层楼板)

13.445

10.550

9.350 (三层楼板)

8.845

7.900

5.600

4.750 (二层楼板)

3.550

1.800

±0.000 (一层楼板)

3010 1715 1050 1715 3010

2400 3900 2400

10500

① ② ③ ④

0 1 2 4m

清华大学建筑学院	清华大学天文台复原图	测绘人	孙悦琦　安娜	图纸名称	南立面图
		制图人	孙悦琦		
		班级	建5	打印日期	2018年7月13日
		指导教师	刘亦师	图纸编号	10-4

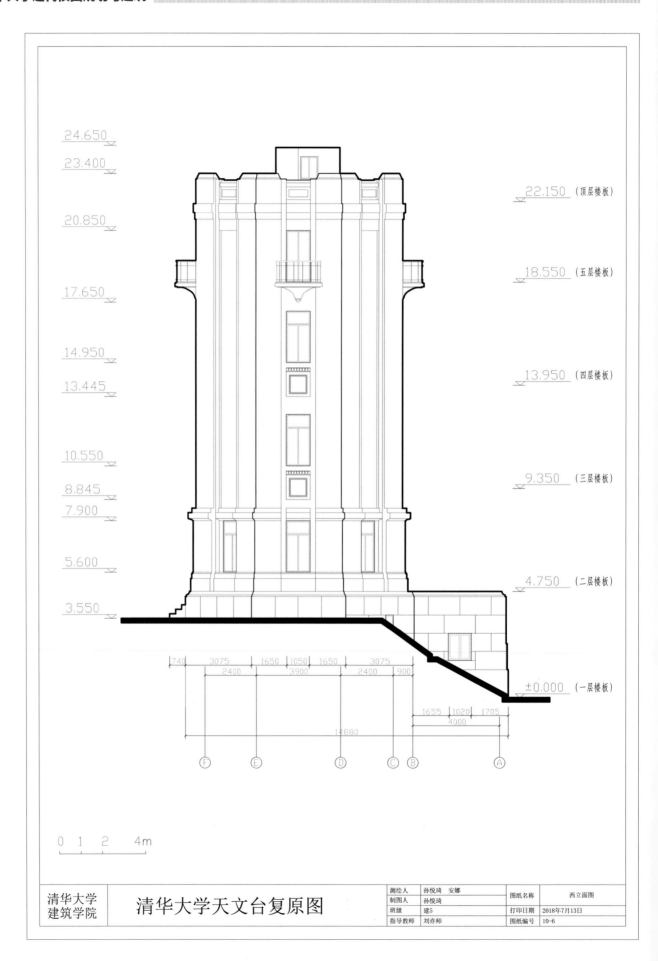

测绘人	孙悦琦 安娜	图纸名称	西立面图
制图人	孙悦琦		
班级	建5	打印日期	2018年7月13日
指导教师	刘亦师	图纸编号	10-6

清华大学
建筑学院

清华大学天文台复原图

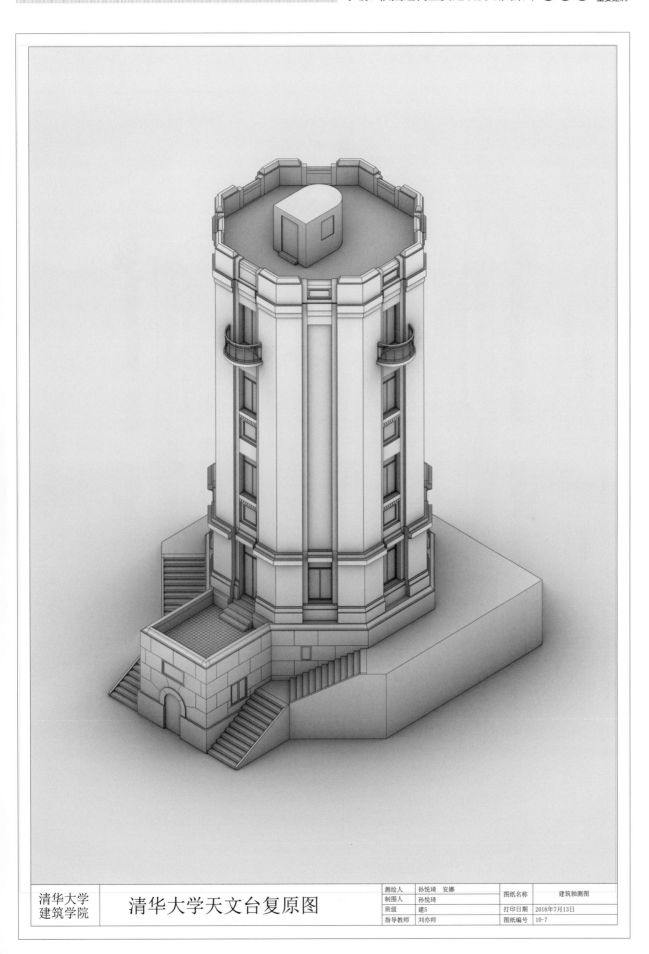

清华大学 建筑学院	清华大学天文台复原图	测绘人	孙悦琦　安娜	图纸名称	建筑轴测图
		制图人	孙悦琦		
		班级	建5	打印日期	2018年7月13日
		指导教师	刘亦师	图纸编号	10-7

测绘人	孙悦琦　安娜	图纸名称	建筑轴测分解图
制图人	孙悦琦		
班级	建5	打印日期	2018年7月13日
指导教师	刘亦师	图纸编号	10-8

清华大学
建筑学院

清华大学天文台复原图

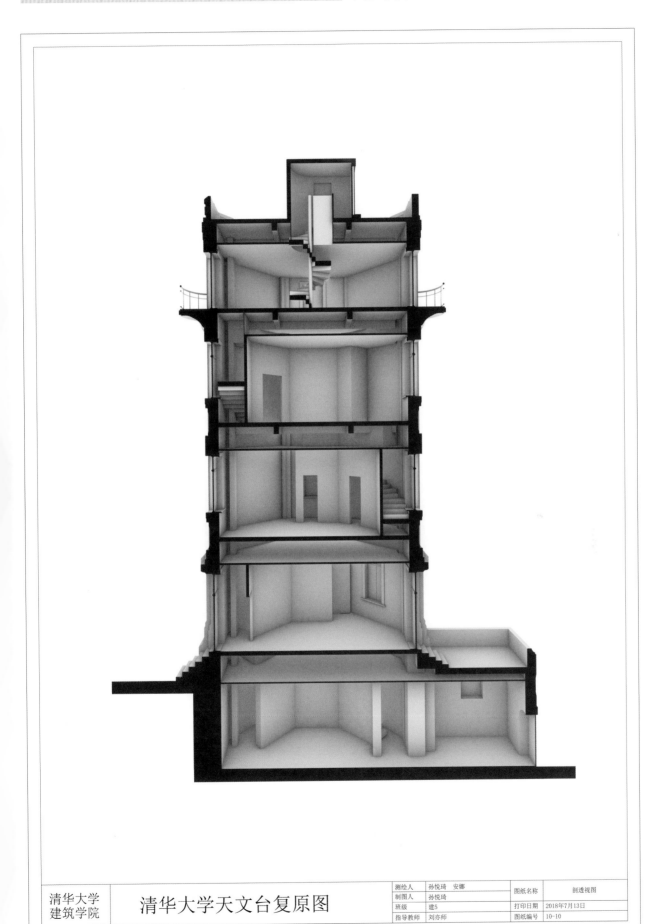

测绘人	孙悦琦　安娜	图纸名称	剖透视图
制图人	孙悦琦		
班级	建5	打印日期	2018年7月13日
指导教师	刘亦师	图纸编号	10-10

清华大学
建筑学院

清华大学天文台复原图

483

12. 老机械馆

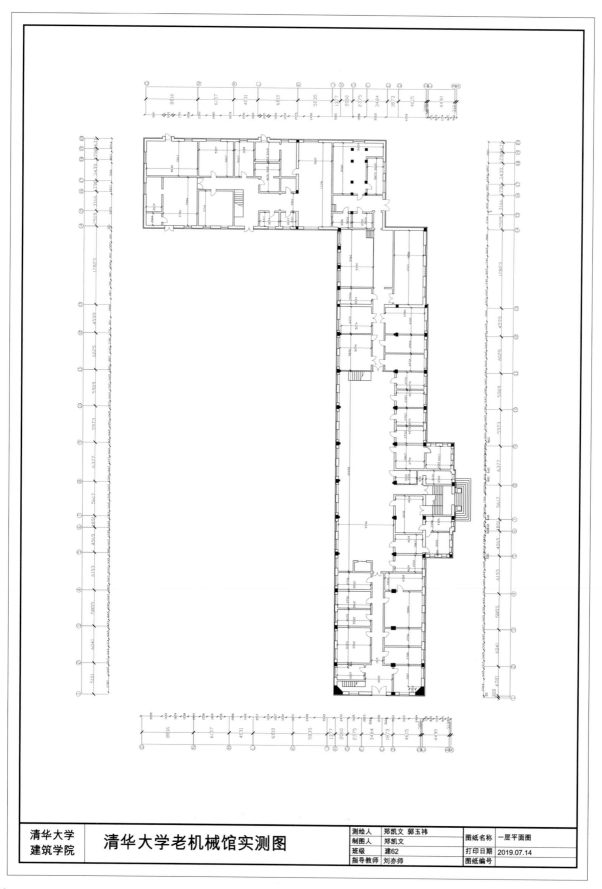

清华大学 建筑学院	清华大学老机械馆实测图	测绘人	郑凯文 郭玉祎	图纸名称	一层平面图
		制图人	郑凯文		
		班级	建62	打印日期	2019.07.14
		指导教师	刘亦师	图纸编号	

484

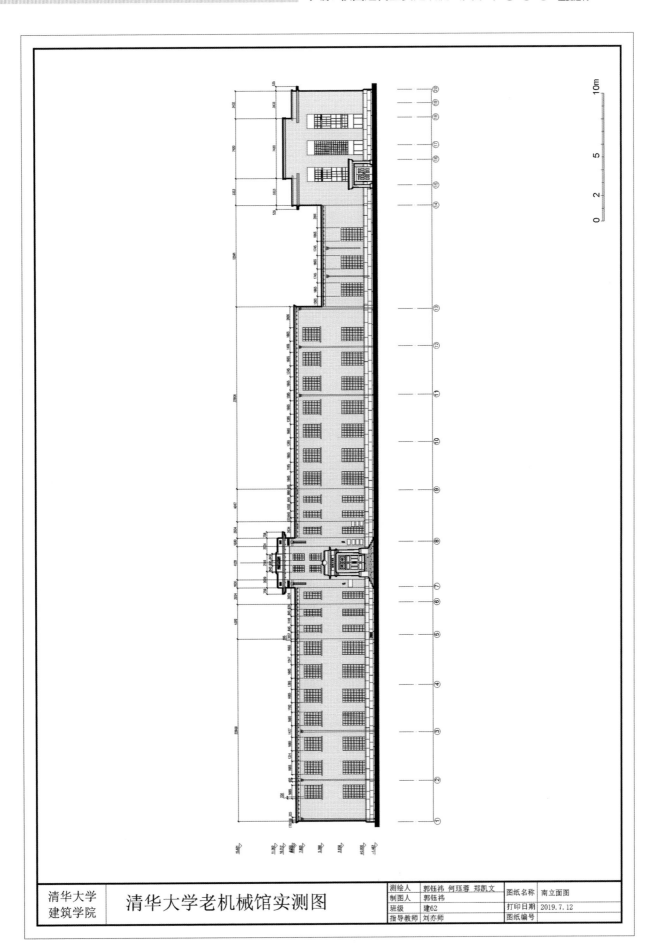

清华大学 建筑学院	清华大学老机械馆实测图	测绘人	郭钰祎 何珏蓉 郑凯文	图纸名称	南立面图
		制图人	郭钰祎		
		班级	建62	打印日期	2019.7.12
		指导教师	刘亦师	图纸编号	

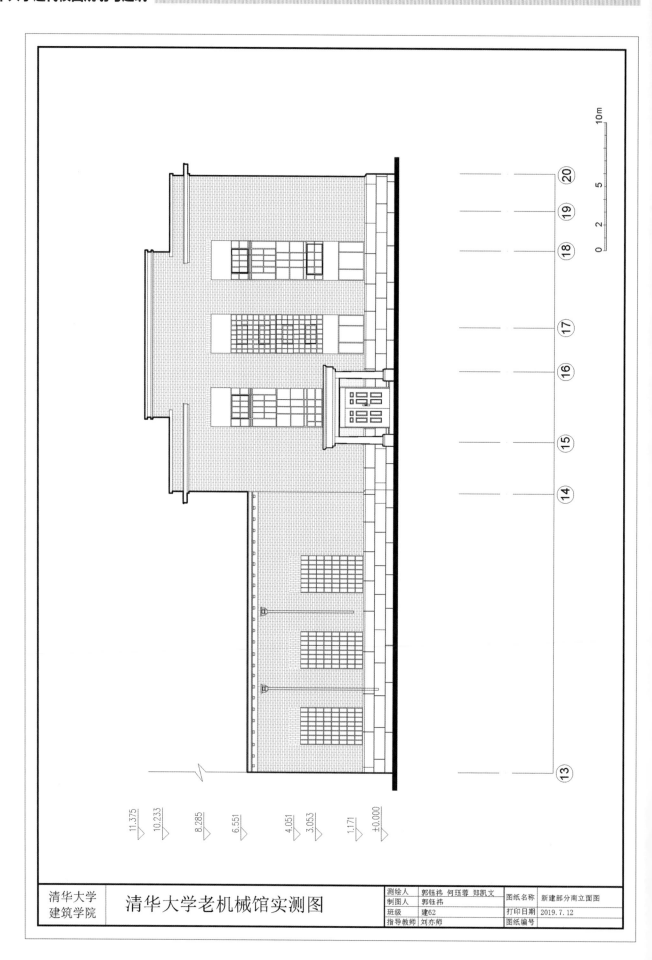

清华大学
建筑学院

清华大学老机械馆实测图

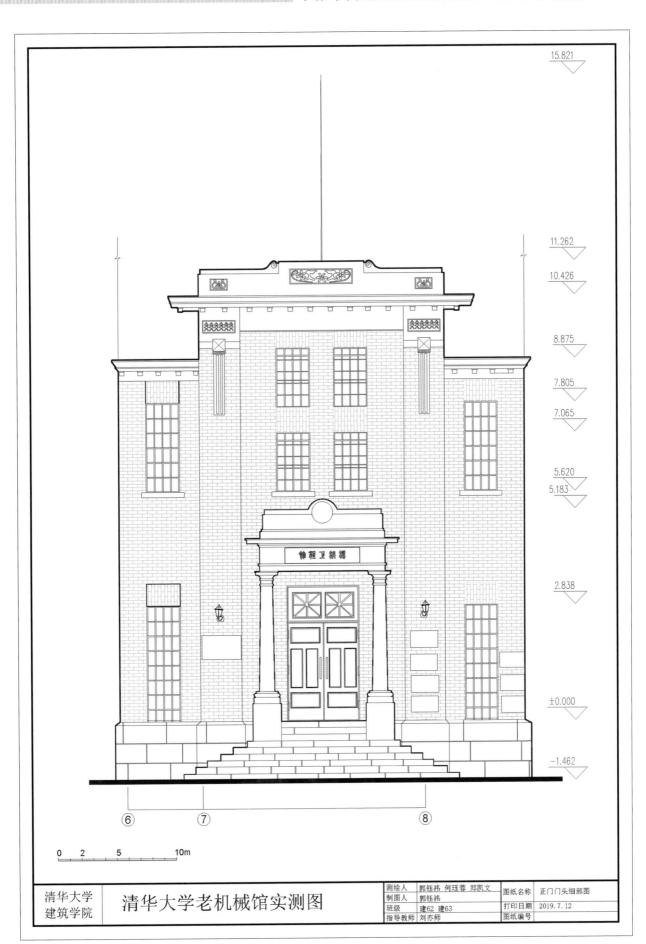

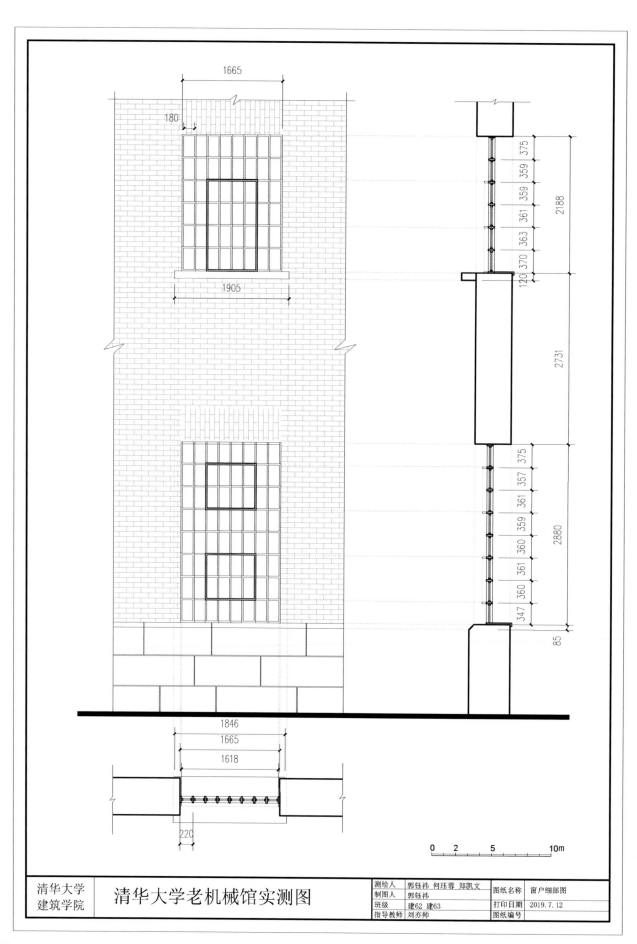

清华大学
建筑学院

清华大学老机械馆实测图

测绘人	郭钰祎 何珏蓉 郑凯文	图纸名称	窗户细部图
制图人	郭钰祎		
班级	建62 建63	打印日期	2019.7.12
指导教师	刘亦师	图纸编号	

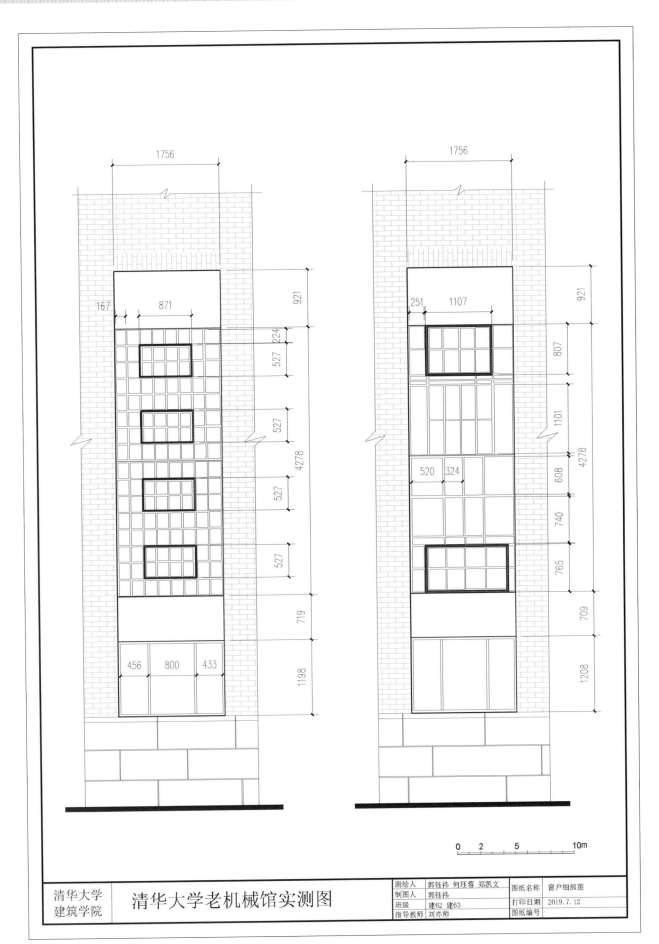

清华大学 建筑学院	**清华大学老机械馆实测图**	

测绘人	郭钰祎 何珏蓉 郑凯文	图纸名称	窗户细部图
制图人	郭钰祎		
班级	建62 建63	打印日期	2019.7.12
指导教师	刘亦师	图纸编号	

		测绘人	郭钰祜 何珏蓉 郑凯文	图纸名称	建筑轴测图
		制图人	何珏蓉		
		班级	建63	打印日期	2019.7.12
		指导教师	刘亦师	图纸编号	测绘人

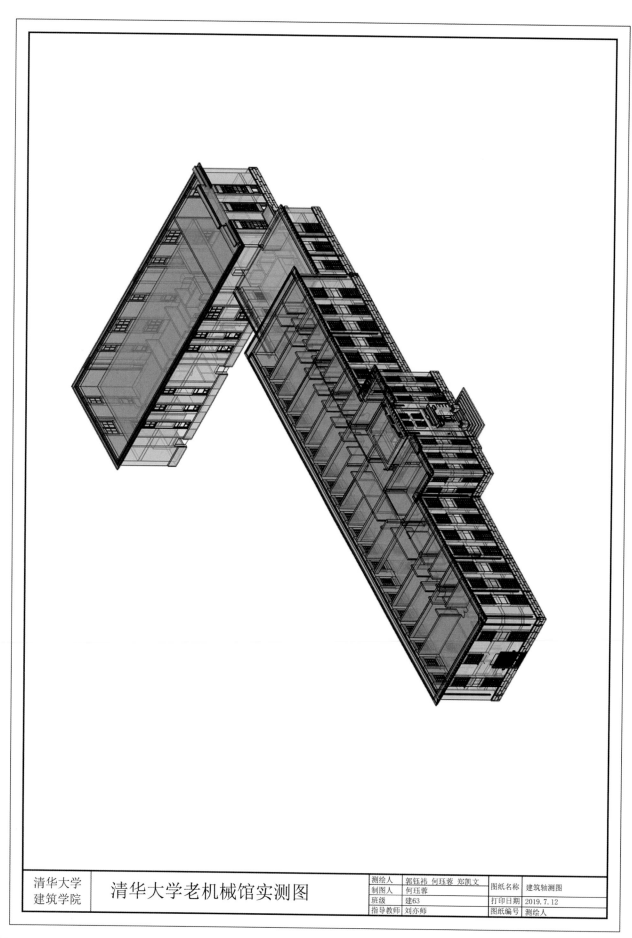

清华大学 建筑学院	清华大学老机械馆实测图	测绘人	郭钰祜 何珏蓉 郑凯文	图纸名称	建筑轴测图
		制图人	何珏蓉		
		班级	建63	打印日期	2019.7.12
		指导教师	刘亦师	图纸编号	测绘人

13. 老水利馆

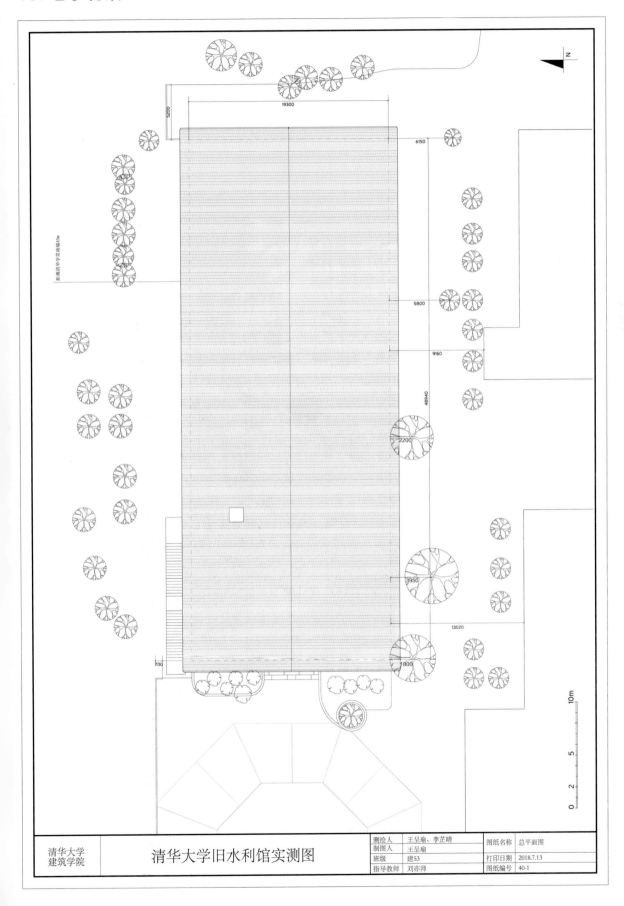

清华大学 建筑学院	清华大学旧水利馆实测图	测绘人	王呈瑜、李芷晴	图纸名称	总平面图
		制图人	王呈瑜		
		班级	建53	打印日期	2018.7.13
		指导教师	刘亦师	图纸编号	40-1

491

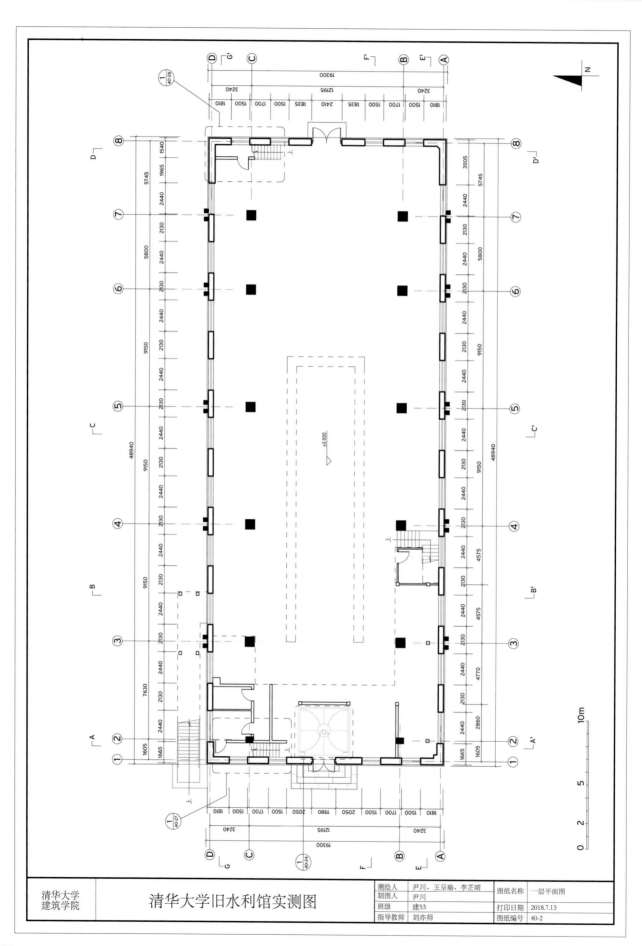

测绘人	尹川、王呈瑜、李芷晴	图纸名称	一层平面图
制图人	尹川		
班级	建53	打印日期	2018.7.13
指导教师	刘亦师	图纸编号	40-2

清华大学
建筑学院

清华大学旧水利馆实测图

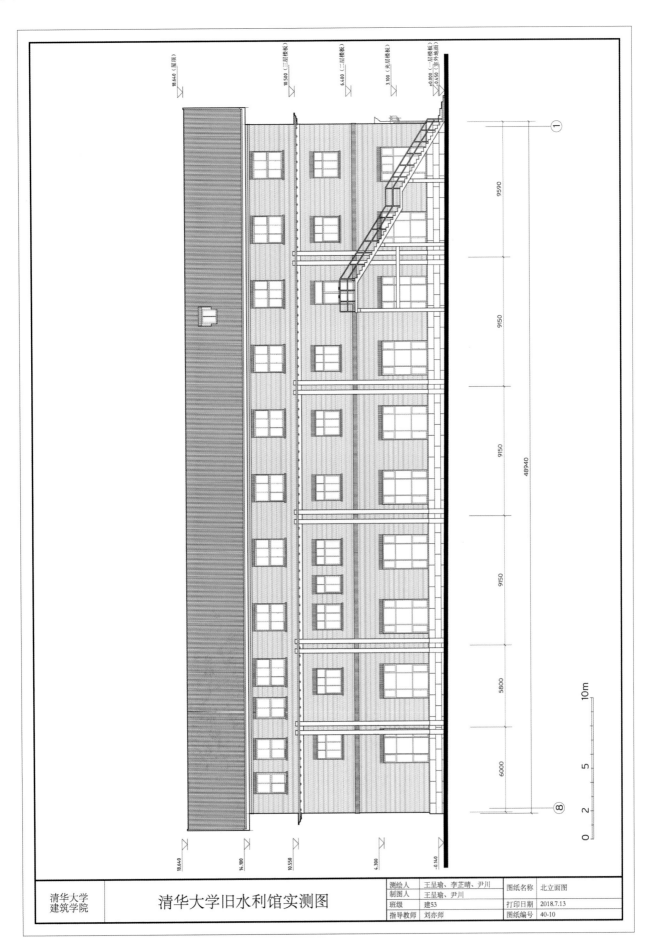

18.640（屋顶）

10.500（三层楼板）

6.400（二层楼板）

3.100（夹层楼板）

±0.000（一层楼板）
-0.450（室外地面）

①

9590

9150

48940

9150

9150

5800

6000

⑧

18.640

14.100

10.550

4.300

-0.140

10m

5

2

0

<table>
<tr><td rowspan="4">清华大学
建筑学院</td><td rowspan="4">清华大学旧水利馆实测图</td><td>测绘人</td><td>王呈瑜、李芷晴、尹川</td><td>图纸名称</td><td>北立面图</td></tr>
<tr><td>制图人</td><td>王呈瑜、尹川</td><td></td><td></td></tr>
<tr><td>班级</td><td>建53</td><td>打印日期</td><td>2018.7.13</td></tr>
<tr><td>指导教师</td><td>刘亦师</td><td>图纸编号</td><td>40-10</td></tr>
</table>

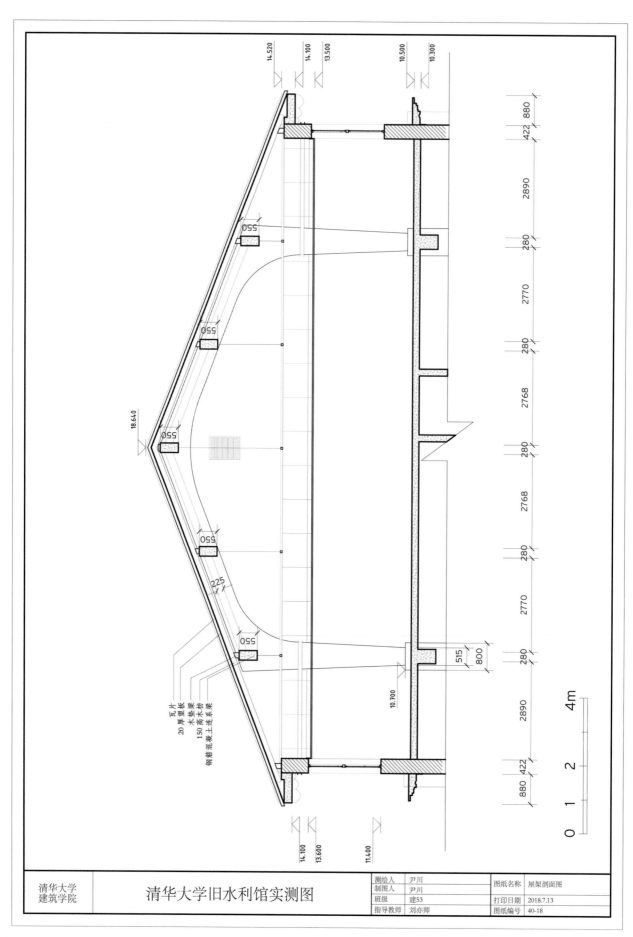

瓦片
20 厚望板
木垫条
150 南木坊梁
钢筋混凝土连系梁

清华大学
建筑学院

清华大学旧水利馆实测图

测绘人	尹川		图纸名称	屋架剖面图
制图人	尹川			
班级	建53		打印日期	2018.7.13
指导教师	刘亦师		图纸编号	40-18

494

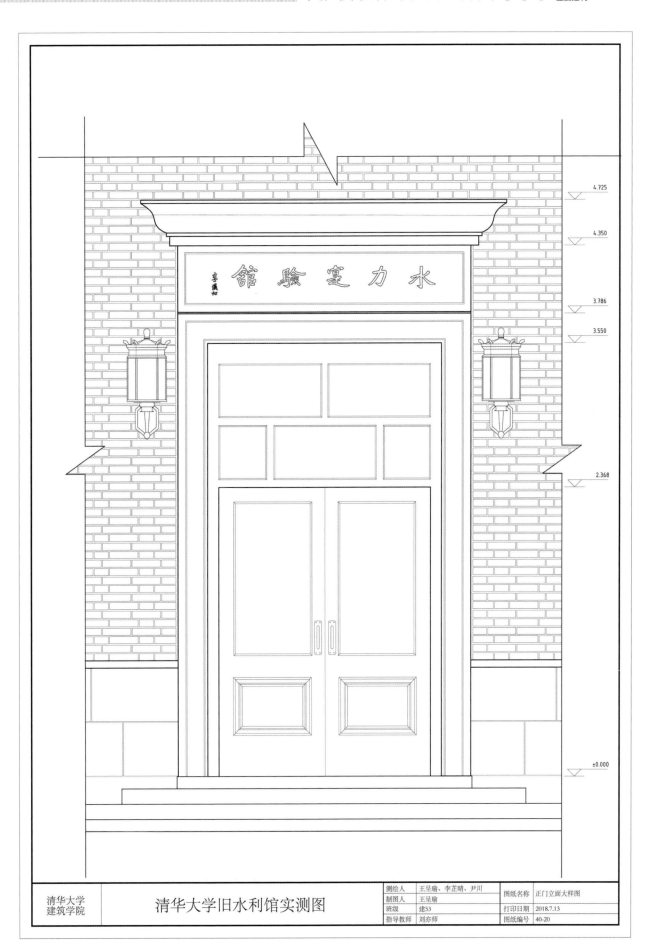

清华大学
建筑学院

清华大学旧水利馆实测图

测绘人	王呈瑜、李芷晴、尹川	图纸名称	正门立面大样图
制图人	王呈瑜		
班级	建53	打印日期	2018.7.13
指导教师	刘亦师	图纸编号	40-20

495

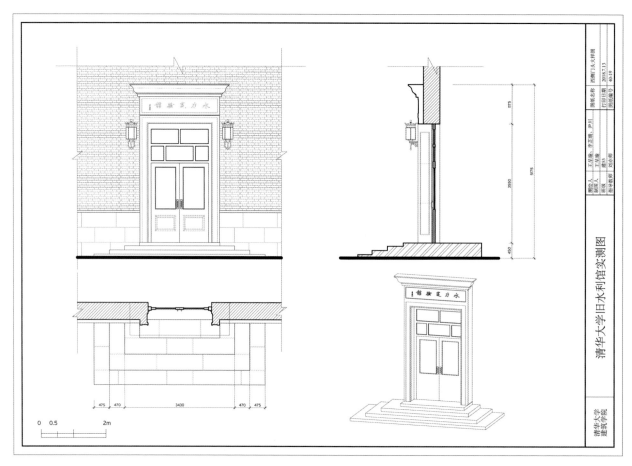

清华大学旧水利馆实测图

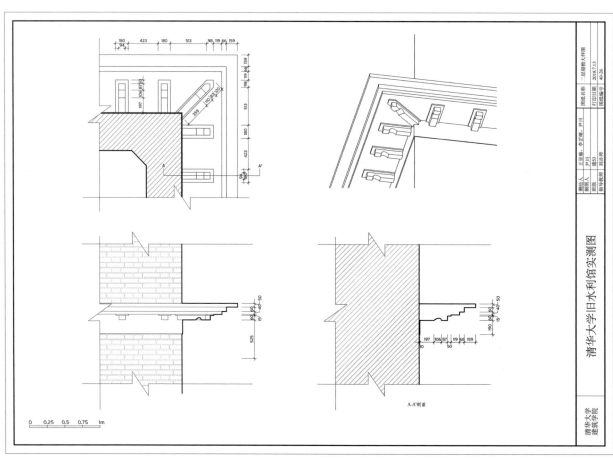

清华大学旧水利馆实测图

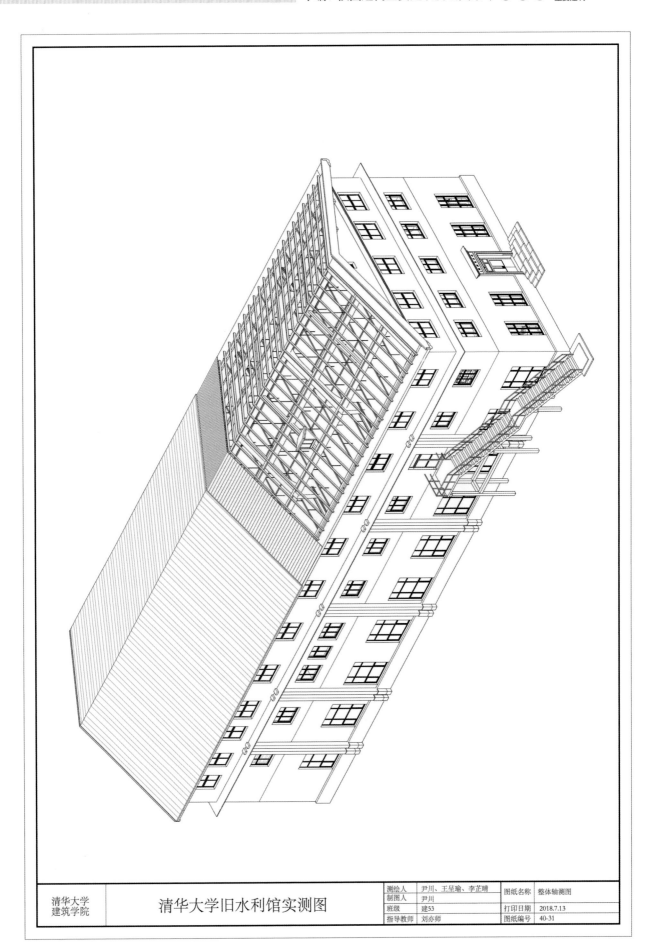

测绘人	尹川、王呈瑜、李芷晴		图纸名称	整体轴测图
制图人	尹川			
班级	建53		打印日期	2018.7.13
指导教师	刘亦师		图纸编号	40-31

清华大学建筑学院

清华大学旧水利馆实测图

测绘人	尹川、王呈瑜、李芷晴	图纸名称	整体结构分析图
制图人	尹川		
班级	建53	打印日期	2018.7.13
指导教师	刘亦师	图纸编号	40-32

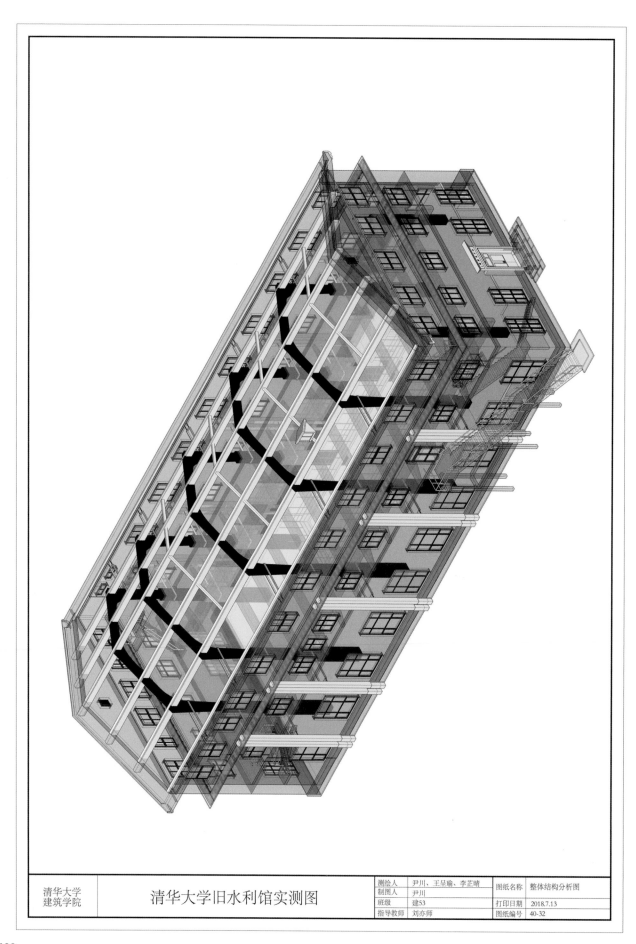

清华大学
建筑学院

清华大学旧水利馆实测图

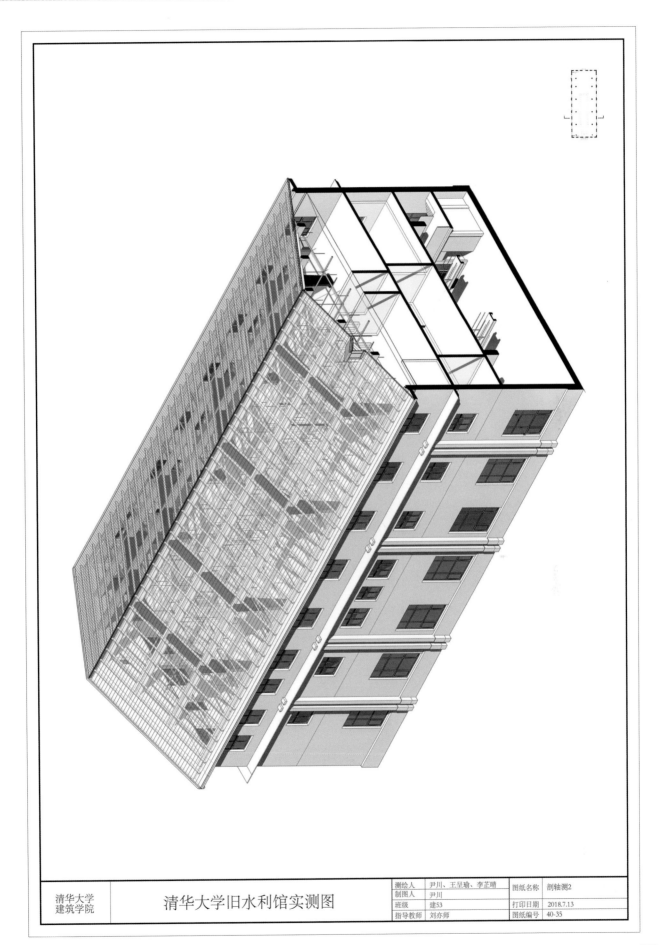

测绘人	尹川、王呈瑜、李芷晴	图纸名称	剖轴测2
制图人	尹川		
班级	建53	打印日期	2018.7.13
指导教师	刘亦师	图纸编号	40-35

清华大学
建筑学院

清华大学旧水利馆实测图

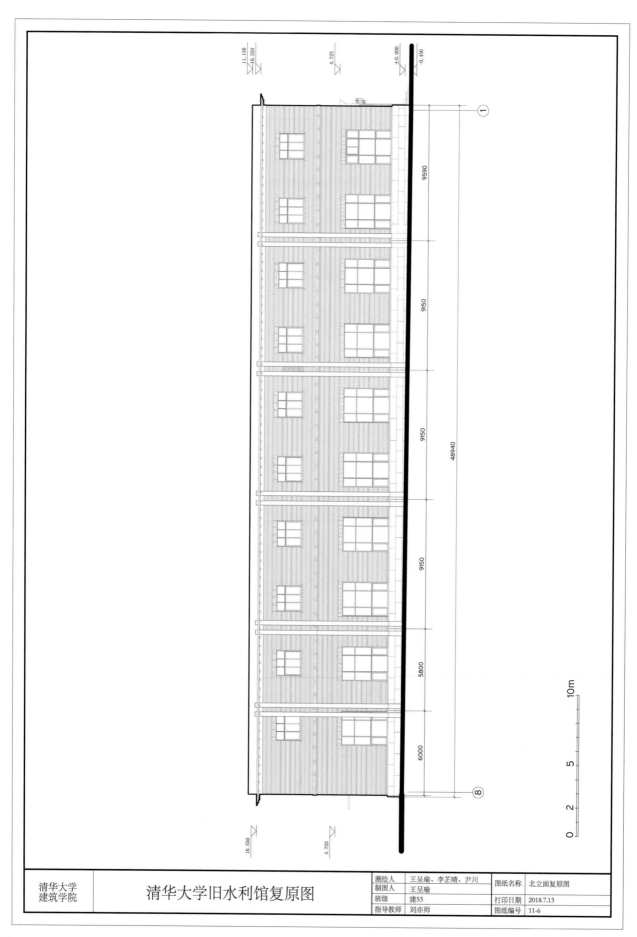

清华大学旧水利馆复原图

清华大学 建筑学院	清华大学旧水利馆复原图	测绘人	王呈瑜、李芷晴、尹川	图纸名称	北立面复原图
		制图人	王呈瑜		
		班级	建53	打印日期	2018.7.13
		指导教师	刘亦师	图纸编号	11-6

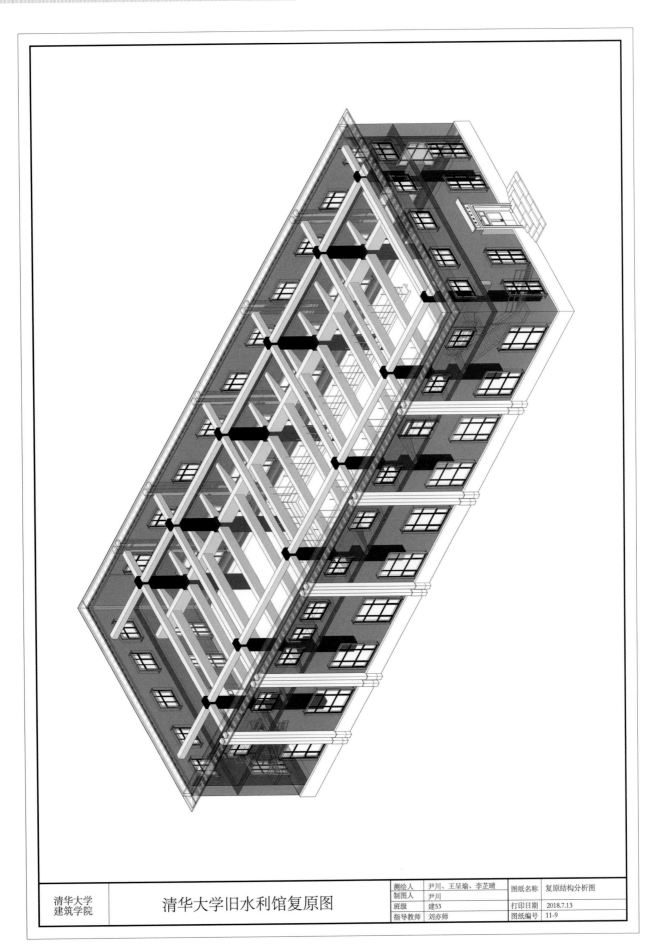

测绘人	尹川、王呈瑜、李芷晴	图纸名称	复原结构分析图
制图人	尹川		
清华大学 建筑学院	清华大学旧水利馆复原图	班级	建53	打印日期	2018.7.13
		指导教师	刘亦师	图纸编号	11-9

	测绘人	尹川、王呈瑜、李芷晴	图纸名称	复原轴测图（西北视角）
	制图人	尹川		

清华大学 建筑学院	清华大学旧水利馆复原图	测绘人	尹川、王呈瑜、李芷晴	图纸名称	复原轴测图（西北视角）
		制图人	尹川		
		班级	建53	打印日期	2018.7.13
		指导教师	刘亦师	图纸编号	11-7

14. 老电机馆

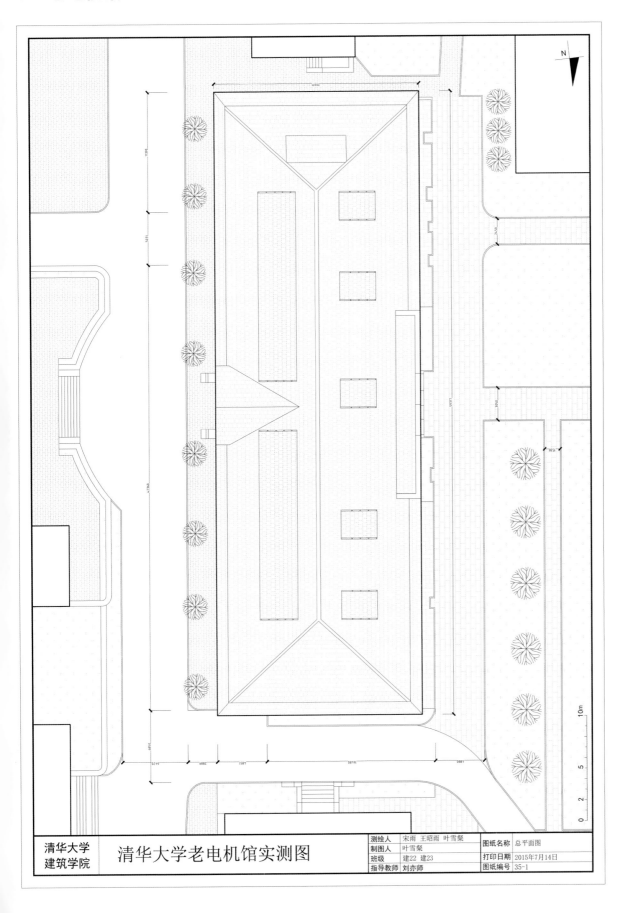

		测绘人	宋雨 王昭雨 叶雪粲	图纸名称	总平面图
清华大学 建筑学院	清华大学老电机馆实测图	制图人	叶雪粲		
		班级	建22 建23	打印日期	2015年7月14日
		指导教师	刘亦师	图纸编号	35-1

503

测绘人	王昭雨 宋雨 叶雪燊	图纸名称	一层平面图
制图人	王昭雨 宋雨		
班级	建22	打印日期	2015年7月14日
指导教师	刘亦师	图纸编号	35-2

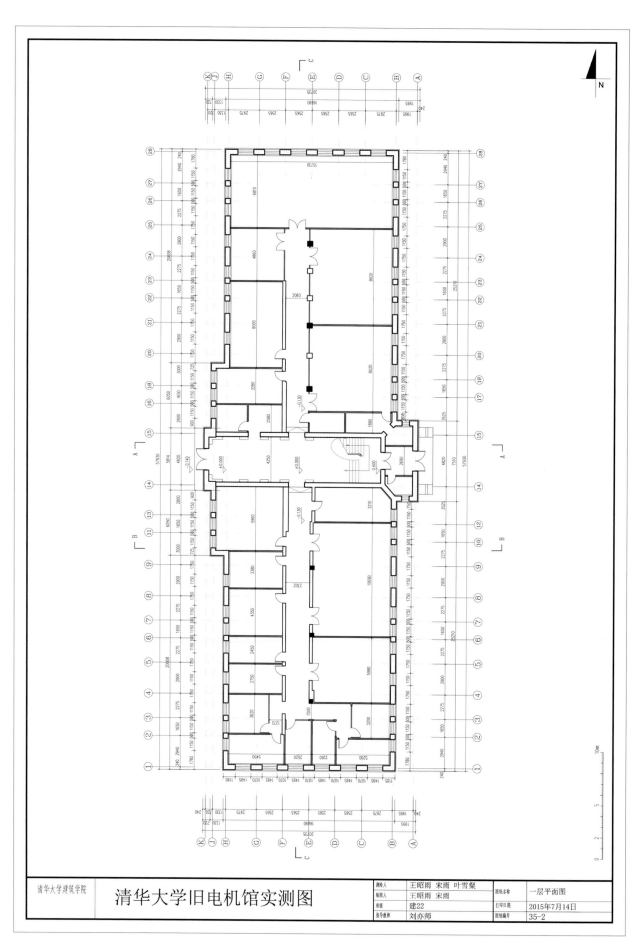

清华大学建筑学院　清华大学旧电机馆实测图

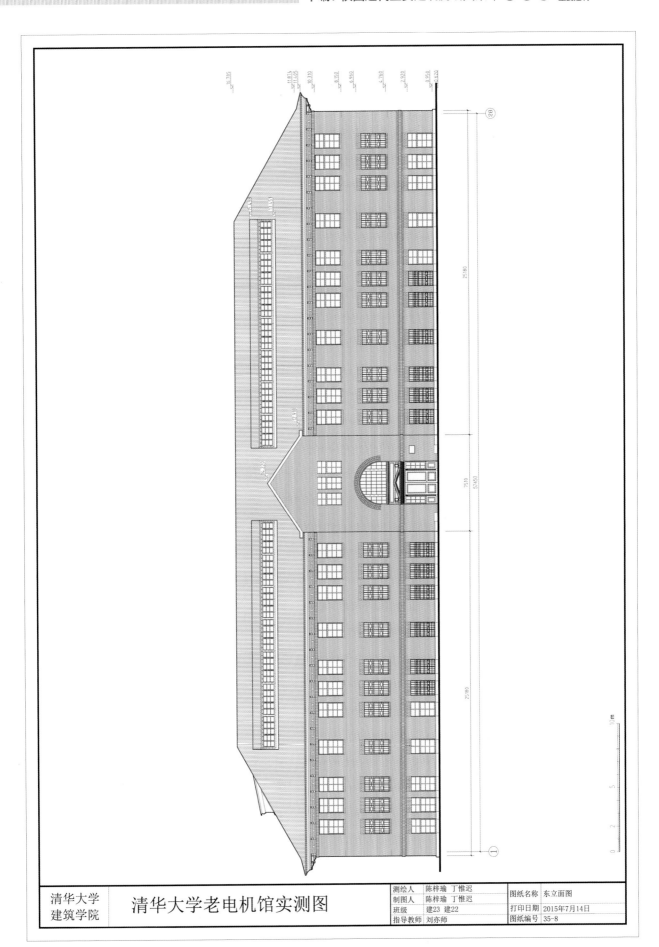

清华大学 建筑学院	清华大学老电机馆实测图	测绘人	陈梓瑜 丁惟迟	图纸名称	东立面图
		制图人	陈梓瑜 丁惟迟	打印日期	2015年7月14日
		班级	建23 建22		
		指导教师	刘亦师	图纸编号	35-8

505

电机工程馆

实业救命
学地不足

清华大学 建筑学院	清华大学老电机馆实测图	测绘人	陈梓瑜		图纸名称	电机馆东立面门大样图
		制图人	陈梓瑜			
		班级	建23		打印日期	2015年7月14日
		指导教师	刘亦师		图纸编号	35-23

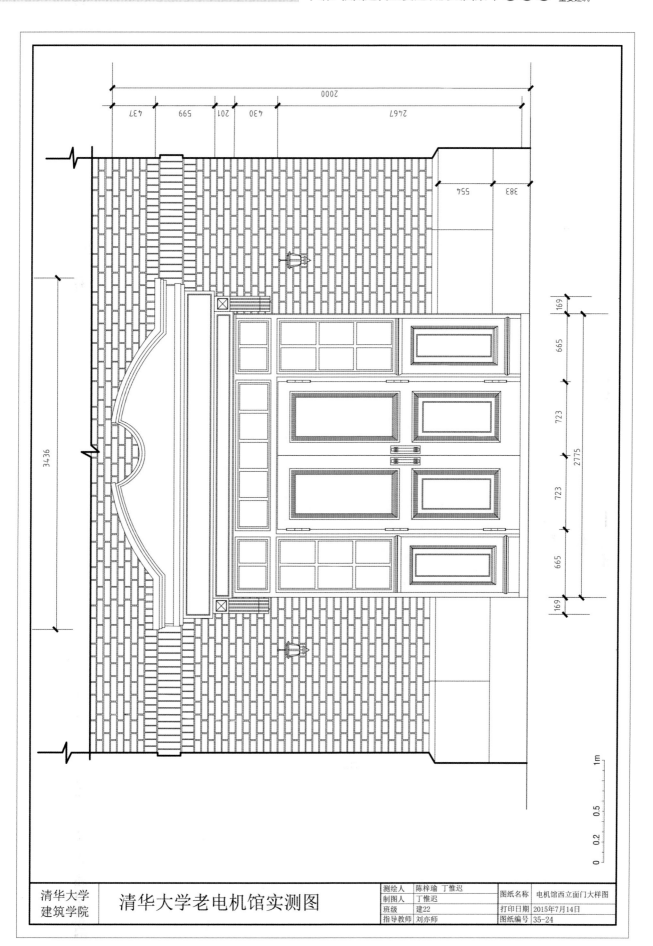

清华大学建筑学院	清华大学老电机馆实测图	测绘人	陈梓瑜 丁惟迟	图纸名称	电机馆西立面门大样图
		制图人	丁惟迟		
		班级	建22	打印日期	2015年7月14日
		指导教师	刘亦师	图纸编号	35-24

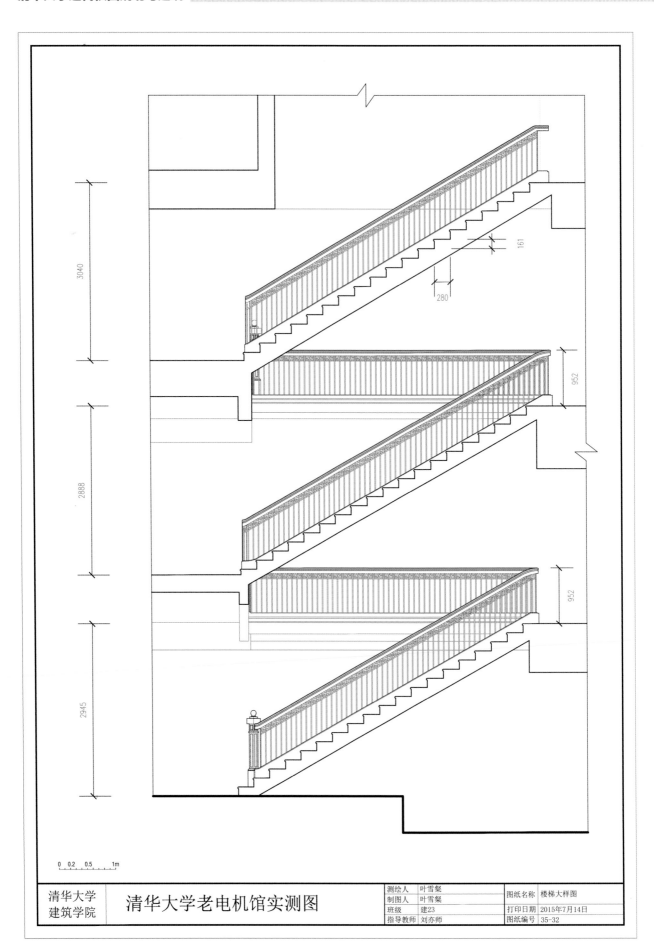

	测绘人	叶雪粲	图纸名称	楼梯大样图
清华大学 建筑学院	制图人	叶雪粲		
	班级	建23	打印日期	2015年7月14日
	指导教师	刘亦师	图纸编号	35-32

清华大学老电机馆实测图

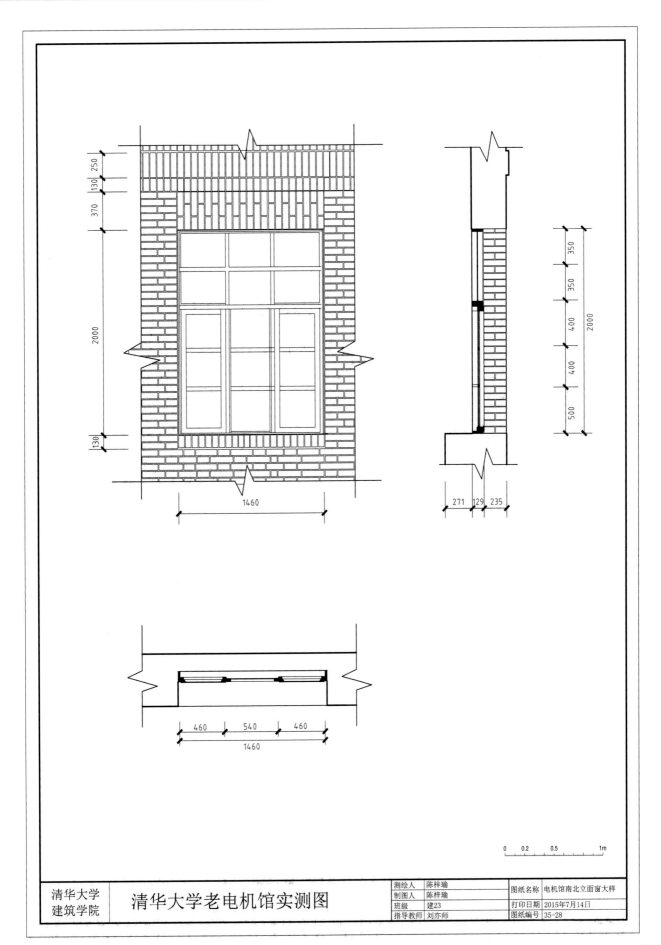

测绘人	陈梓瑜		图纸名称	电机馆南北立面窗大样
制图人	陈梓瑜			
班级	建23		打印日期	2015年7月14日
指导教师	刘亦师		图纸编号	35-28

清华大学
建筑学院

清华大学老电机馆实测图

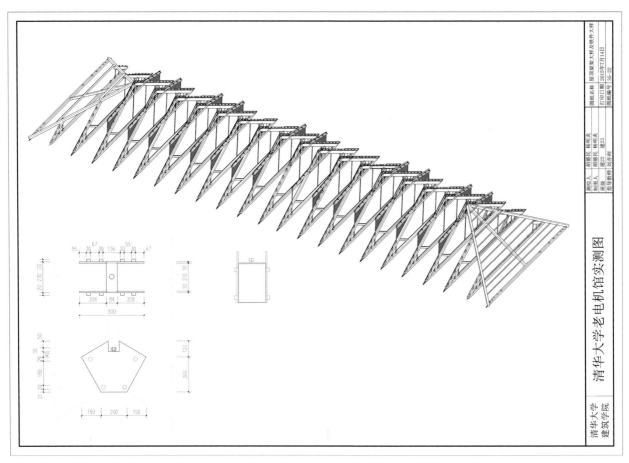

特殊砌砖法与牛腿立面

45度角屋檐处牛腿立面

牛腿大样图

长牛腿大样图

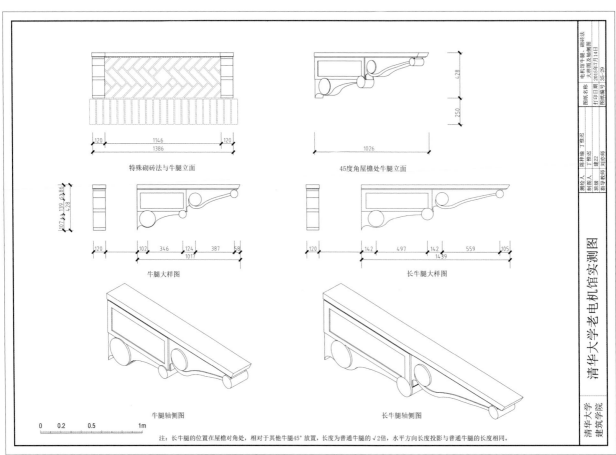

牛腿轴侧图

长牛腿轴侧图

0 0.2 0.5 1m

注：长牛腿的位置在屋檐对角处，相对于其他牛腿45°放置，长度为普通牛腿的√2倍，水平方向长度投影与普通牛腿的长度相同。

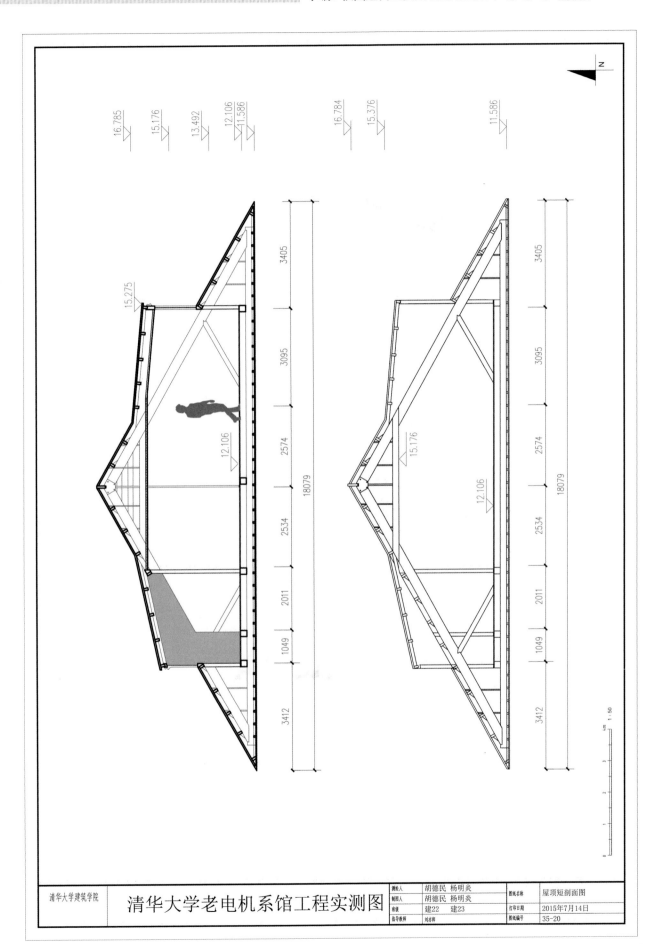

清华大学建筑学院	清华大学老电机系馆工程实测图				
		测绘人	胡德民 杨明炎	图纸名称	屋顶短剖面图
		制图人	胡德民 杨明炎		
		班级	建22 建23	打印日期	2015年7月14日
		指导教师	刘志辉	图纸编号	35-20

	清华大学建筑学院	清华大学老电机馆实测图	测绘人	胡德民　杨明炎	图纸名称	建筑轴测图
			制图人	胡德民　杨明炎		
			班级	建22　建23	打印日期	2015年7月14日
			指导教师	刘亦师	图纸编号	

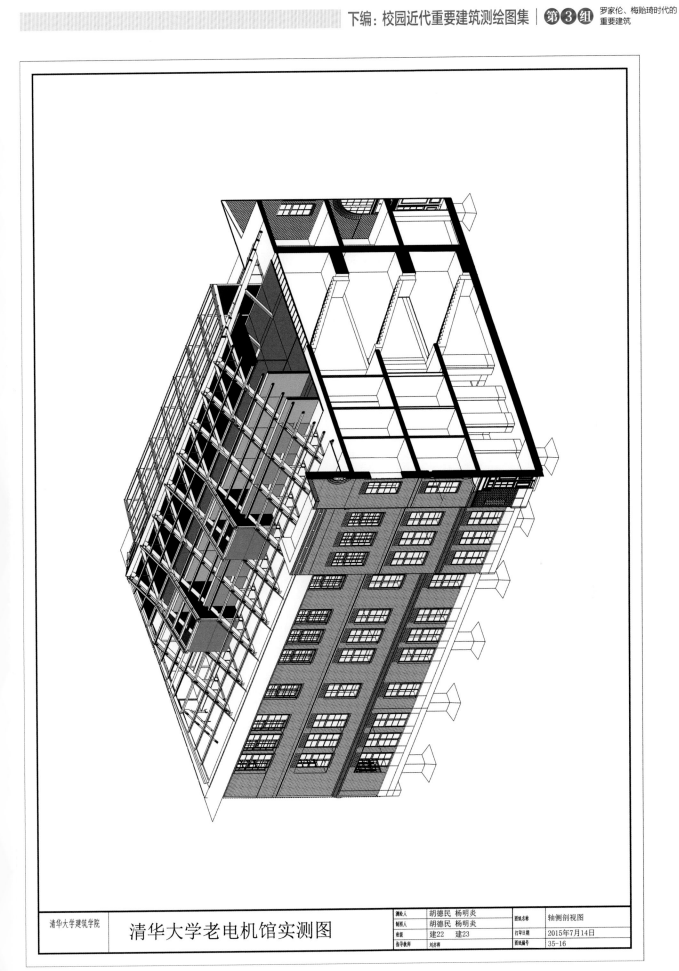

测绘人	胡德民 杨明炎	图纸名称	轴侧剖视图
制图人	胡德民 杨明炎		
班级	建22 建23	打印日期	2015年7月14日
指导教师	刘志勇	图纸编号	35-16

清华大学建筑学院

清华大学老电机馆实测图

15. 老化学馆

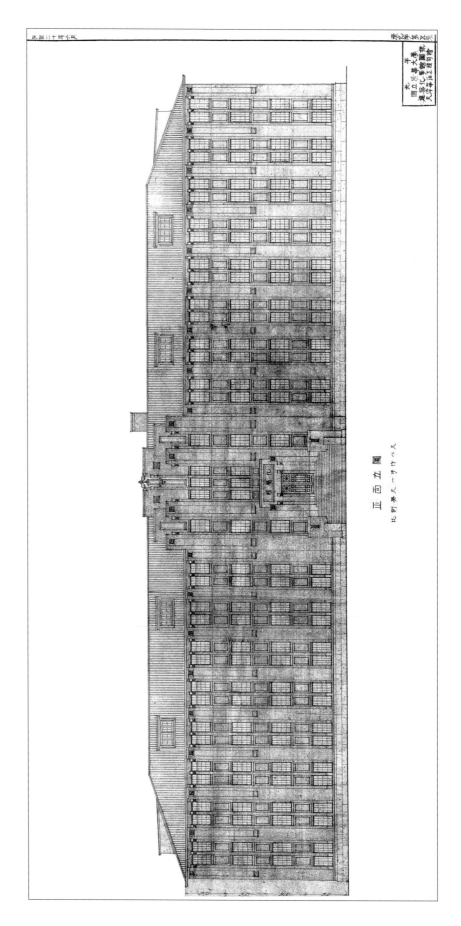

沈理源之化学馆立面设计图

来源：清华大学档案馆

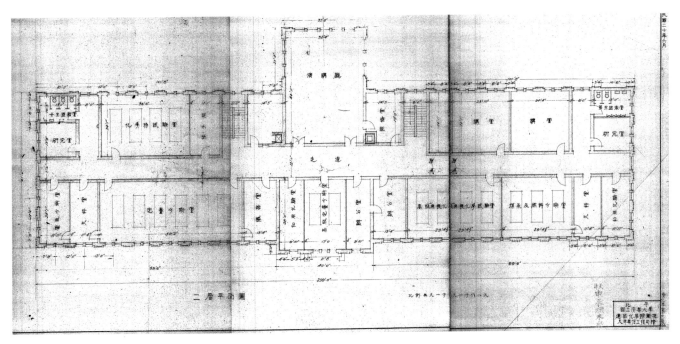

沈理源之化学馆平面设计图
来源：清华大学档案馆

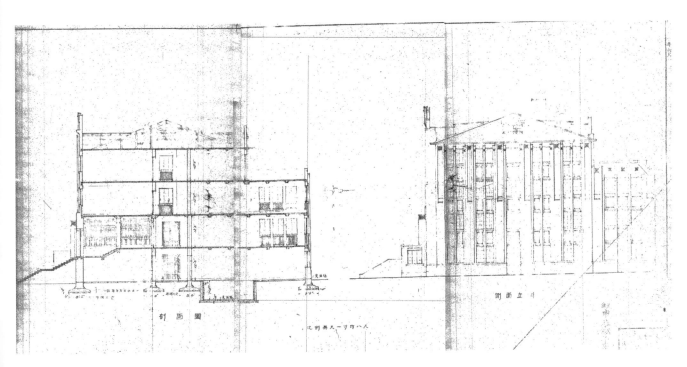

沈理源之化学馆剖面设计图
来源：清华大学档案馆

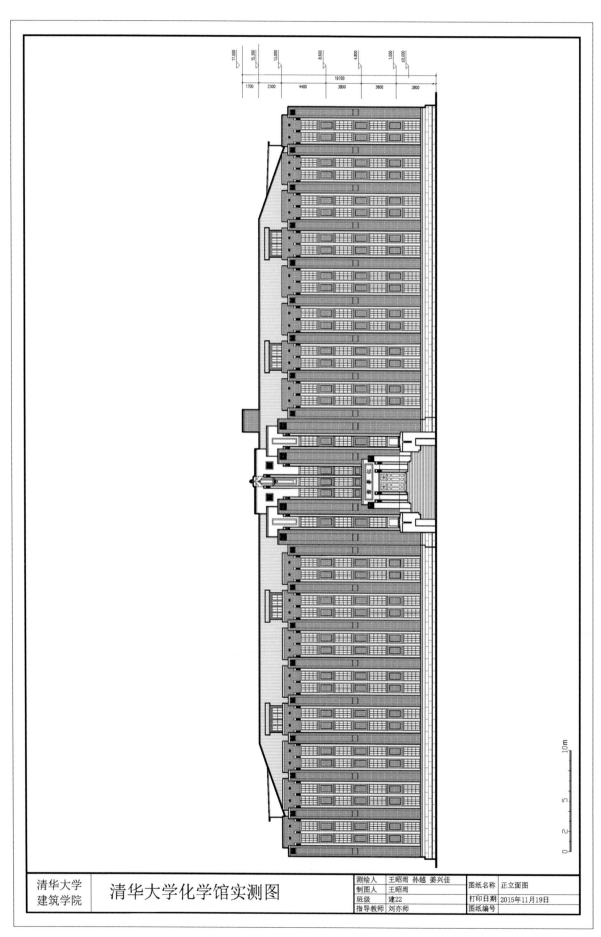

测绘人	王昭雨 孙越 姜兴佳	图纸名称	正立面图
制图人	王昭雨		
班级	建22	打印日期	2015年11月19日
指导教师	刘亦师	图纸编号	

清华大学
建筑学院

清华大学化学馆实测图

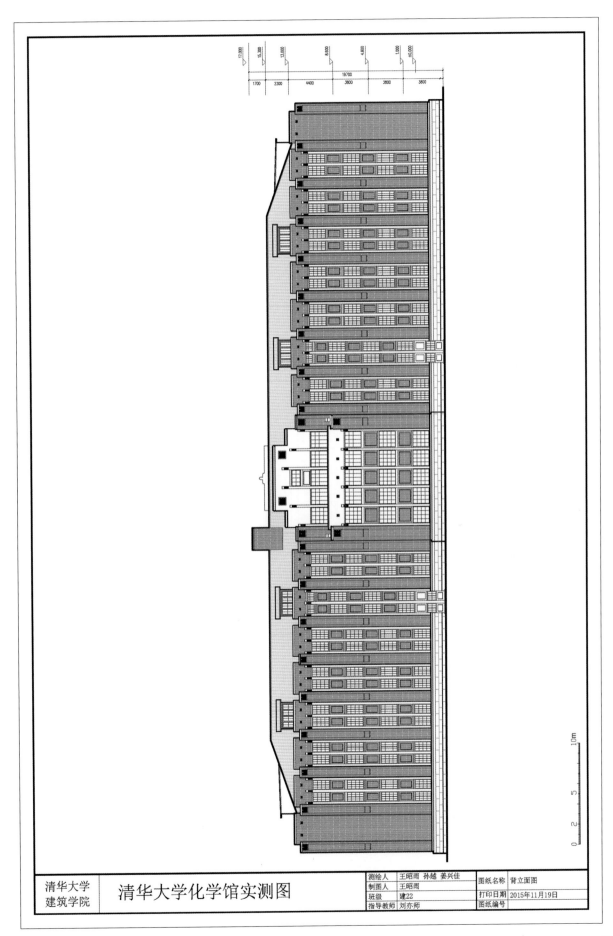

清华大学 建筑学院	清华大学化学馆实测图	测绘人	王昭雨 孙越 姜兴佳	图纸名称	背立面图	
		制图人	王昭雨	打印日期	2015年11月19日	
		班级	建22	图纸编号		
		指导教师	刘亦师			

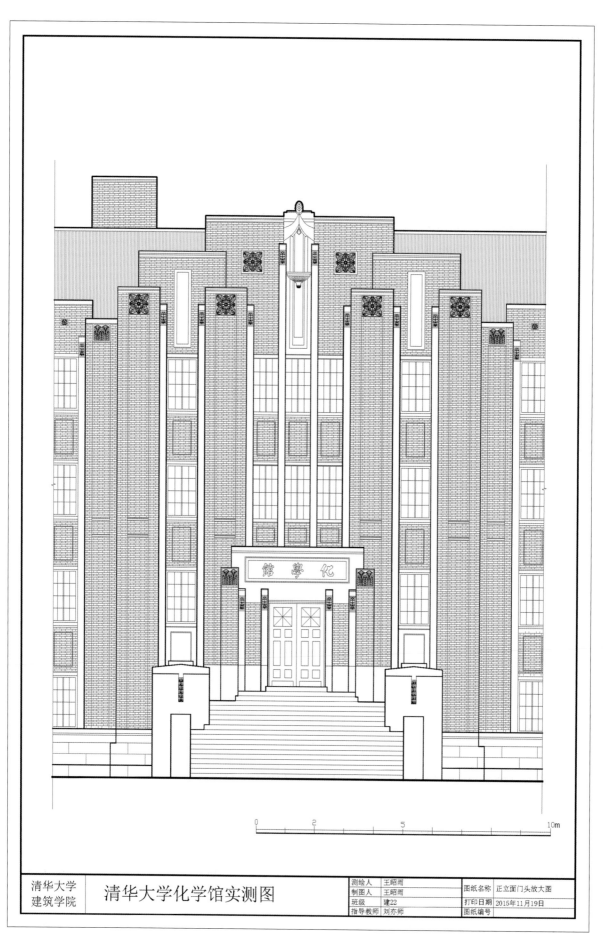

化學館

清华大学 建筑学院	清华大学化学馆实测图	测绘人	王昭雨	图纸名称	正立面门头放大图
		制图人	王昭雨		
		班级	建22	打印日期	2015年11月19日
		指导教师	刘亦师	图纸编号	

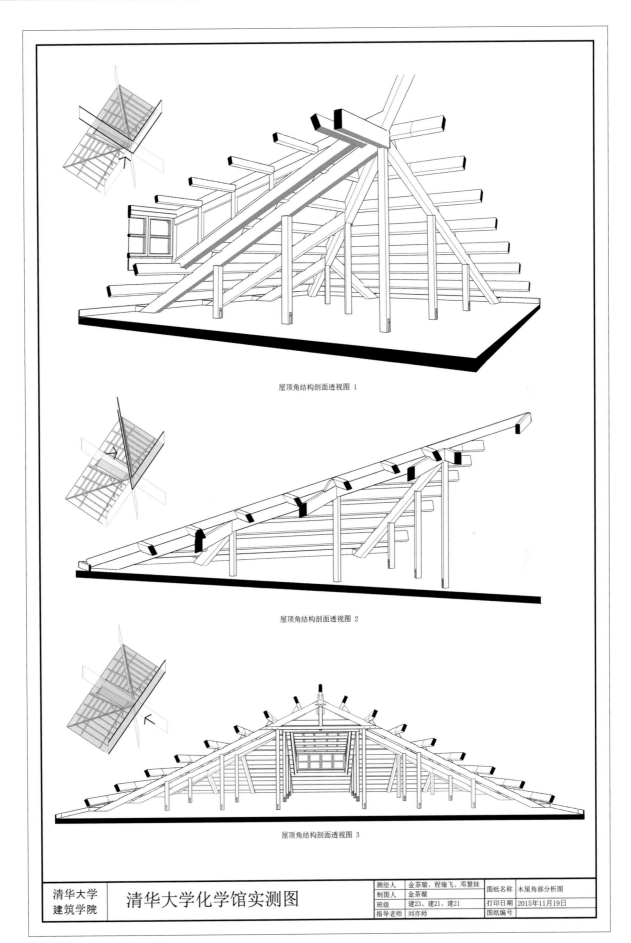

屋顶角结构剖面透视图 1

屋顶角结构剖面透视图 2

屋顶角结构剖面透视图 3

清华大学化学馆实测图	测绘人	金茶璇、程瑜飞、邓慧妹	图纸名称	木屋角部分析图
	制图人	金茶璇		
	班级	建23、建21、建21	打印日期	2015年11月19日
	指导老师	刘亦师	图纸编号	

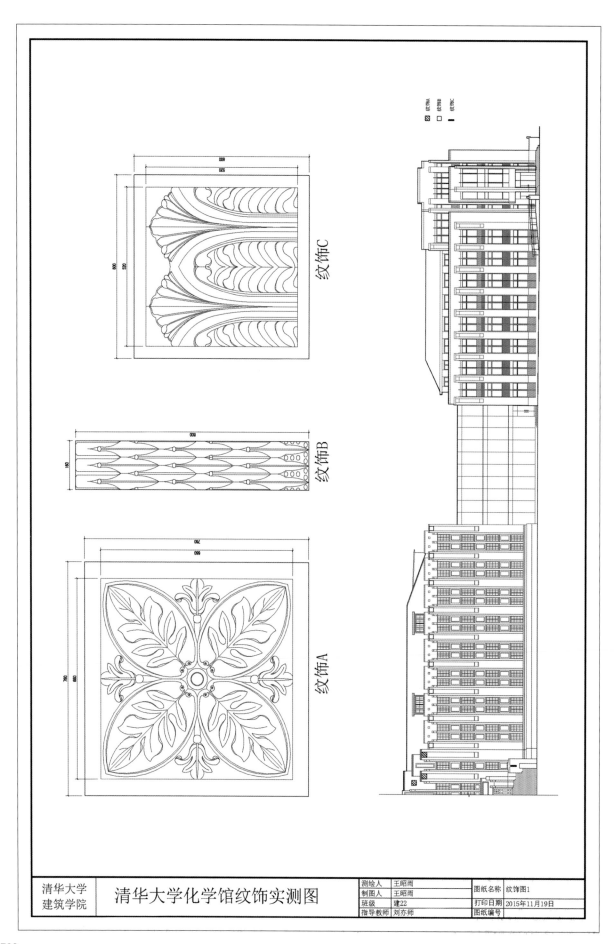

纹饰C

纹饰B

纹饰A

清华大学 建筑学院	清华大学化学馆纹饰实测图	测绘人	王昭雨	图纸名称	纹饰图1
		制图人	王昭雨		
		班级	建22	打印日期	2015年11月19日
		指导教师	刘亦师	图纸编号	

第④组　学生宿舍

16. 明斋

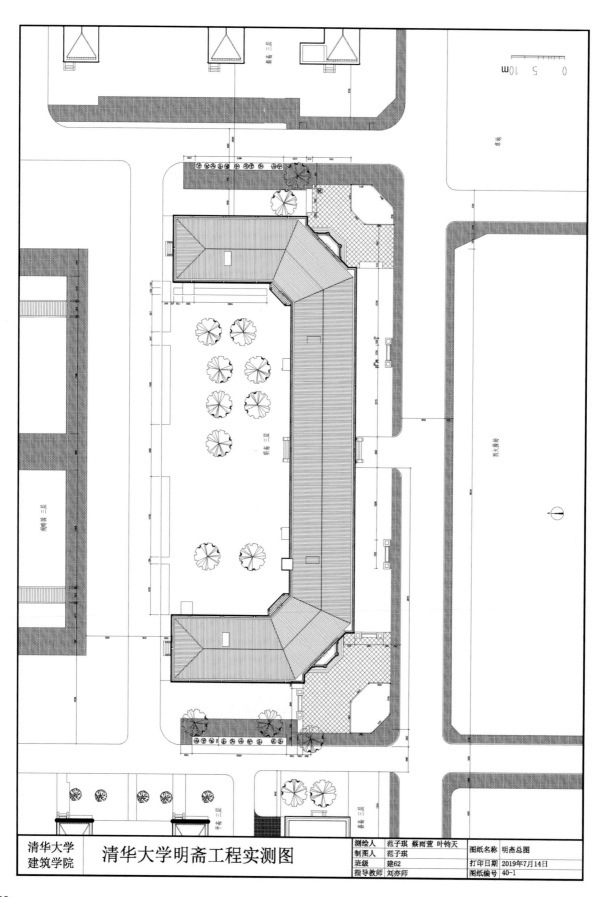

清华大学 建筑学院	清华大学明斋工程实测图	测绘人	范子琪 蔡雨萱 叶钧天	图纸名称	明斋总图
		制图人	范子琪		
		班级	建62	打印日期	2019年7月14日
		指导教师	刘亦师	图纸编号	40-1

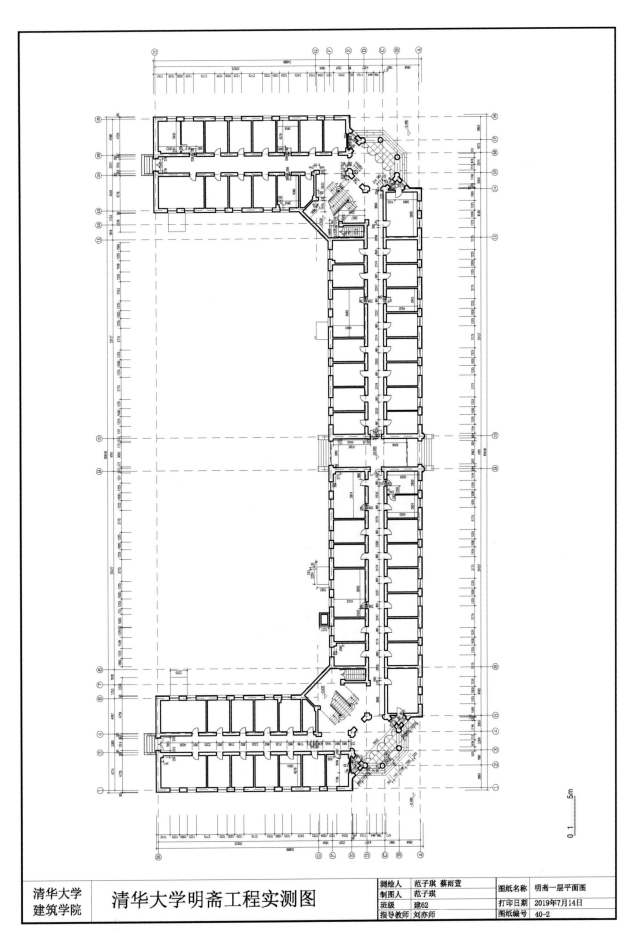

0 1 5m

清华大学
建筑学院

清华大学明斋工程实测图

测绘人	范子琪 蔡雨萱	图纸名称	明斋一层平面图
制图人	范子琪		
班级	建62	打印日期	2019年7月14日
指导教师	刘亦师	图纸编号	40-2

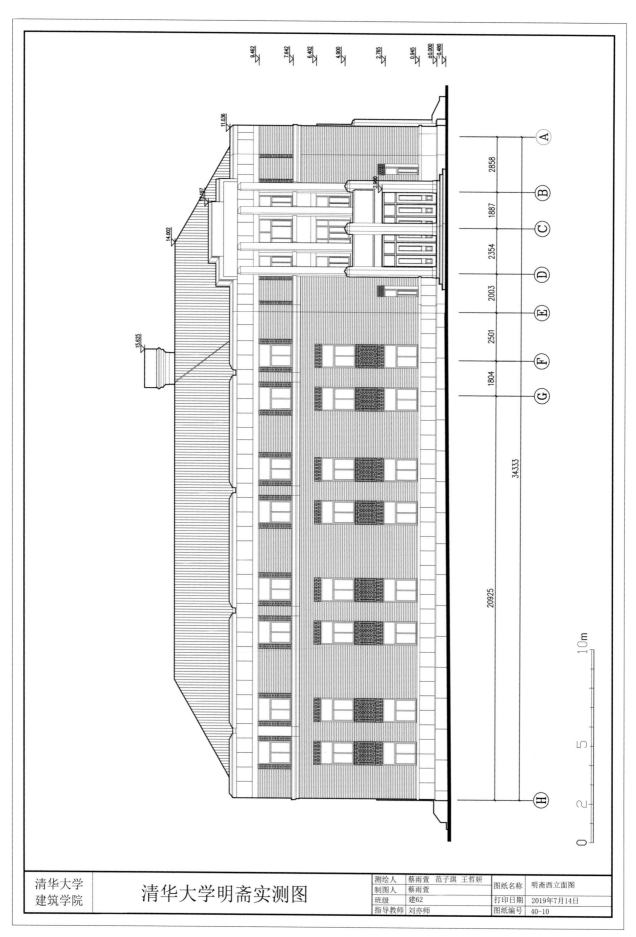

测绘人	蔡雨萱 范子琪 王哲妍	图纸名称	明斋西立面图
制图人	蔡雨萱		
班级	建62	打印日期	2019年7月14日
指导教师	刘亦师	图纸编号	40-10

清华大学
建筑学院

清华大学明斋实测图

14.000

11.036

9.462

7.642

6.443

4.253

2.700

2.425

0.980

±0.000

-0.245

-0.657

2404　220　577　225　　2960　　1202　　2385

测绘人	王哲妍 叶钧天	图纸名称	明斋南立面门头大样
制图人	王哲妍 范子琪		
班级	建62	打印日期	2019年7月14日
指导教师	刘亦师	图纸编号	40-17

清华大学
建筑学院

清华大学明斋实测图

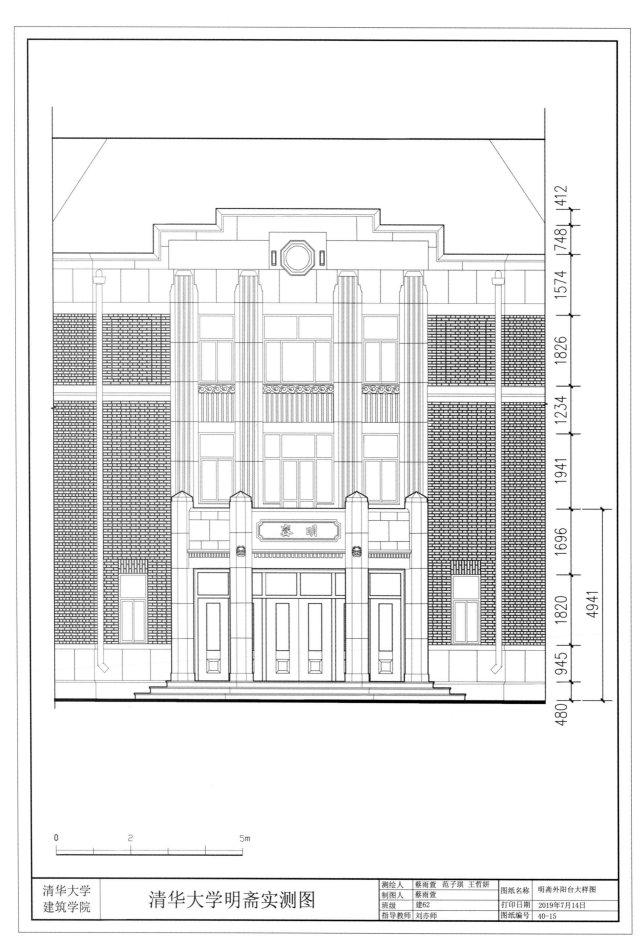

0   2   5m

清华大学 建筑学院	清华大学明斋实测图	测绘人	蔡雨萱 范子琪 王哲妍	图纸名称	明斋外阳台大样图
		制图人	蔡雨萱		
		班级	建62	打印日期	2019年7月14日
		指导教师	刘亦师	图纸编号	40-15

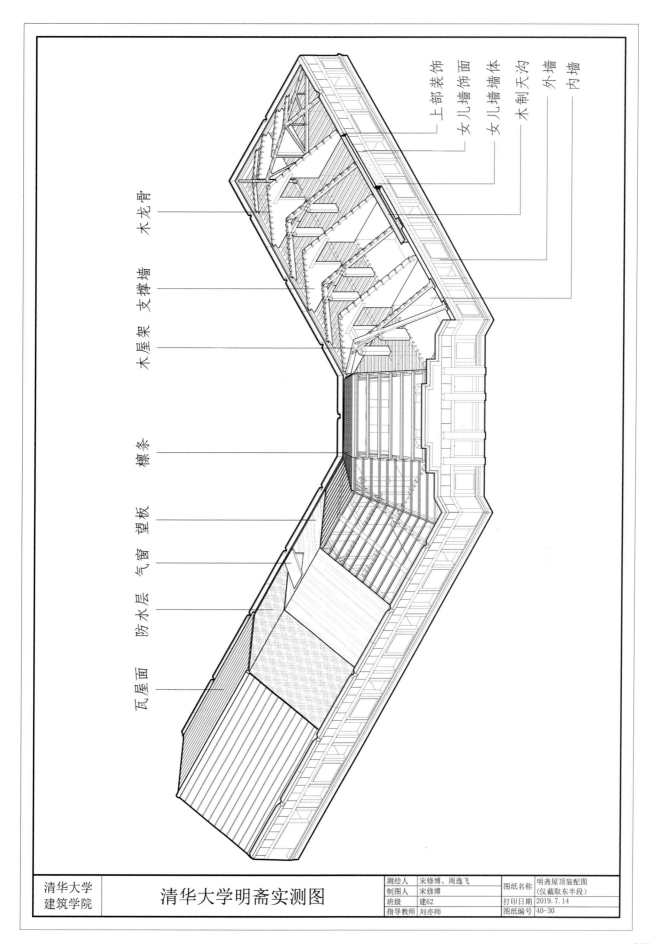

上部装饰
女儿墙饰面
女儿墙墙体
木制天沟
外墙
内墙

木龙骨

支撑墙

木屋架

檩条

望板

气窗

防水层

瓦屋面

清华大学
建筑学院

清华大学明斋实测图

测绘人	宋修博、周逸飞	图纸名称	明斋屋顶装配图
制图人	宋修博		（仅截取东半段）
班级	建62	打印日期	2019.7.14
指导教师	刘亦师	图纸编号	40-30

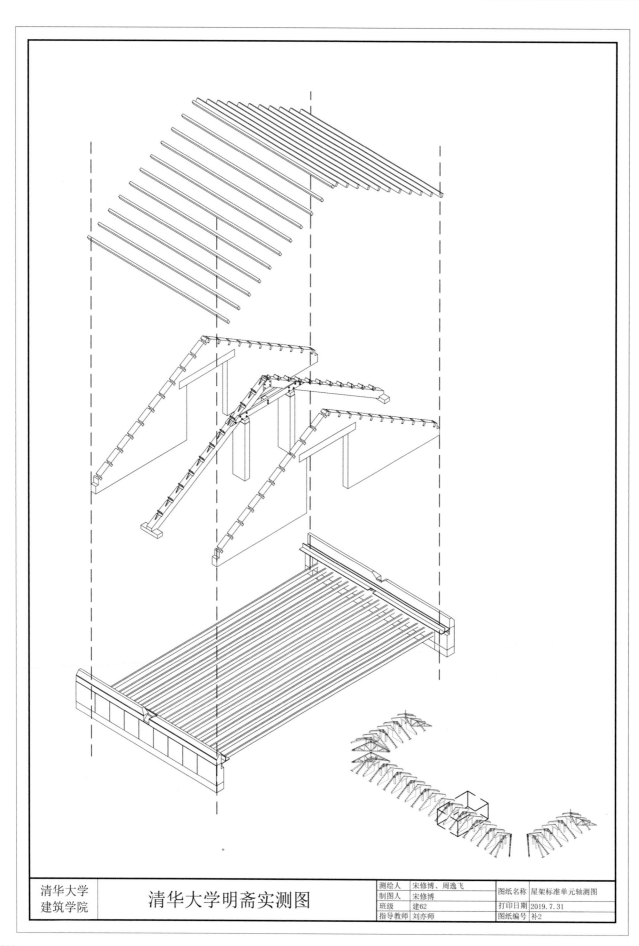

清华大学
建筑学院

清华大学明斋实测图

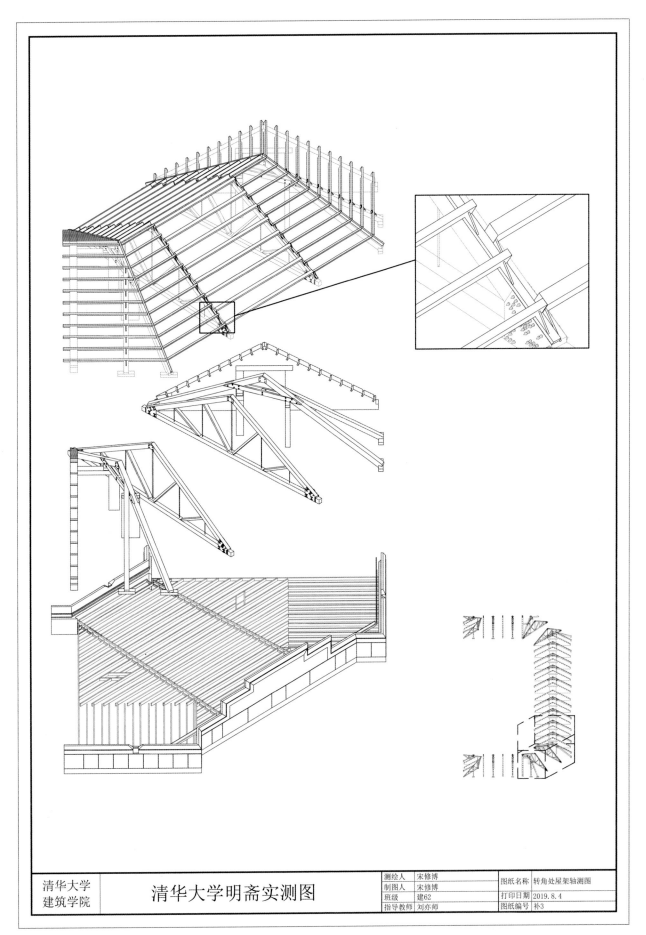

测绘人	宋修博		图纸名称	转角处屋架轴测图
制图人	宋修博			
班级	建62		打印日期	2019.8.4
指导教师	刘亦师		图纸编号	补3

清华大学
建筑学院

清华大学明斋实测图

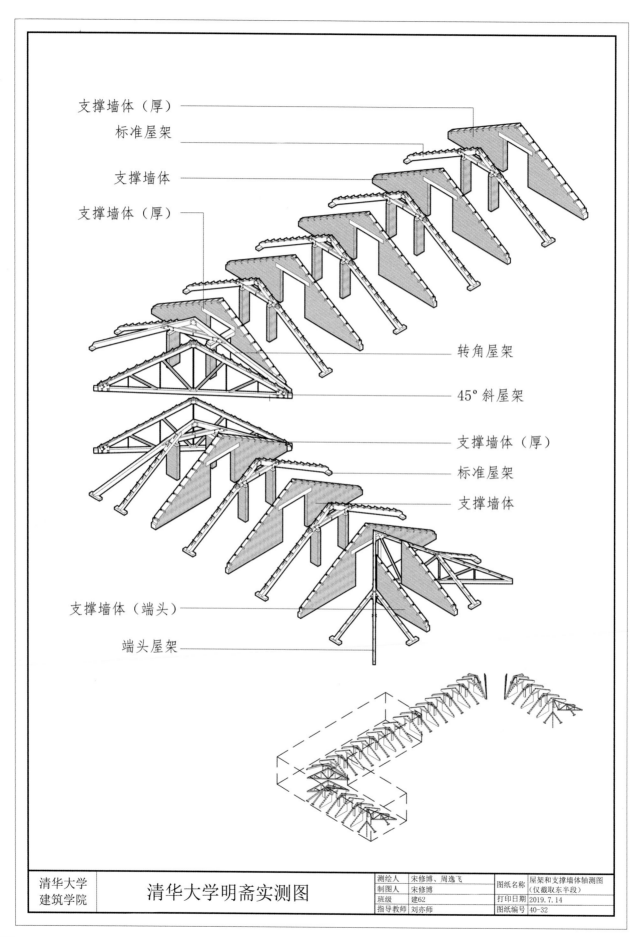

支撑墙体（厚）

标准屋架

支撑墙体

支撑墙体（厚）

转角屋架

45°斜屋架

支撑墙体（厚）

标准屋架

支撑墙体

支撑墙体（端头）

端头屋架

清华大学 建筑学院	清华大学明斋实测图	测绘人	宋修博、周逸飞	图纸名称	屋架和支撑墙体轴测图 （仅截取东半段）
		制图人	宋修博		
		班级	建62	打印日期	2019.7.14
		指导教师	刘亦师	图纸编号	40-32

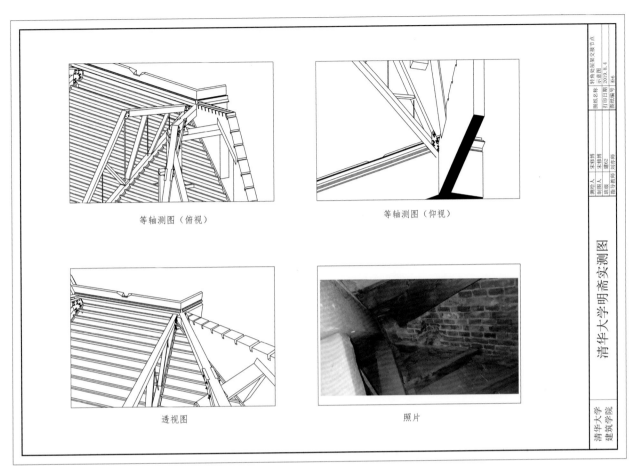

等轴测图（俯视）　　等轴测图（仰视）

透视图　　照片

图纸名称　转角处檩架交接节点 示意图
打印日期　2019.8.4
图纸编号　W6

测绘人　宋修博
制图人　宋修博
班级　建62
指导教师　刘亦师

清华大学明斋实测图

清华大学
建筑学院

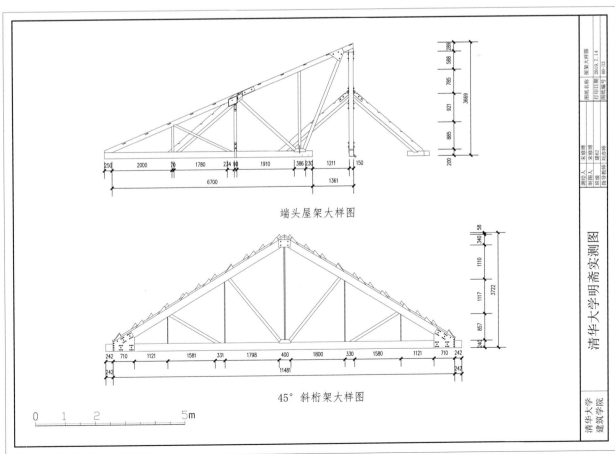

端头屋架大样图

45° 斜桁架大样图

图纸名称　屋架大样图
打印日期　2019.7.14
图纸编号　40-33

测绘人　宋修博
制图人　宋修博
班级　建62
指导教师　刘亦师

清华大学明斋实测图

0　1　2　　　　　5m

清华大学
建筑学院

测图人	宋修博	图纸名称	明斋整体轴测图
制图人	宋修博		（去除屋面）
班级	建62	打印日期	2019.7.31
指导教师	刘亦师	图纸编号	补1

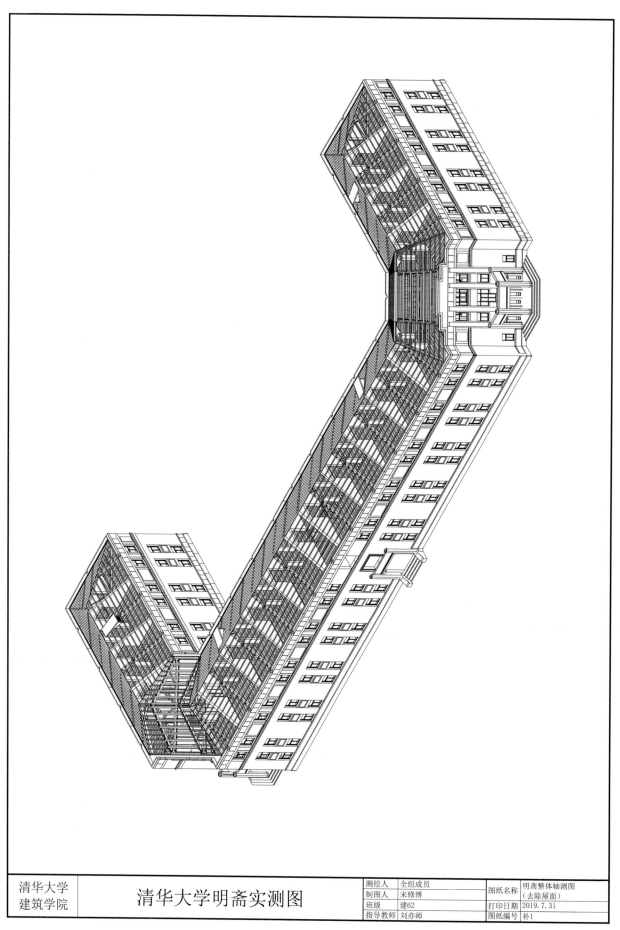

清华大学 建筑学院	清华大学明斋实测图	测绘人	全组成员	图纸名称	明斋整体轴测图
		制图人	宋修博		（去除屋面）
		班级	建62	打印日期	2019.7.31
		指导教师	刘亦师	图纸编号	补1

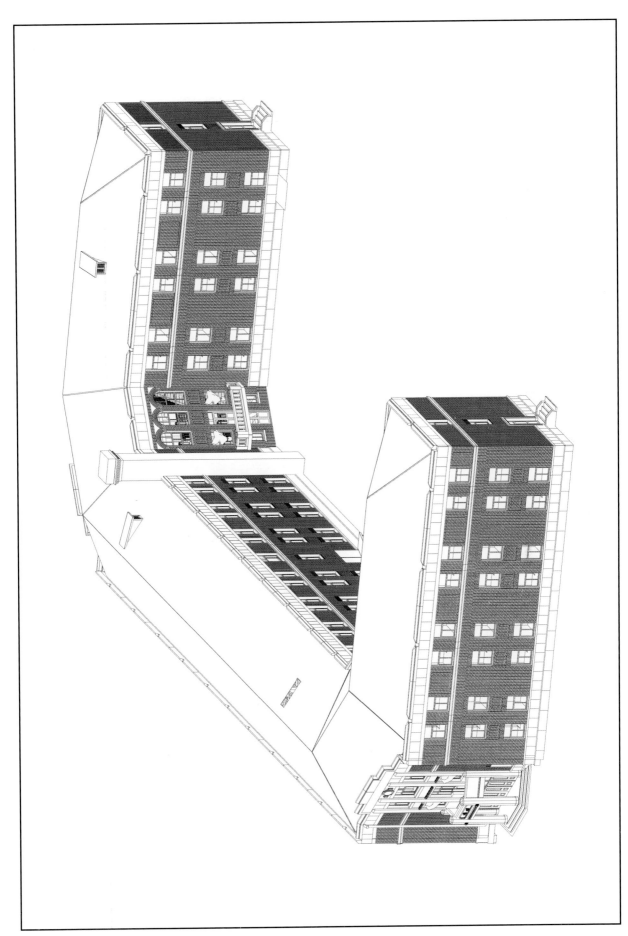

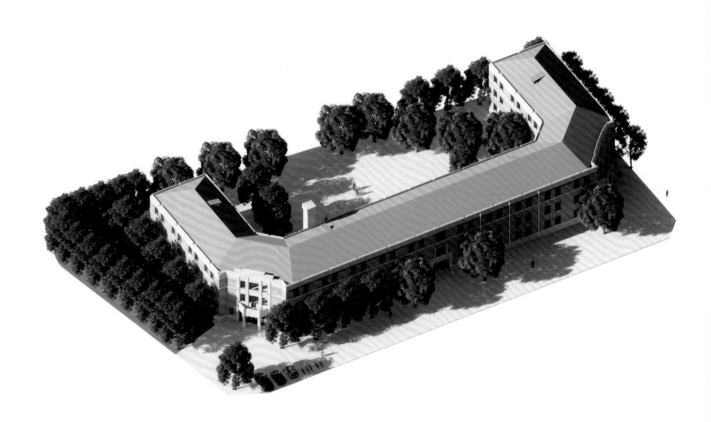

17. 善斋

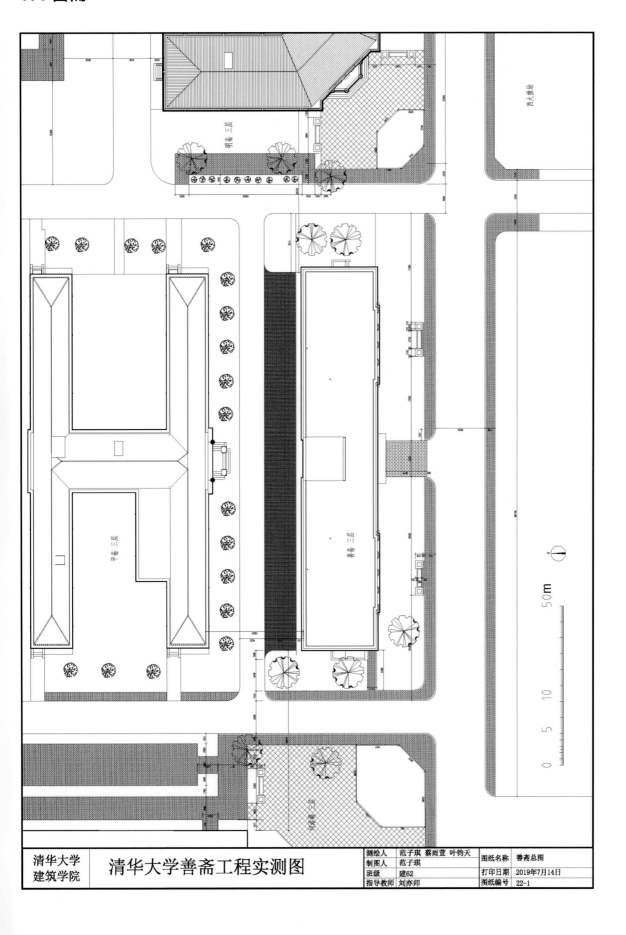

50m

清华大学 建筑学院	清华大学善斋工程实测图	测绘人	范子琪 蔡雨萱 叶钧天	图纸名称	善斋总图
		制图人	范子琪	打印日期	2019年7月14日
		班级	建62		
		指导教师	刘亦师	图纸编号	22-1

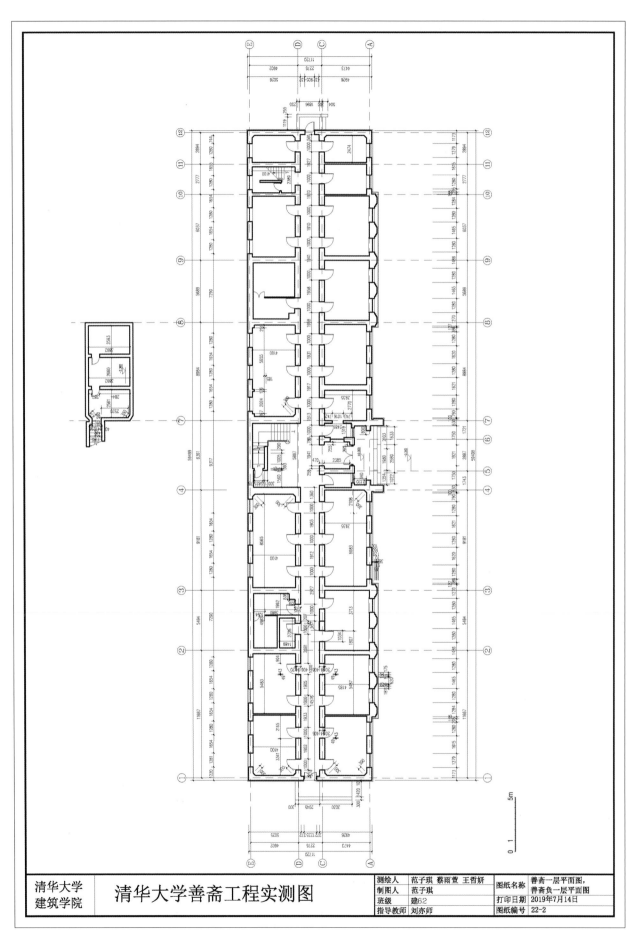

清华大学
建筑学院

清华大学善斋工程实测图

测绘人	范子琪 蔡雨萱 王哲妍	图纸名称	善斋一层平面图, 善斋负一层平面图
制图人	范子琪		
班级	建62	打印日期	2019年7月14日
指导教师	刘亦师	图纸编号	22-2

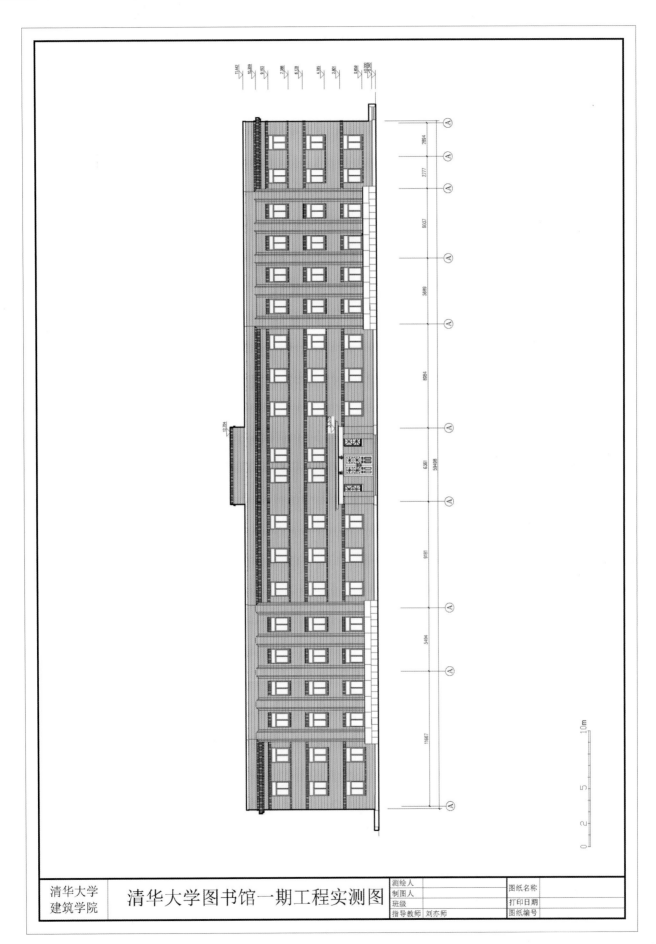

清华大学
建筑学院

清华大学图书馆一期工程实测图

测绘人		图纸名称	
制图人		打印日期	
班级		图纸编号	
指导教师	刘亦师		

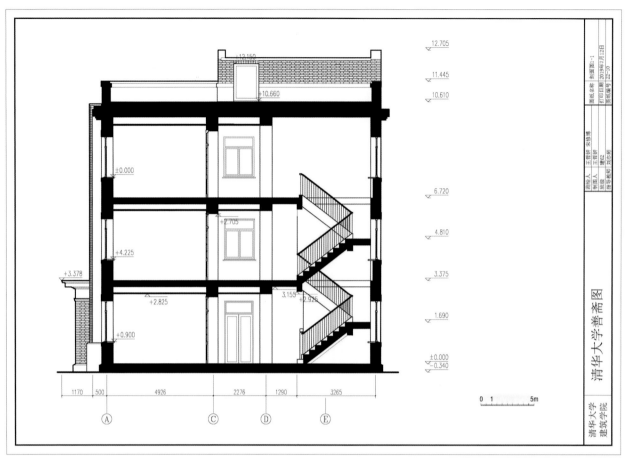

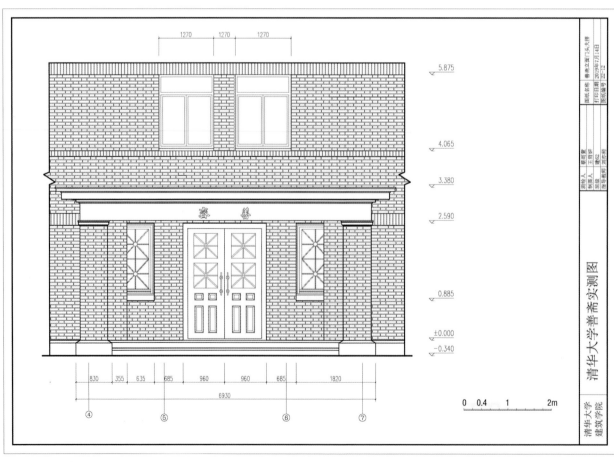

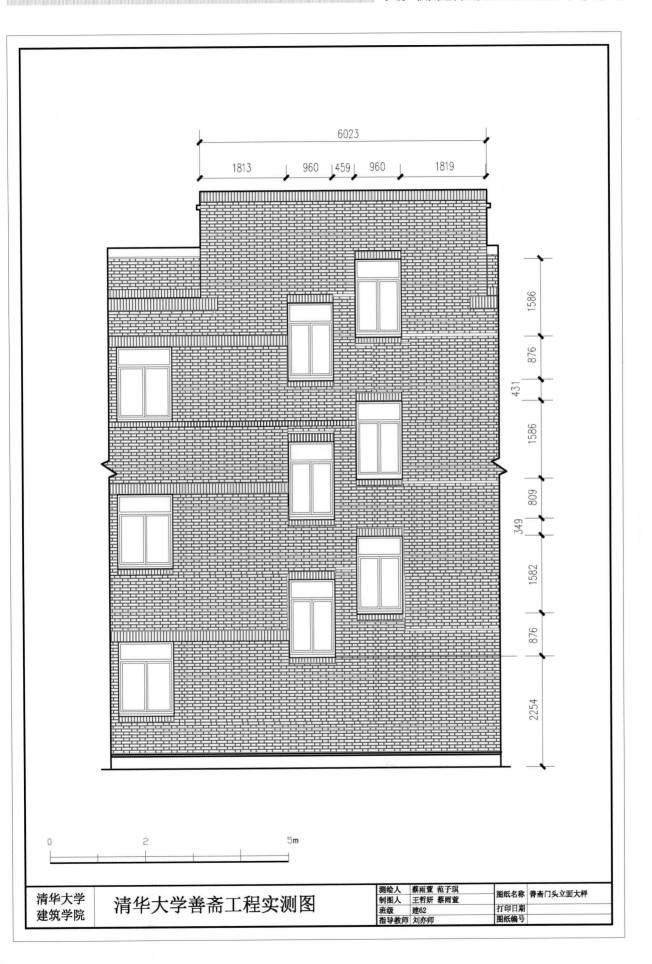

测绘人	蔡雨萱 范子琪	图纸名称	善斋门头立面大样
制图人	王哲妍 蔡雨萱		
班级	建62	打印日期	
指导教师	刘亦师	图纸编号	

清华大学
建筑学院

清华大学善斋工程实测图

539

清华大学 建筑学院	清华大学善斋实测图	测绘人	蔡雨萱 范子琪 王哲妍	图纸名称	善斋立面扶壁柱大样
		制图人	王哲妍 蔡雨萱		
		班级	建62	打印日期	2019年7月14日
		指导教师	刘亦师	图纸编号	22-14

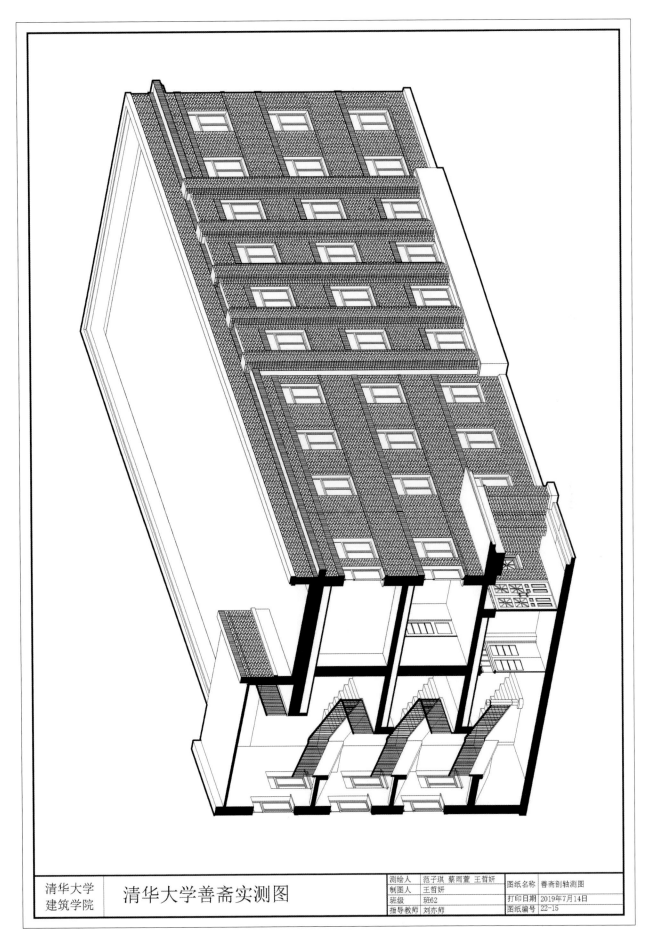

测绘人	范子琪 蔡雨萱 王哲妍	图纸名称	善斋剖轴测图
制图人	王哲妍		
班级	班62	打印日期	2019年7月14日
指导教师	刘亦师	图纸编号	22-15

清华大学
建筑学院　　　清华大学善斋实测图

18. 新斋

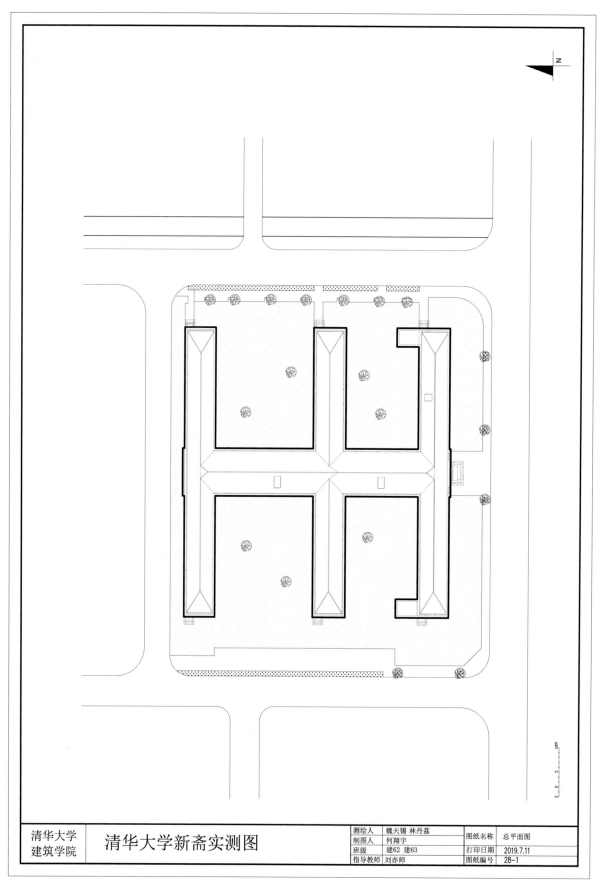

测绘人	魏天锡 林丹荔	图纸名称	总平面图
制图人	何翔宇		
班级	建62 建63	打印日期	2019.7.11
指导教师	刘亦师	图纸编号	28-1

清华大学
建筑学院　　清华大学新斋实测图

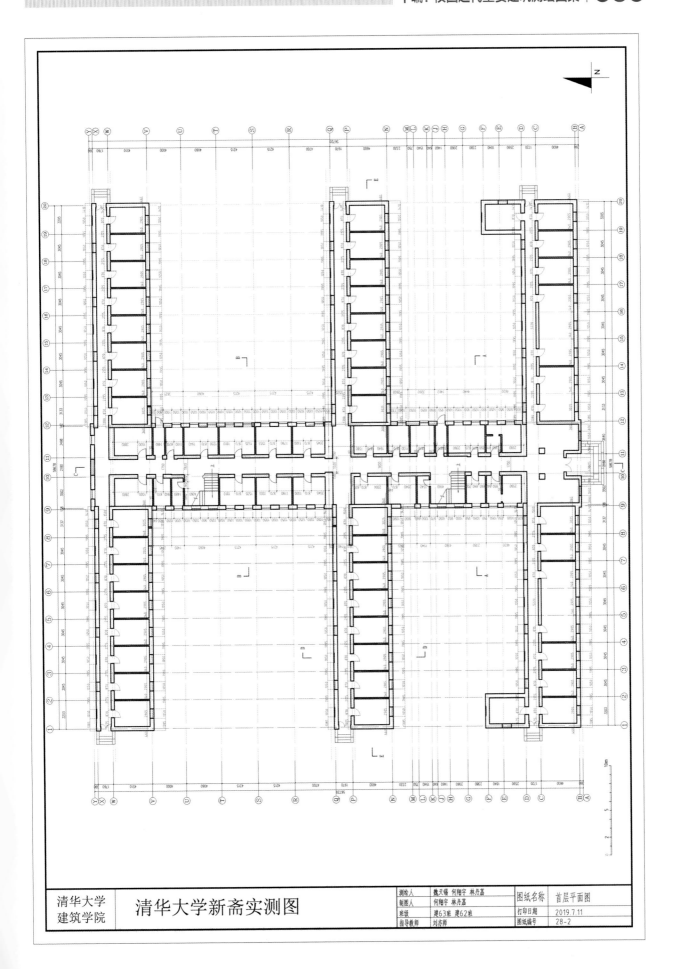

清华大学
建筑学院

清华大学新斋实测图

测绘人	戴天骥 何翔宇 林丹荔	图纸名称	首层平面图
制图人	何翔宇 林丹荔		
班级	建63班 建62班	打印日期	2019.7.11
指导教师	刘亦师	图纸编号	28-2

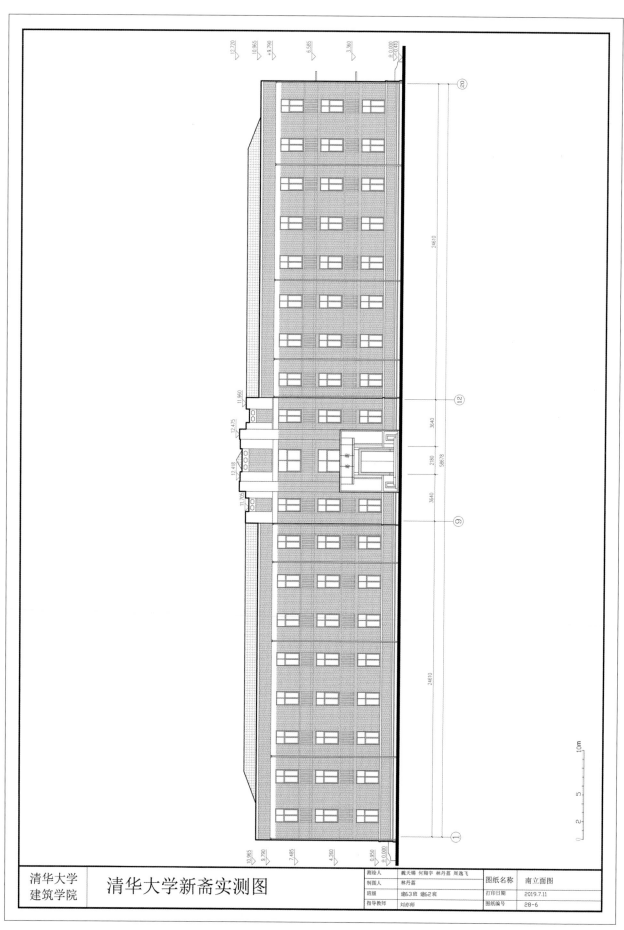

清华大学
建筑学院　　清华大学新斋实测图

测绘人	魏天锡 何翔宇 林丹蕴 周逸飞	图纸名称	南立面图
制图人	林丹蕴	打印日期	2019.7.11
班级	建63班 建62班		
指导教师	刘亦师	图纸编号	28-6

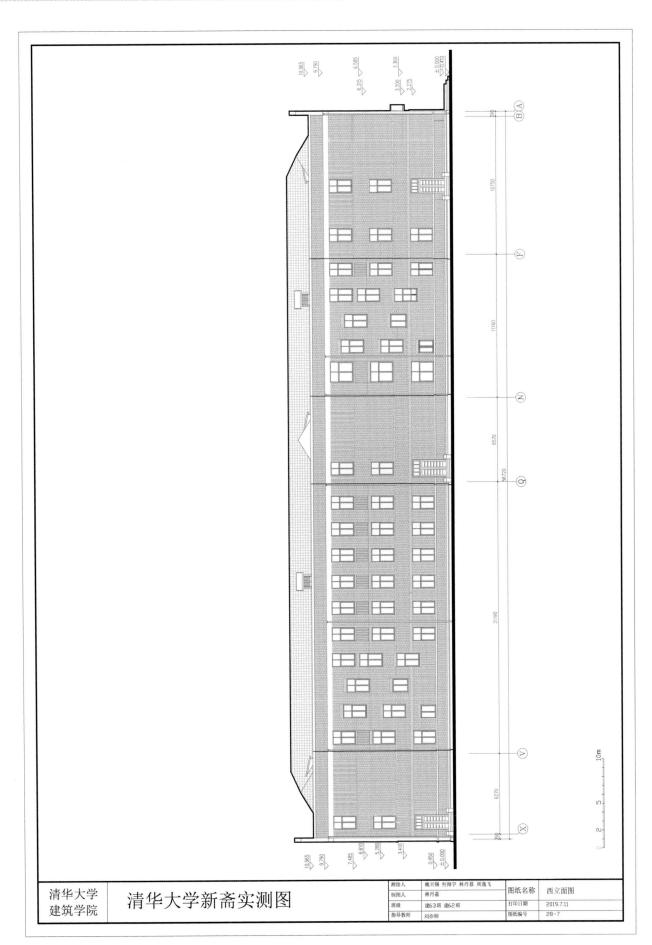

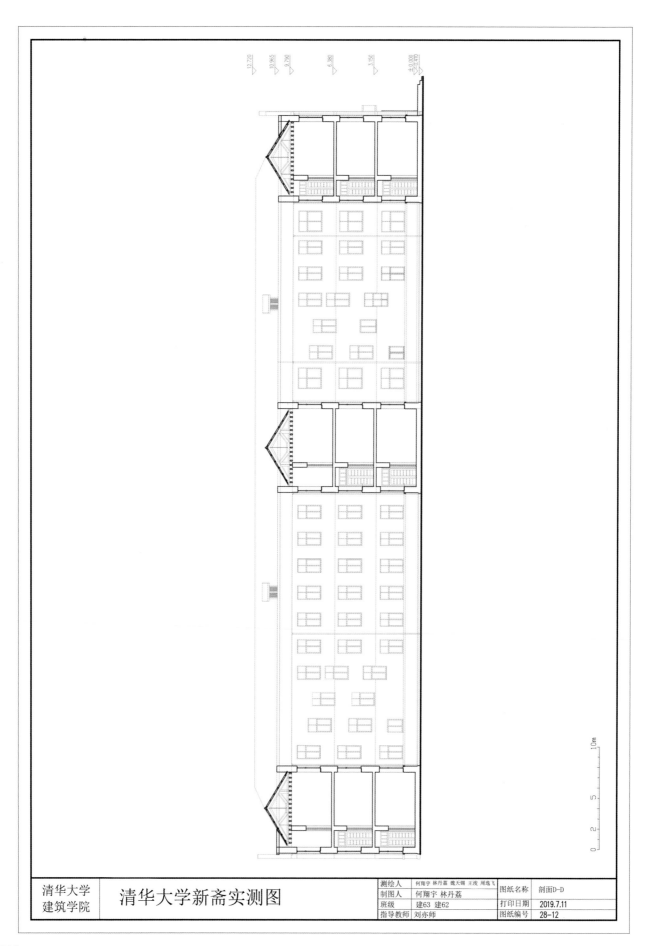

测绘人	何翔宇 林丹荔 魏天锡 王浚 周逸飞	图纸名称	剖面D-D
制图人	何翔宇 林丹荔		
班级	建63 建62	打印日期	2019.7.11
指导教师	刘亦师	图纸编号	28-12

清华大学
建筑学院

清华大学新斋实测图

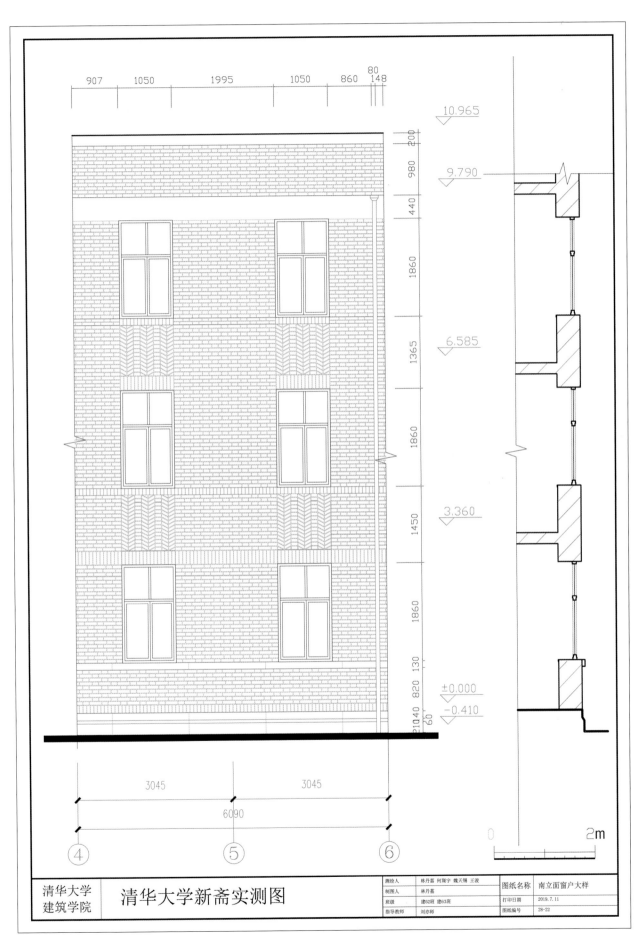

测绘人	林丹磊 何翔宇 魏天锡 王凌	图纸名称	南立面窗户大样
制图人	林丹磊		
班级	建62班 建63班	打印日期	2019.7.11
指导教师	刘亦师	图纸编号	28-22

清华大学
建筑学院　　清华大学新斋实测图

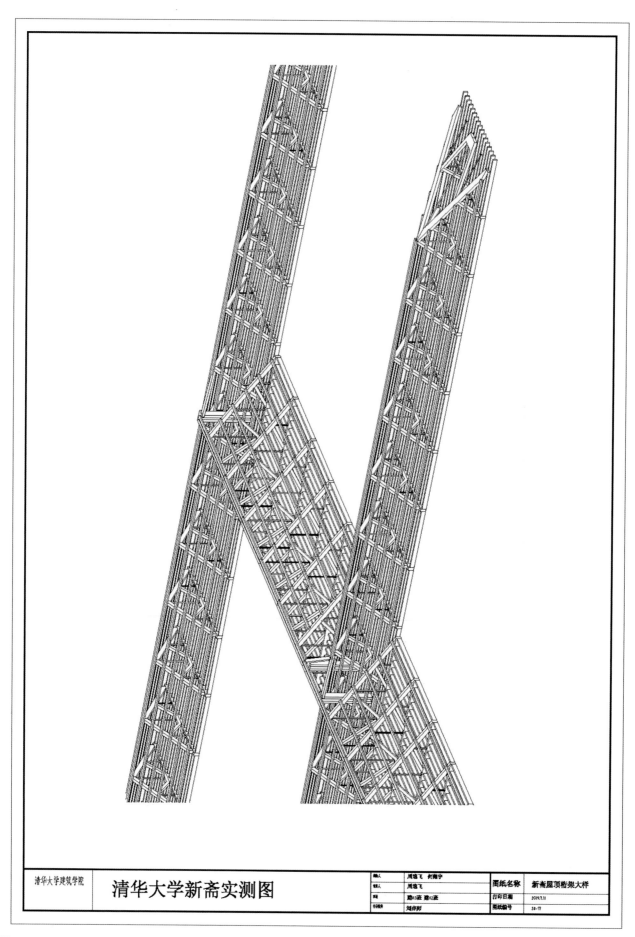

清华大学建筑学院　清华大学新斋实测图

制图人	周堉飞 何陶宇	图纸名称	新斋屋顶桁架大样
审核人	周堉飞		
比例	建43班 建42班	打印日期	2019.7.11
指导教师	刘亦师	图纸编号	24-17

测绘人	魏天赐 何翔宇 林丹蕴 王浚 周逸飞	图纸名称	轴测图
制图人	林丹蕴		
班级	建63班 建62班	打印日期	2019.7.11
指导教师	刘亦师	图纸编号	28-27

清华大学
建筑学院

清华大学新斋实测图

| 测绘人 | 魏大锡 林丹荔 何翔宇 | 图纸名称 | 实测照片 |

清华大学 建筑学院	清华大学新斋实测图	测绘人	魏大锡 林丹荔 何翔宇	图纸名称	实测照片
		制图人	何翔宇		
		班级	建62 建63	打印日期	2019.7.11
		指导教师	刘亦师	图纸编号	—

19. 平斋

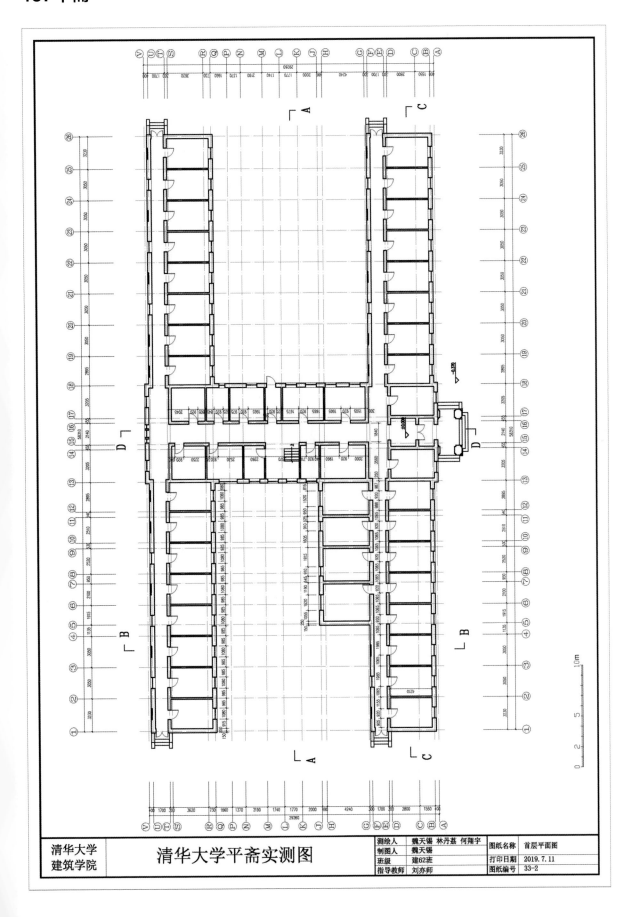

清华大学
建筑学院

清华大学平斋实测图

测绘人	魏天锡 林丹荔 何翔宇	图纸名称	首层平面图
制图人	魏天锡		
班级	建62班	打印日期	2019.7.11
指导教师	刘亦师	图纸编号	33-2

551

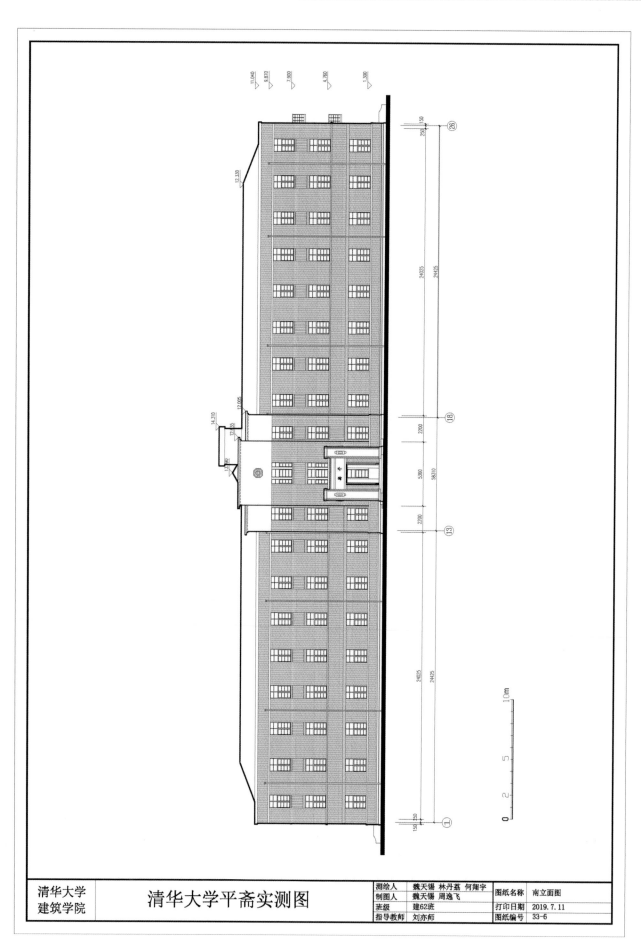

清华大学 建筑学院	清华大学平斋实测图	测绘人	魏天锡 林丹荔 何翔宇	图纸名称	南立面图
		制图人	魏天锡 周逸飞		
		班级	建62班	打印日期	2019.7.11
		指导教师	刘亦师	图纸编号	33-6

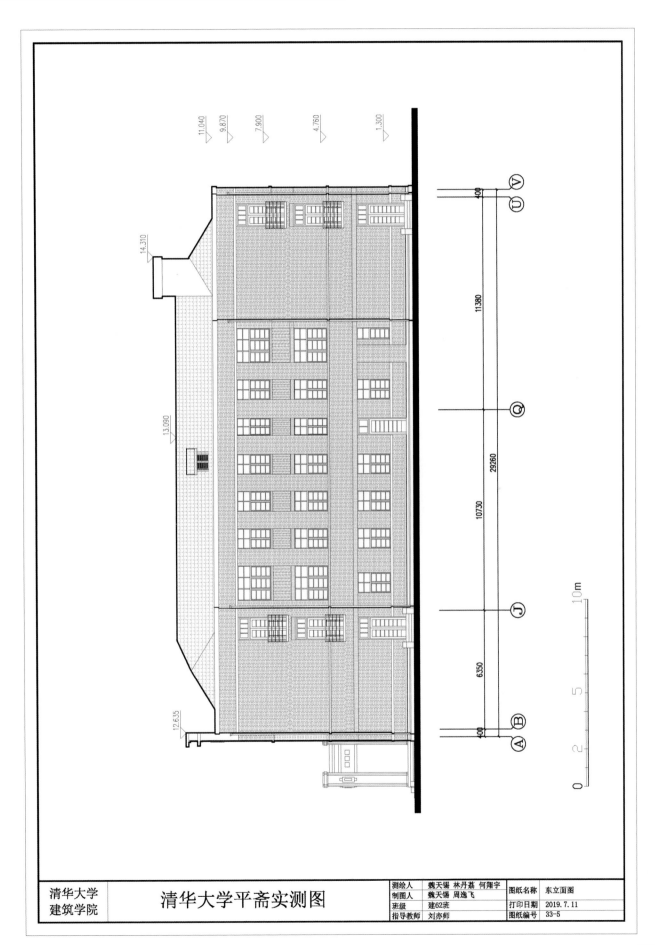

测绘人	魏天锡 林丹荔 何翔宇	图纸名称	东立面图
制图人	魏天锡 周逸飞		
班级	建62班	打印日期	2019.7.11
指导教师	刘亦师	图纸编号	33-5

清华大学
建筑学院

清华大学平斋实测图

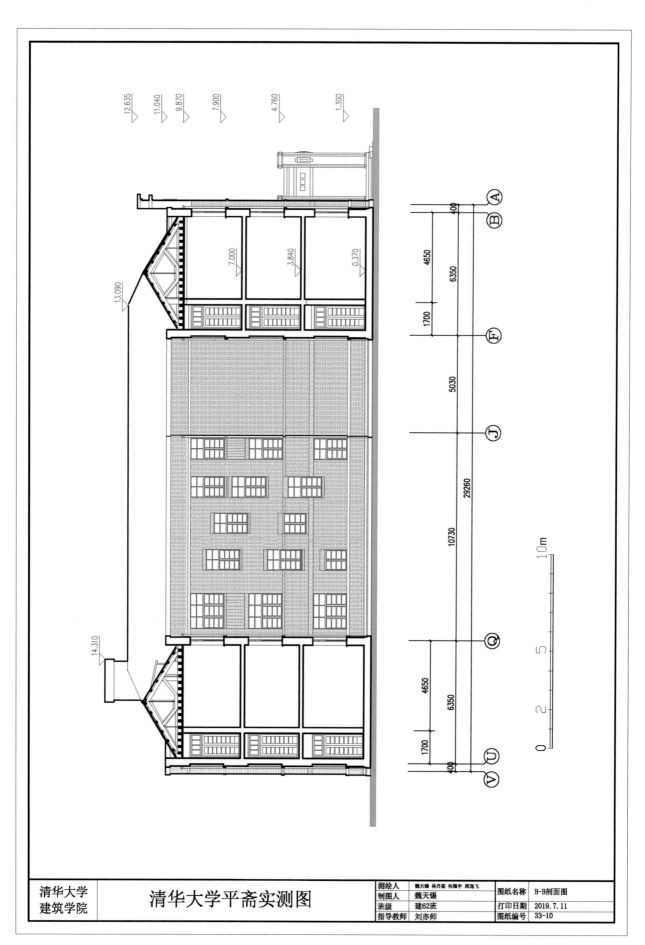

清华大学
建筑学院

清华大学平斋实测图

测绘人	魏天锡 林丹嘉 何海宇 周逸飞	图纸名称	B-B剖面图
制图人	魏天锡	打印日期	2019.7.11
班级	建62班	图纸编号	33-10
指导教师	刘亦师		

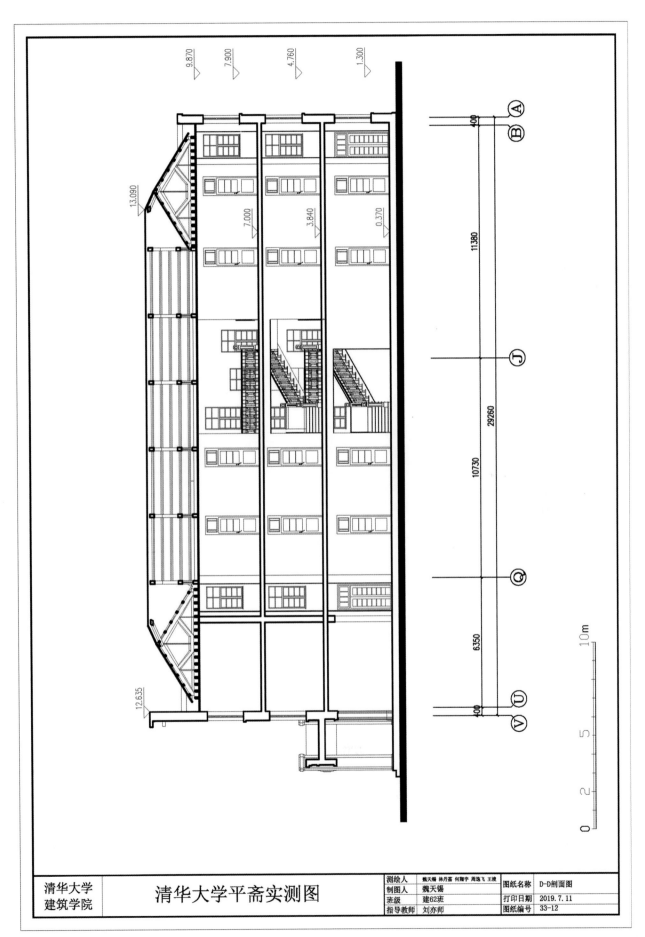

清华大学 建筑学院	清华大学平斋实测图	测绘人	魏天锡 林方磊 何翔宇 周逸飞 王璞	图纸名称	D-D剖面图
		制图人	魏天锡		
		班级	建62班	打印日期	2019.7.11
		指导教师	刘亦师	图纸编号	33-12

测绘人	林丹磊 何翔宇 魏天锡	图纸名称	南立面大样
制图人	魏天锡 林丹磊		
班级	建62班 建63班	打印日期	2019.7.11
指导教师	刘亦师	图纸编号	33-17

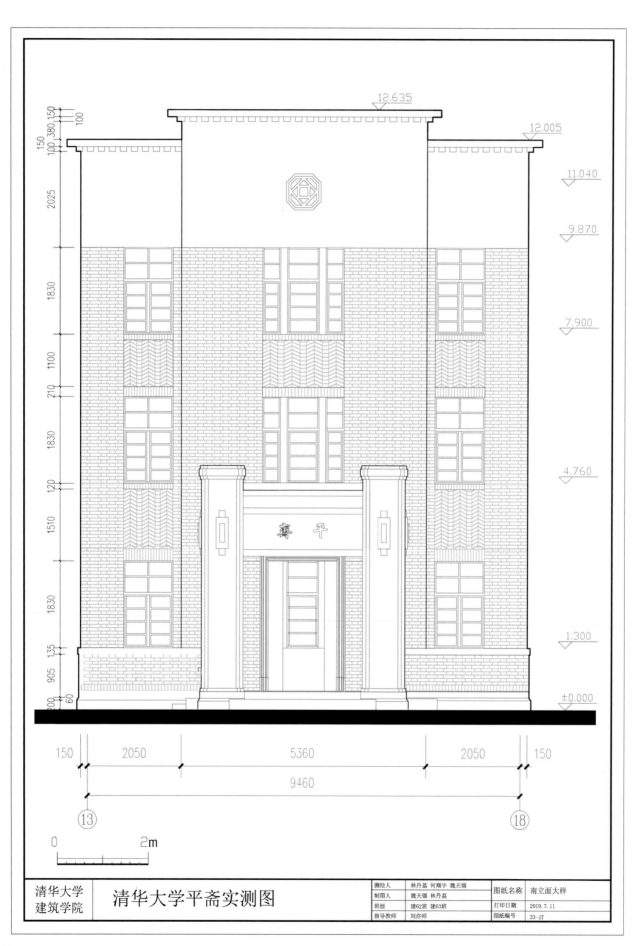

12.635

12.005

11.040

9.870

7.900

4.760

1.300

±0.000

齋 平

150 · 2050 · 5360 · 2050 · 150

9460

⑬ ⑱

0   2m

清华大学 建筑学院	清华大学平斋实测图	测绘人	林丹磊 何翔宇 魏天锡	图纸名称	南立面大样
		制图人	魏天锡 林丹磊		
		班级	建62班 建63班	打印日期	2019.7.11
		指导教师	刘亦师	图纸编号	33-17

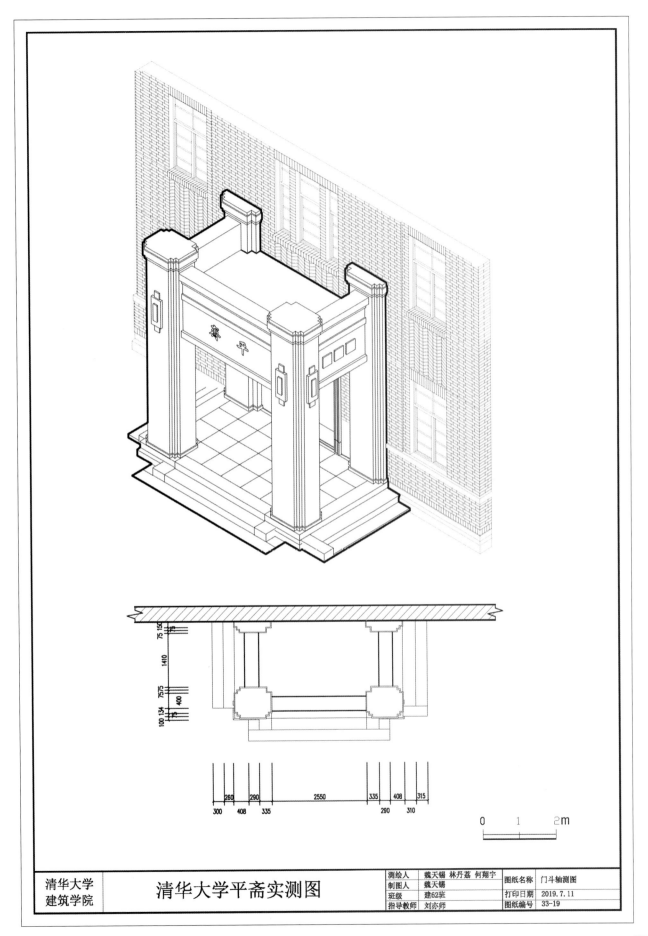

清华大学平斋实测图

测绘人	魏天锡 林丹荔 何翔宇	图纸名称	门斗轴测图
制图人	魏天锡		
班级	建62班	打印日期	2019.7.11
指导教师	刘亦师	图纸编号	33-19

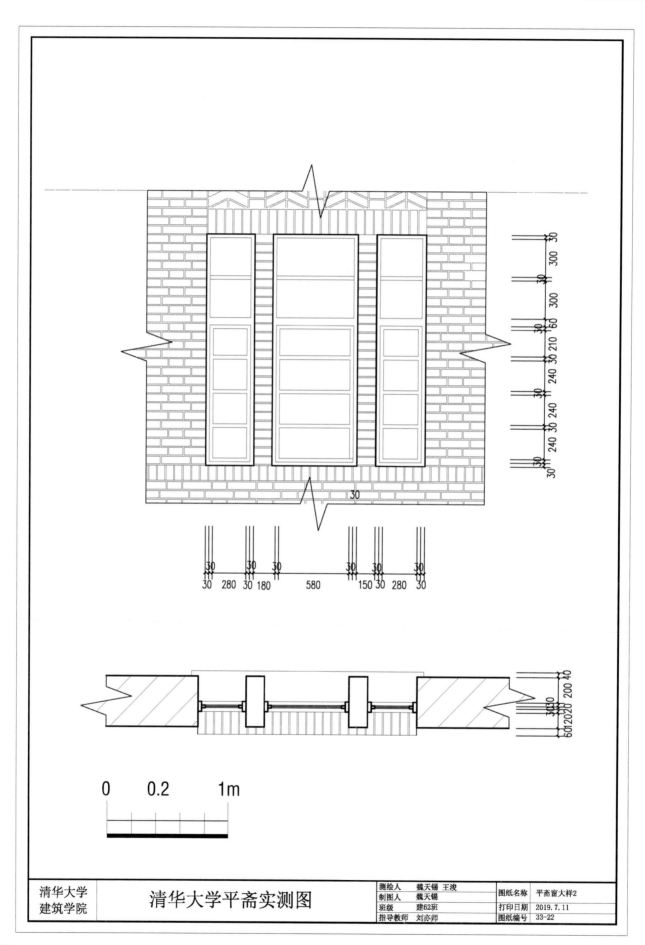

清华大学
建筑学院

清华大学平斋实测图

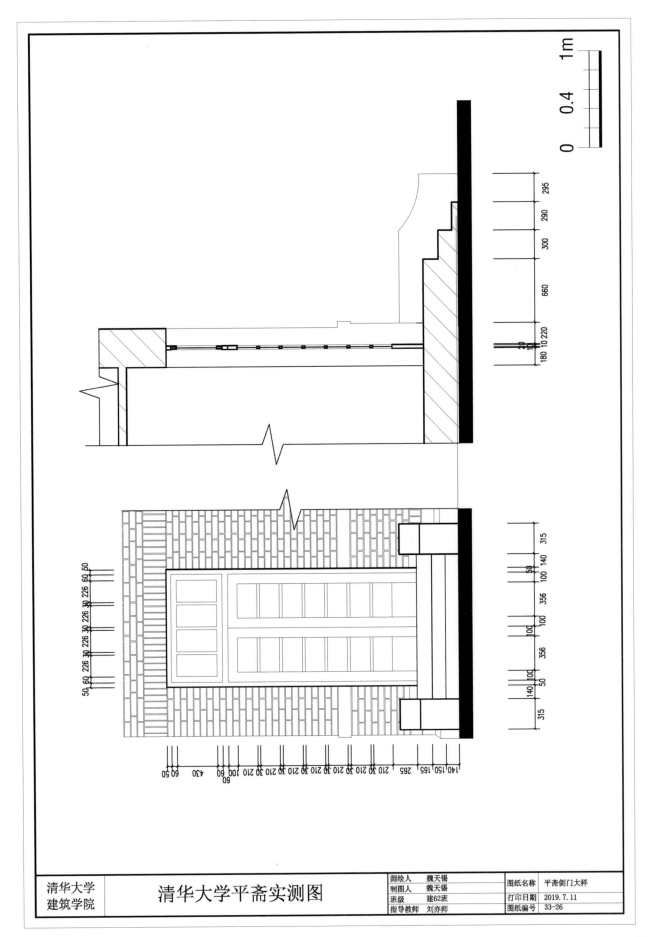

清华大学平斋实测图

测绘人	魏天锡	图纸名称	平斋侧门大样
制图人	魏天锡		
班级	建62班	打印日期	2019.7.11
指导教师	刘亦师	图纸编号	33-26

清华大学
建筑学院

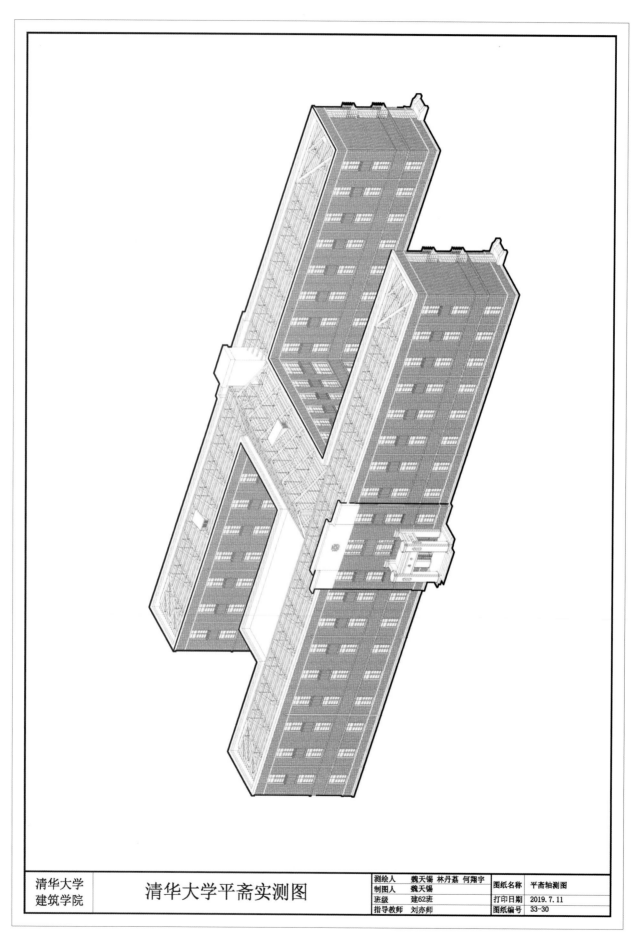

清华大学 建筑学院	清华大学平斋实测图	测绘人	魏天锡 林丹荔 何翔宇	图纸名称	平斋轴测图
		制图人	魏天锡		
		班级	建62班	打印日期	2019.7.11
		指导教师	刘亦师	图纸编号	33-30

清华大学 建筑学院	清华大学平斋实测图	测绘人	魏天锦 林丹荔 何翔宇	图纸名称	实测照片
		制图人	何翔宇		
		班级	建62 建63	打印日期	2019.7.11
		指导教师	刘亦师	图纸编号	附1

20.1~4号学生公寓 （1~4号学生宿舍建成于1954年，已非近代时期，但其测绘图亦收录于本编。）

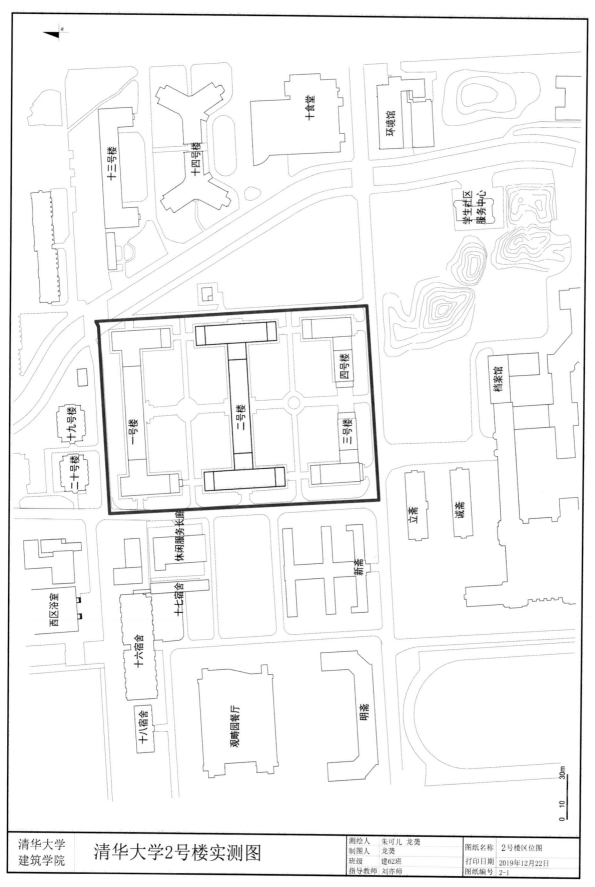

测绘人	朱可儿 龙荄		图纸名称	2号楼区位图
清华大学 建筑学院	制图人	龙荄		
	班级	建62班	打印日期	2019年12月22日
	指导教师	刘亦师	图纸编号	2-1

清华大学2号楼实测图

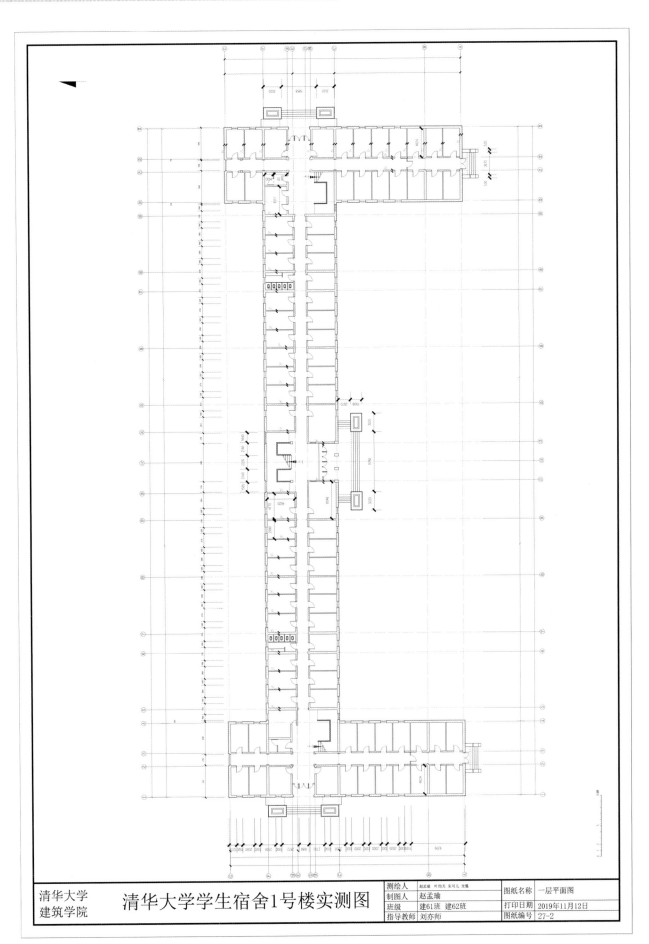

清华大学
建筑学院

清华大学学生宿舍1号楼实测图

测绘人	赵孟瑜 叶均天 朱可儿 龙馨	图纸名称	一层平面图
制图人	赵孟瑜		
班级	建61班 建62班	打印日期	2019年11月12日
指导教师	刘亦师	图纸编号	27-2

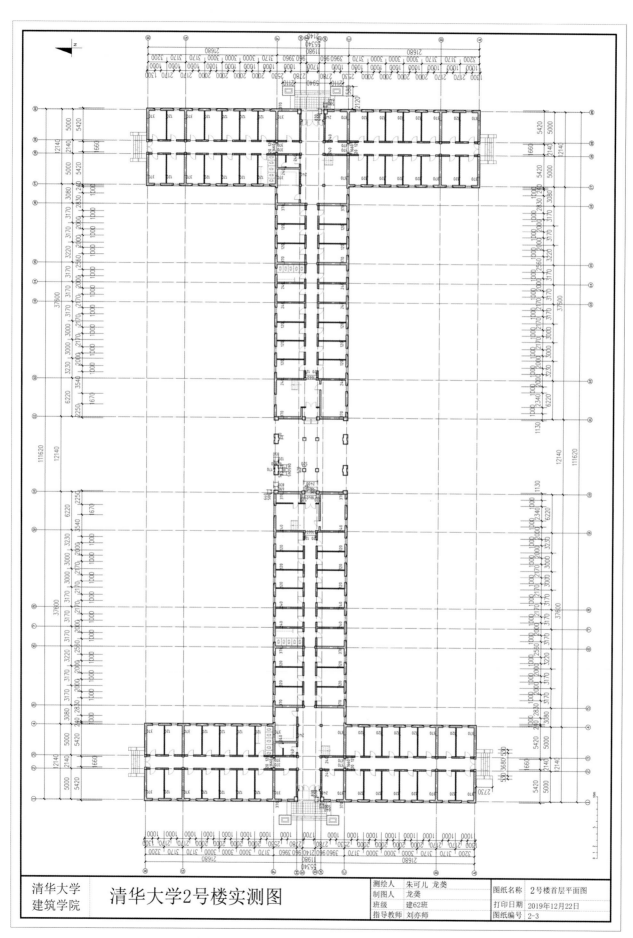

测绘人	朱可儿 龙葵
制图人	龙葵
班级	建62班
指导教师	刘亦师

图纸名称	2号楼首层平面图
打印日期	2019年12月22日
图纸编号	2-3

清华大学
建筑学院

清华大学2号楼实测图

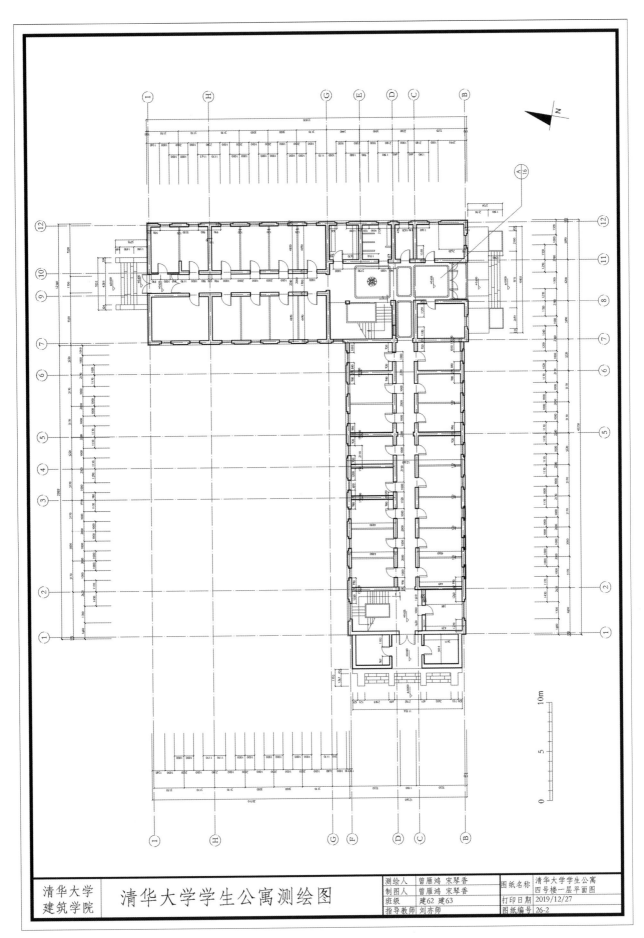

测绘人	曾雁鸿 宋琴香	图纸名称	清华大学学生公寓 四号楼一层平面图	
制图人	曾雁鸿 宋琴香			
班级	建62 建63	打印日期	2019/12/27	
指导教师	刘亦师	图纸编号	26-2	

清华大学 建筑学院

清华大学学生公寓测绘图

测绘人　朱可儿 龙奕
制图人　朱可儿
指导教师　刘亦师
图纸名称　二号楼北立面图
图纸编号　2-8

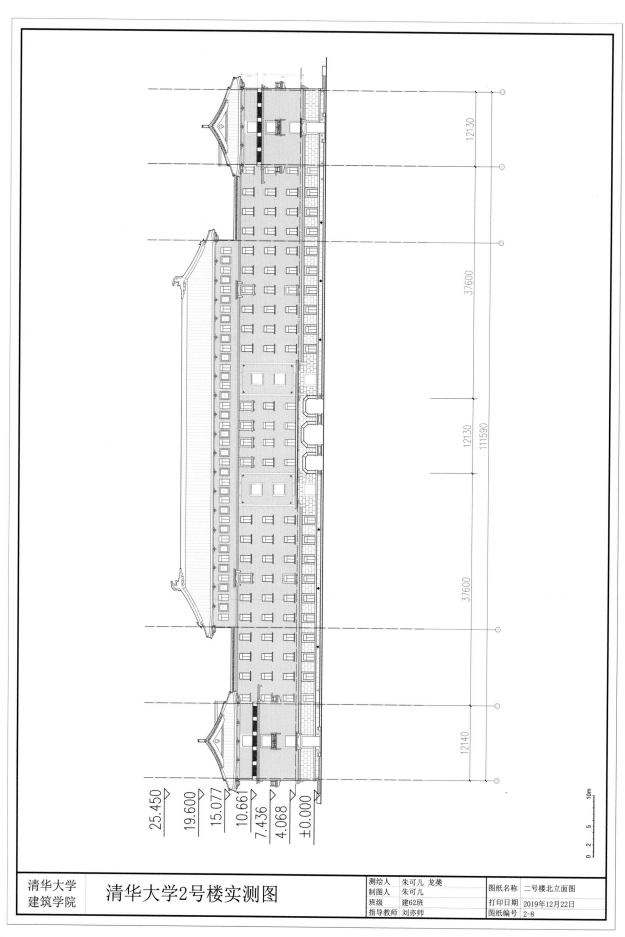

清华大学建筑学院	清华大学2号楼实测图

测绘人	朱可儿 龙奕	图纸名称	二号楼北立面图
制图人	朱可儿		
班级	建62班	打印日期	2019年12月22日
指导教师	刘亦师	图纸编号	2-8

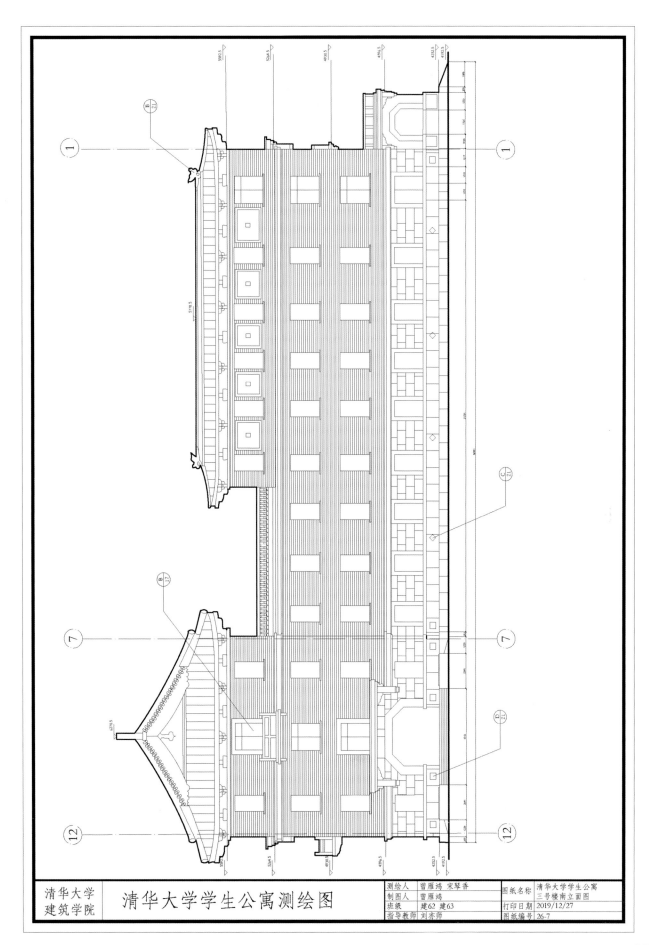

清华大学
建筑学院

清华大学学生公寓测绘图

测绘人	曾雁鸿 宋琴香	图纸名称	清华大学学生公寓三号楼南立面图
制图人	曾雁鸿		
班级	建62 建63	打印日期	2019/12/27
指导教师	刘亦师	图纸编号	26-7

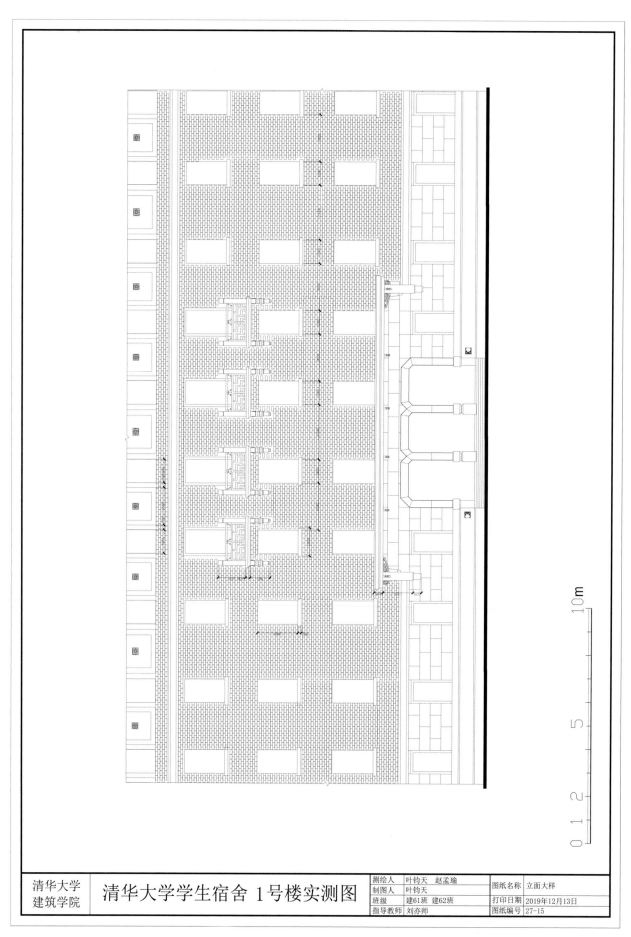

10m

5

2
1
0

测绘人	叶钧天　赵孟瑜	图纸名称	立面大样
制图人	叶钧天		
班级	建61班　建62班	打印日期	2019年12月13日
指导教师	刘亦师	图纸编号	27-15

清华大学
建筑学院

清华大学学生宿舍 1 号楼实测图

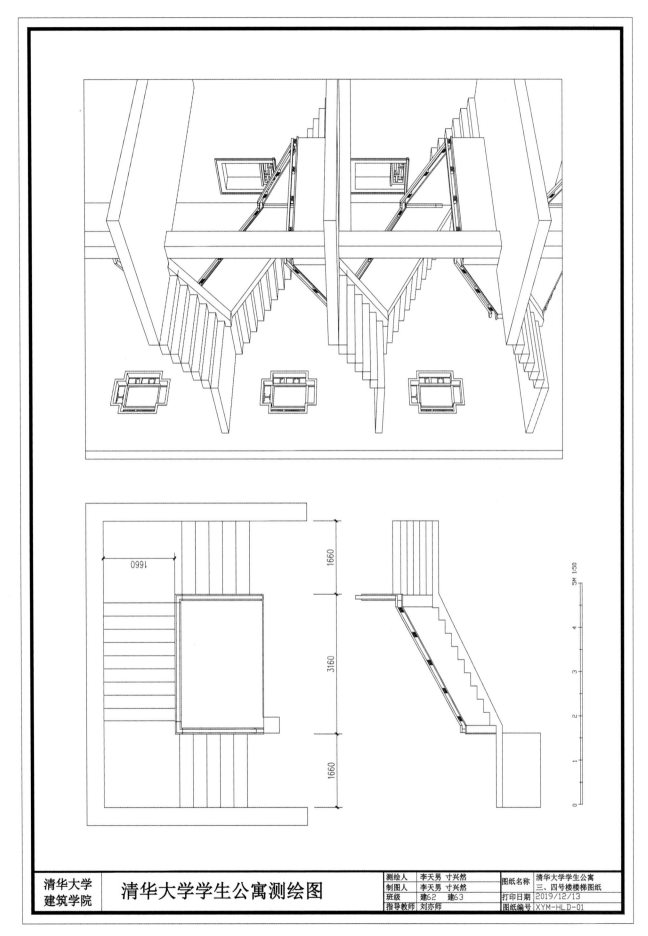

测绘人	李天男 寸兴然	图纸名称	清华大学学生公寓
制图人	李天男 寸兴然		三、四号楼楼梯图纸
班级	建62 建63	打印日期	2019/12/13
指导教师	刘亦师	图纸编号	XYM-HLD-01

清华大学
建筑学院

清华大学学生公寓测绘图

测绘人	叶钧天 赵孟瑜		图纸名称	楼梯平立面及轴测
制图人	赵孟瑜			
班级	建61班 建62班		打印日期	2019年11月12日
指导教师	刘亦师		图纸编号	27-21

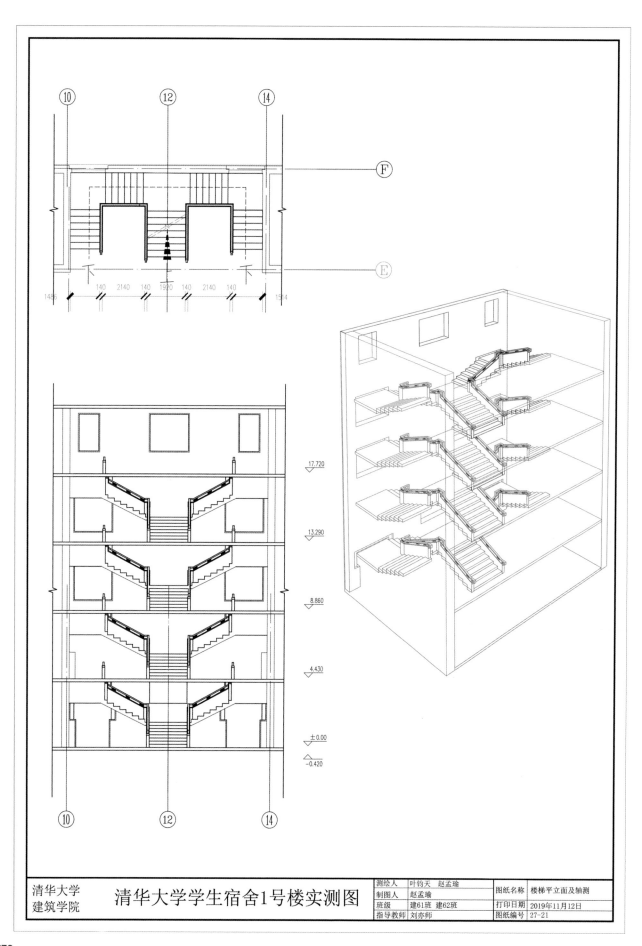

清华大学
建筑学院

清华大学学生宿舍1号楼实测图

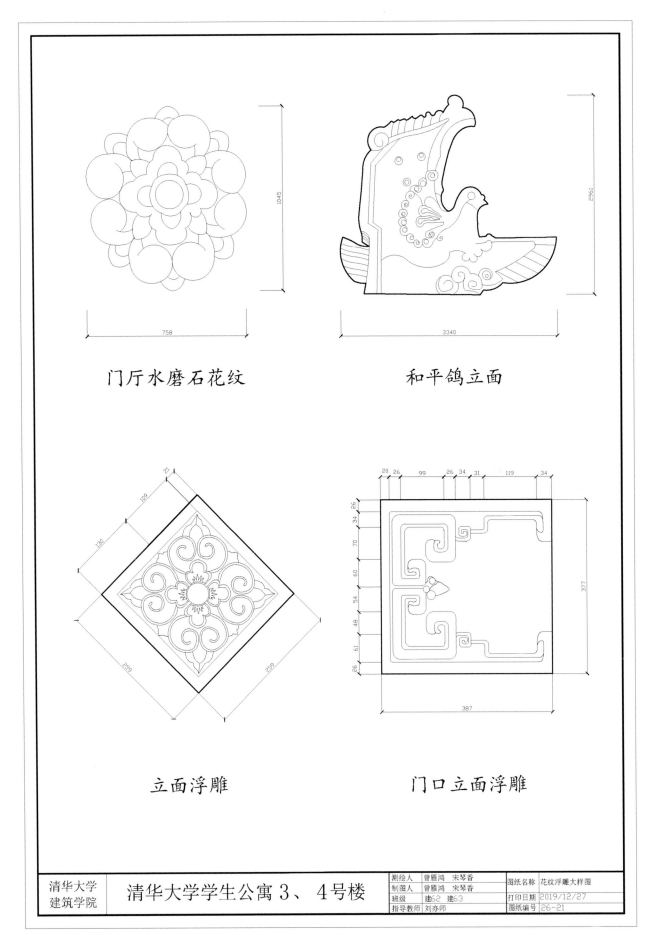

门厅水磨石花纹

和平鸽立面

立面浮雕

门口立面浮雕

测绘人	曾雁鸿　宋琴香	图纸名称	花纹浮雕大样图
制图人	曾雁鸿　宋琴香		
班级	建62　建63	打印日期	2019/12/27
指导教师	刘亦师	图纸编号	26-21

清华大学
建筑学院

清华大学学生公寓 3、4号楼

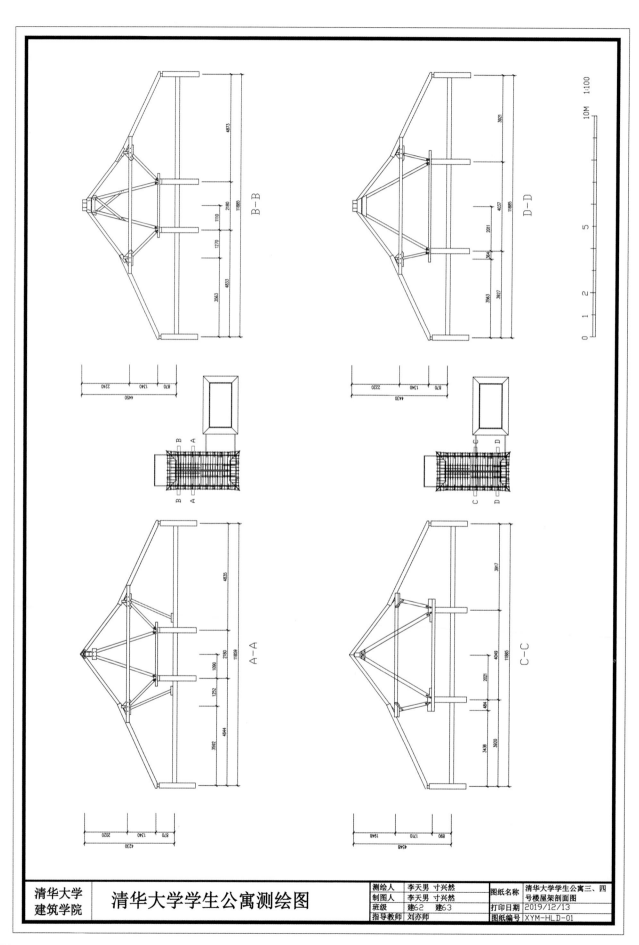

清华大学 建筑学院	清华大学学生公寓测绘图	测绘人	李天男 寸兴然	图纸名称	清华大学学生公寓三、四号楼屋架剖面图
		制图人	李天男 寸兴然		
		班级	建62 建63	打印日期	2019/12/13
		指导教师	刘亦师	图纸编号	XYM-HLD-01

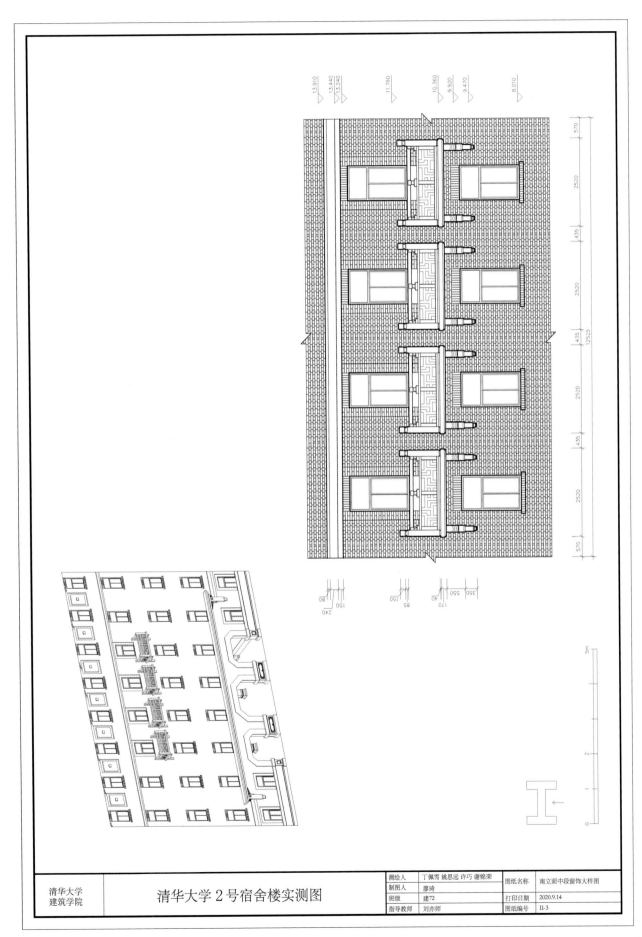

清华大学 建筑学院	清华大学 2 号宿舍楼实测图	测绘人	丁佩雪 姚思远 许巧 谢锦荣	图纸名称	南立面中段窗饰大样图
		制图人	廖琦	打印日期	2020.9.14
		班级	建72		
		指导教师	刘亦师	图纸编号	II-3

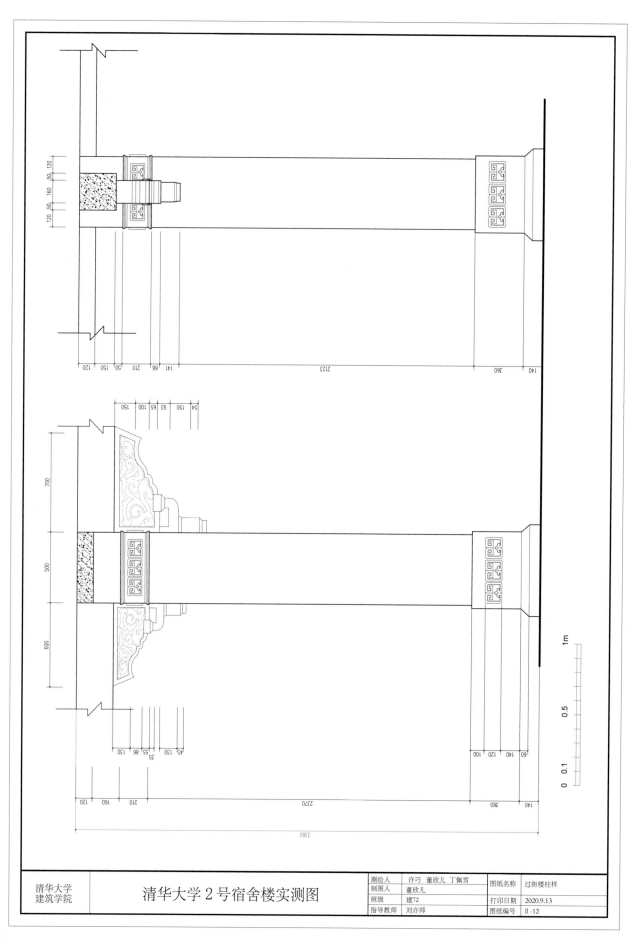

测绘人	许巧 董欣儿 丁佩雪	图纸名称	过街楼柱样
制图人	董欣儿		
班级	建72	打印日期	2020.9.13
指导教师	刘亦师	图纸编号	II -12

清华大学
建筑学院

清华大学 2 号宿舍楼实测图

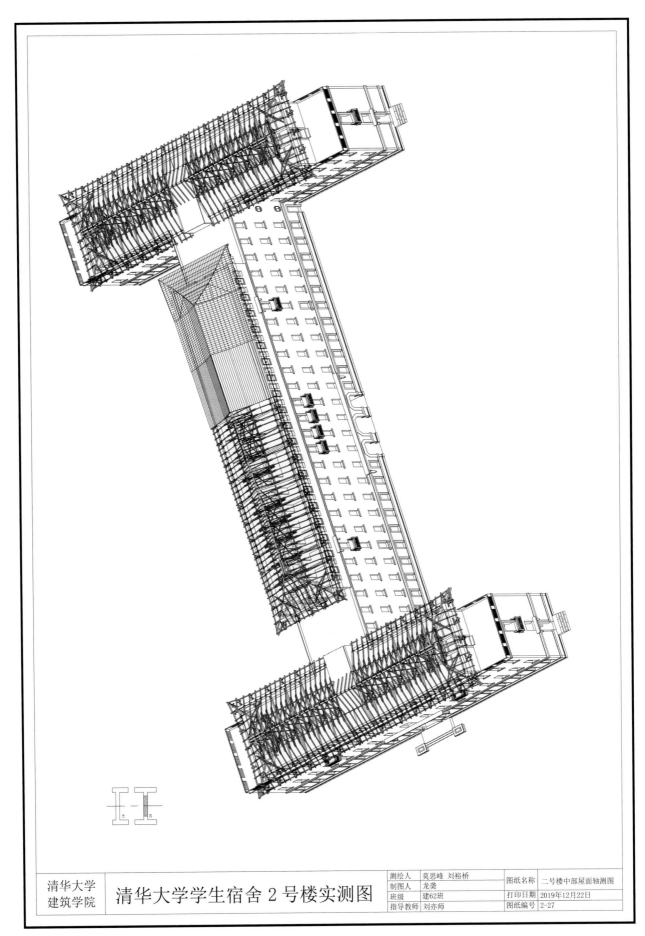

清华大学
建筑学院　　清华大学学生宿舍 2 号楼实测图

测绘人	莫思峰 刘裕桥	图纸名称	二号楼中部屋面轴测图
制图人	龙葵		
班级	建62班	打印日期	2019年12月22日
指导教师	刘亦师	图纸编号	2-27

清华大学二校门
91·11·9·

　　我第一次进清华园是 2001 年夏天。那时我从武汉到北京参加 GRE 的暑假培训班，住在北京科技大学（原北京钢铁学院）。有一天早晨骑自行车从清华南门进入学校，沿路找到图书馆，不经意间抬头看到大礼堂的后身。大礼堂的穹顶在夏日晨曦中雄沉巍峨的样子给我留下了深刻印象。

　　2003 年我考入清华大学建筑学院建筑历史研究所，选择近代建筑史作为研究方向。当时秦佑国教授是建筑学院院长，给我们上课时讲到全国只有 4 所大学的校园被列入国家级文物保护单位：清华大学、燕京大学（北京大学）、武汉大学和原东北大学。这说的是 2003 年前后的事情，后来陆续被列入国家级保护单位的大学校址增加不少。秦先生当时课上常用清华和燕大的校园作比较，二者都是同一个美国建筑师——墨菲设计，但建筑风格迥异：清华用的是经济实用的西洋样式，而燕大的建筑都被覆以中国式大屋顶，建筑细部也截然不同。秦先生的解释，两校的建筑风格分别反映了庚子事变之后中国和西方政府的不同政策：清华采用西洋建筑样式代表了清政府学习西方的决心，而作为教会大学的燕大则倾向于尊重中国文化，避免再度激发民愤。

　　这些观点在我当时听来很新颖，颇受启发。读硕士那几年住在清华校园内，师从张复合先生学习近代建筑史，虽然常从大礼堂、明斋（靠近食堂）等建筑旁经过，但当时在听课、参加讲座和做硕士论文的研究之外再无余力，根本没想过要去研究本校的这些近代建筑。之后到美国读博士，五年半的时间基本接续硕士论文的研究，也无暇旁顾其他，只是在 2012 年年初拿到博士学位以后到耶鲁大学游玩时，抽空去翻阅了一下耶鲁神学院中与中国教会学校以及清华有关的资料。

　　2012 年我回清华做博士后，于是年 11 月前往美国弗吉尼亚州杰斐逊国际研究中心（International Center for Jefferson Studies）做为期一个月的访问研究，有机会实地考察弗吉尼亚大学的大草坪区与清华的相似与差别。等访问期快结束时，偶然碰到利用业余时间研究弗吉尼亚大学圆厅图书馆（Rotunda）穹顶结构的执业建筑师 Doug Harnsberger，他向我介绍了关斯塔维诺穹顶建造体系（Guastavino Dome & Ribs）。

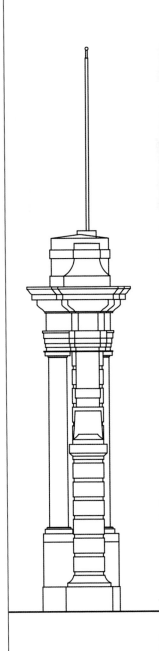

我听后大为吃惊。清华大礼堂虽然和圆厅图书馆外观有类似之处，但明显与后者不同的是，大礼堂穹顶没有开采光顶。我当时就在想：清华大礼堂的穹顶结构到底是怎样的呢？

为了解决这个问题，我改变行程，于2012年12月特地绕道去耶鲁大学试着查找相关资料。最初在耶鲁大学神学院没有找到，后来在耶鲁大学主图书馆一层的手稿与档案室中发现墨菲档案（Murphy Papers）内就有一盒清华学校的设计资料。很幸运，我发现了大礼堂的平面、立面和剖面设计图，剖面上且标注拼写有错误的"Guastavino Dome & Ribs"，简直如获至宝。我回京后立即与清华房管处联系，最终在2013年7月对大礼堂进行了全面的测绘。以测绘图与原始设计图比较，能清晰地反映出穹顶结构以至整体设计的差别。这一工作很快发表在当年11月的《建筑学报》上，也即为本书第3章3.5节的部分内容。

对大礼堂的这一研究经历使我深有感触，也对我个人的研究取法和学术走向产生了很大影响。首先，我在此前的研究中没怎么用过真正意义上的原始档案，更不用说经测量新绘的图像史料。这一次的经历使我感到爬梳原始档案和未经广泛使用的历史文献，并且积极组织测绘取得第一手资料，前景广阔，大有可为。我后来又去过一次耶鲁大学，在欧柏林学院（Oberlin College）的档案馆也找到一些与墨菲相关的资料。我梳理了这些档案，又利用《清华周刊》等记载，开始了针对清华早期校园建设的研究。与此同时，在清华房管处的支持、协调下，2014年和2015年我们完成了对墨菲设计的"四大工程"的测绘工作，此后将测绘对象的范围比年扩大到20世纪20年代和30年代及至50年代时期的建筑。

其次，在研究大礼堂的具体设计、建造及清华早期的校园建设时，我发现了清华校园的主要建筑之所以选择西方样式，既有宏大理想的擘画（将清华建设为北洋政府的"中央大学"），也有立足现实的考量（出于造价与便于维护的经济考虑放弃大屋顶样式），而这在白纸黑字的会谈记录和往来函件中都能找到确凿的证据。找到这些文件时我是真的高兴——时隔多年，在当初秦佑国先生宏观式的解释之外，终于找到史料上的依据了。这些新发现推动了清华校史的研究，加深了对近代建筑发展的认识。"务博综、尚实证"，历史研究不正是这样进步的吗？

我在清华大学从事近代建筑史的研究和教学，自感把清华近代校园和建筑的研究做好、做深责无旁贷。由于还有别的工作同时进行，爱将清华的研究分成几个历史时期分别开展，使之既能独立成文也相互联系。这一工程从2013年开始，时断时续。最先进行的是和墨菲有关的部分，包括研究和走访他在设计清华之前的那些校园设计（卢弥斯学校和雅礼大学），顺带着把1928年以前的清华建设情况梳理了一遍。此后花了不少时间考察20世纪50年代的校园建设、访谈健在的亲历者，整理出1948—1966年清华的几次规划和具体建设，包括1～4号学生公寓和主楼等个案研究。然后才回过头来清理1928—1948年间的清华校园的规划和建设情况。通过这些研究，大体上掌握了清华校园从1909年建校到20世纪60年代的发展线索和基本史实。这就是本书上编的内容。

这期间，校园建筑的测绘工作一直在进行。每年测绘都安排在暑假小学期，即6月底或7月初春季学期期末考试完后的头一周，动员清华大学建筑学院大三的学生参加。这往往是北京最闷热、暴晒最甚的时候。除使用全站仪测量高点外，我们用的全部是手工测量方式，而且攀高爬下，要克服畏难、恐高情绪，条件相对艰苦。尤其每次测绘的重点总是积满灰尘、长年无人进入的屋顶，从仅容身过的检修口爬上去后，要随时注意脚底不要踩空摔下，且每到下午闷热难当，我深怕有人中暑；爬出屋顶外测量关键数据时，瓦面又常烫得不能着手。万幸这许多年来，没有发生事故。但 2013 年测绘时，孙旭东同学为量取更多数据便于建模，一脚踩漏吊顶，险状犹在眼前。因为是教学计划（测绘实习）的一部分而且地点又在校内，除能提供每人一瓶矿泉水外，再无任何经济或物质补助，但这似乎也不影响为时两周的测绘。这些年来，我看到 20 来岁的同学们朝气勃发，以饱满的热情进行测量和绘图工作。他们深信这些图纸终会有利于后人了解这些老建筑，因此投入了额外的心血，其"作业"往往超出技术图纸的要求。我始终认为，近代建筑测绘很好地体现了清华同学的专业素质和精神面貌。在将这些测绘图纸结集出版之际，我对参与过校内近代建筑测绘的同学们表示由衷的谢意和敬意。他们的名字列在各自所绘的图纸上。

2020 年年初新冠疫情爆发，整个春季学期只能在家中通过网络上课。为响应学校线上教学改革的号召，我在这次《中国近代建筑史》课程的作业中，采取自愿报名方式，让部分选课同学把此前已测绘的部分清华近代建筑，根据测绘图重新建模。此外，也安排他们分别录入和翻译了一些与近代清华校园建设有关的英文文献，其"作业"收入到本书中编。他们的名字也列在各自翻译的文稿之后。这一工作将来还要进行下去。

2013 年夏开始的清华校园近代建筑测绘，是在当时清华房管处郁鼎文处长和连彦青、王玉英两位老师的大力支持下开展的；此后历年的测绘工作均得到房管处的积极配合。这种校内跨部门的合作是这项研究的起点和支撑，作者对此机缘铭记在心。文中不少部分曾与金富军、杨国华两位好友反复讨论，得益匪浅。这些年来这些研究常需到清华大学档案馆查阅资料，朱俊鹏、王俊熙两位老师提供了很多便利。本书在研究和撰写过程中得到国家自然科学基金、清华大学自主科研经费、清华大学建筑学院"双一流"共建经费资助，并且清华大学宣传部和校史馆诸多师友多方帮助、设法解决了出版中的各种问题；本书交稿后，一次因建筑学院院史访谈高冀先生时，高先生得知此事，慷慨将其所绘"清华十景"钢笔画见赠，作者一并肃志谢悃。我尤其感谢开辟了清华校园规划与建设研究方向的前辈罗森教授，先生奖掖后进、有问必答、热心解惑，更对本书书稿逐章讨论，提示疏漏、校正错失，笔者获益良多。但是限于时间精力，一些重要问题只能待之以后，如各时期的营造厂和具体施工情况，同时还未系统梳理台湾地区所藏档案，等等。总之，书中疵谬、遗憾甚多，笔者独负其责，幸祈海内外诸君子有以教之，以便将来改正。

2020 年 6 月 28 日

附图：清华大学校园近代建筑分布图（底图为1959年清华地图）

清 华 大 学 平 面 图

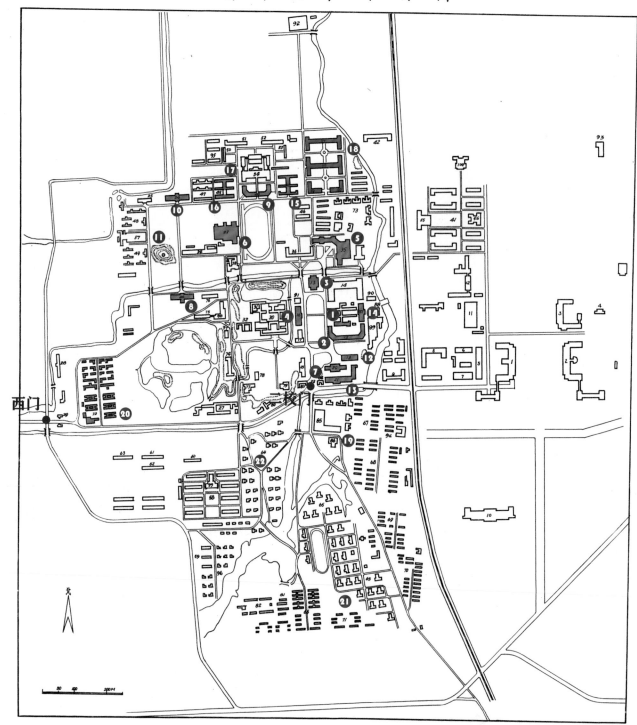

西门

校门

北

50 100 200M

图注（图中号码代表之建筑名称）：

1 同方部
2 清华学堂（一院，亦称"高等科"）
3 大礼堂
4 科学馆
5 图书馆一期及二期
6 西体育馆（"罗斯福纪念体育馆"）及后馆
7 土木馆（亦称"工艺馆"）
8 生物学馆

9 明斋
10 化学馆
11 气象台（今天文台）
12 老水利馆（水力实验室）
13 机械工程馆
14 电机工程馆
15 新斋
16 善斋

17 平斋
18 学生宿舍1～4号楼
19 南院住宅（照澜院）
20 西院住宅（包括新西院）
21 新林院住宅
22 胜因院

底图来源：《清华大学一览》编辑室. 清华大学一览[A]. 北京：清华大学，1959.

清华大学历史建筑地图
Historic Buildings at Tsinghua University

  为推进我国高校历史建筑的保护和宣传工作，清华大学房地产管理处暨文物保护办公室和清华大学建筑学院合作编制了《清华大学历史建筑地图》。以图文并茂的形式，按清华发展的三个历史时期，即清华学堂（1911—1912）—清华学校（1912—1928）—清华大学（1928—），简述了17座近代建成的国家级文保建筑的位置、特征及现状。

策划：清华大学房地产管理处、清华大学建筑学院
摄影：郭海军、刘亦师   版面设计：张晓莉
历史资料来源：墨菲档案、《清华周刊》《清华风物志》等

清华大学时期——四大

建筑结构：钢混结构
建筑年代：1930年
建筑面积：4221.00m²
占地面积：1055.30m²
设 计 人：杨廷宝

1933

8 气象台

  又名"天文台"，"四大建筑"之一，也是当时清华园里最高的建筑。气象台位于生物馆校河西北侧小山坡上，高五层，约24米（90英尺），呈八角形，内设螺旋钢梯，是清华理学院重要设施之一，原属地学系，其设备均为当时国际先进水平。

1931